Noise-Controlling Casings

Due to technological development, noise and vibration are ubiquitous in the human environment and belong to the most significant threats to man's wellbeing.

Noise-Controlling Casings offers a range of feasible noise-controlling strategies for different kinds of devices generating excessive noise. Depending on the required performance and the availability of energy sources, three solution categories are presented: passive (no external energy is needed, but performance is limited), semi-active (little energy is needed, but performance achieves higher values) and active (best performance, but an external energy source is needed). Two very important benefits of these proposed solutions are global noise reduction (in an entire enclosure or the surrounding space) and compact technology (contrary to other active noise control solutions requiring a large number of secondary sources and distributed sensors). Many of the solutions presented are original approaches by the authors, their own developed concepts and new elements and designs that have gained recognition in prestigious journals. The book provides a theoretical background to the research, looking at system configurations, mathematical modelling, signal processing implementation and numerical analysis.

The proposed ideas can be applied to any devices provided they have casings of thin walls or they can be enclosed by casings of thin walls. Applications include industrial devices, household appliances, vehicle or aircraft cabins and more.

This book will be of interest to professionals and students in the fields of acoustics, vibration, signal processing, control, automotive and aircraft engineering.

Noise-Controlling Casings

Edited by
Marek Pawelczyk
Stanislaw Wrona

CRC Press
Taylor & Francis Group
Boca Raton London New York

CRC Press is an imprint of the
Taylor & Francis Group, an **informa** business

First edition published 2023
by CRC Press
6000 Broken Sound Parkway NW, Suite 300, Boca Raton, FL 33487-2742

and by CRC Press
4 Park Square, Milton Park, Abingdon, Oxon, OX14 4RN

CRC Press is an imprint of Taylor & Francis Group, LLC

ISBN: 9781032226972 (hbk)
ISBN: 9781032227023 (pbk)
ISBN: 9781003273806 (ebk)

DOI: 10.1201/9781003273806

Typeset in NimbusRomNo9L-Regu font
by KnowledgeWorks Global Ltd.

Publisher's note: This book has been prepared from camera-ready copy provided by the authors.

Dedication

To our families and supporters

Contents

Preface

Due to technological development, noise and vibration are ubiquitous in the human environment and belong to the most significant threats to man's wellbeing. The European Environment Agency reports alarmingly that noise pollution is a major environmental health concern in Europe. Some of the most common noise and vibration sources are devices of different kinds, including industrial, household and automotive. Sound-absorbing materials are commonly applied to reduce excessive device noise. However, they are generally ineffective for low frequencies and often are inapplicable due to increase of size and weight of the device and its potential for overheating. Personal hearing protection is not ergonomic, and it is unacceptable for regular living. This motivates scientists to look for new ways to reduce noise, well-fitted to the current challenges and expectations.

The main idea behind the noise-controlling methods proposed by the authors in this book is to act directly at the source of noise. In particular, the idea is to use the casing (also referred to as housing or enclosure) of the noise-generating device to isolate it acoustically from the surrounding environment.

In this book different kinds of casings are considered, starting with casings with rigid frames (to reduce wall interactions), to casings with double panel walls, to casings of flexible structures (with all complicated couplings) and, finally, casings of market-available devices, with sophisticated shapes as well as vibration and noise patterns.

Mathematical, numerical and experimental models of considered casings are developed and presented as they are of particular importance for control methods and for better understanding of existing phenomena, interactions and interpretations of control results.

The book offers a range of feasible noise-controlling strategies for devices generating excessive noise. Depending on the required performance and the availability of energy sources, three solution categories are presented: passive (no external energy is needed, but performance is limited), semi-active (little energy needed, but performance achieves higher values) and active (best performance, but an external energy source is needed). A very important benefit of the proposed solutions is global noise reduction (in an entire enclosure or surrounding space) and compact technology (contrary to other active noise control solutions requiring a large number of secondary sources and sensors distributed). Many of the solutions presented are original approaches by the authors, their own developed concepts and new elements and designs that have gained recognition in prestigious journals.

In order to enable readers to better understand and feel the whole concept of noise-reducing enclosures, and perhaps to replicate their design or even conduct experiments themselves, the authors decided to choose a particular form for this book. It can be regarded as a consistent case study with examples of several casing types of increasing complexity. Therefore, all the material data, construction details, algorithm parameters, practical aspects including system configuration, efficient signal processing implementation, conditions and method of experiments are given. The stages of design, construction and testing are intertwined with the theoretical parts necessary for their execution or better understanding.

The proposed ideas can be applied to any devices provided they have casings of thin walls or they can be enclosed by casings of thin walls. Applications include vehicle or aircraft cabins, and more.

This book is the result of many years of research on noise and vibration control in the Department of Measurements and Control Systems at the Silesian University of Technology in Gliwice, Poland, where the presented concept of noise-reducing device casings was created and developed. The presented results are the result of the project funded by the National Science Centre, Poland, contract number UMO-2017/25/B/ST7/02236. The editors would like to express their deepest thanks to the contractors of this project—employees and PhD students of the Silesian University of Technology, who have made a huge contribution to conducting scientific research, developing the concept of device enclosures and preparing the detailed chapters of this book. They are (alphabetically): Dariusz Bismor (Chapter 6), Anna Chraponska (Chapter 5), Chukwuemeke William Isaac (Chapters 3 and 4), Jerzy Klamka (Chapter 3), Krzysztof Mazur (Chapters 4, 5 and 6), Jaroslaw Rzepecki (Chapters 4 and 5) and Janusz Wyrwal (Chapter 3).

We are particularly grateful for the scientific support provided by eminent scientists from other centres—co-authors of many of our joint studies and publications, the results of which can be found in many places in this book. They are (alphabetically): Jordan Cheer (University of Southampton), Li Cheng (Hong-Kong Polytechnic), Maria de Diego (Polytechnic University of Valencia), Ling Liu (Shanghai Jiao Tong University, Polytechnic University of Milan), Zhushi Rao (Shanghai Jiao Tong University), Xiaojun Qiu (University of Technology Sydney) and Bert Roozen (KU Leuven).

The authors are also very appreciative of all the discussions held during the Department's many seminars and scientific conferences, and the valuable guidance given by reviewers of publications submitted to journals.

The authors are grateful for support in the form of pro-quality programmes within the participation of the Silesian University of Technology in the programme Excellence Initiative—Research University, funded by the Ministry of Education and Science, Poland.

We would like to express our great thanks to the publishers from which some of the material, especially figures and plots, has been used in this book. We have made every effort to provide appropriate acknowledgment.

Our special and heartfelt thanks go to our wonderful families for their incredible support in every way, their understanding and their patience.

We warmly invite all interested researchers, practitioners and industrial partners to cooperate with us, including visiting our laboratory, conducting their own experiments and preparing a joint publication, project or implementation in products. Please contact us at Marek.Pawelczyk@polsl.pl and Stanislaw.Wrona@polsl.pl.

Marek Pawelczyk and Stanislaw Wrona

Gliwice, Poland, 2022

Contributors

Dariusz Bismor
Silesian University of Technology
Gliwice, Poland

Anna Chraponska
Silesian University of Technology
Gliwice, Poland

Chukwuemeke William Isaac
Silesian University of Technology
Gliwice, Poland

Jerzy Klamka
Silesian University of Technology
Gliwice, Poland

Krzysztof Mazur
Silesian University of Technology
Gliwice, Poland

Marek Pawelczyk
Silesian University of Technology
Gliwice, Poland

Jaroslaw Rzepecki
Silesian University of Technology
Gliwice, Poland

Stanislaw Wrona
Silesian University of Technology
Gliwice, Poland

Janusz Wyrwal
Silesian University of Technology
Gliwice, Poland

1 Introduction

1.1 NOISE IN THE MODERN WORLD

Sound accompanies humans in everyday life. First of all, it is used for communication, but it also informs about dangers and warns. Pleasant sounds, such as birds singing, water flow or music, make it easier to relax. People are used to living among sounds. But even in the modern world sounds can also be unpleasant or dangerous. However, a dichotomous classification is difficult and requires taking into account psychoacoustic phenomena and holistic analysis. A sound that in some circumstances may be informative and necessary for doing something, in other condition, for example when one wants to relax, can be disturbing. The degree of annoyance also depends on the mood we are in, our current condition and many other factors. Sound that is unpleasant or disturbing under certain conditions is referred to as noise for the purposes of this study.

Noise is currently one of the most significant hazards to humans in the workplace, making it difficult to communicate by voice, to hear alarms and warnings, and being the source of many ailments and diseases, especially hearing diseases. Noise at home makes it difficult to rest and causes irritation. Long-term exposure to noise at high sound pressure levels can lead to permanent hearing damage. These issues are the subject of many studies and are reflected in defined standards.

The most important sources of noise include equipment - both industrial, domestic and transportation. Noise is most often produced by rotating and reciprocating machinery, as well as by vibration or friction of materials. The source of the noise determines its character. We distinguish between narrowband noise, including tone or multi-tone noise, and broadband noise. In many cases, both types of noise occur together, with varying contribution. Particularly troublesome is impulsive noise associated with a single or series of acoustic events of short duration, usually less than 1 s.

1.2 NOISE REDUCTION METHODS

Humans can try to protect themselves against noise in various ways. To some extent, noise can be reduced by improving the design of the equipment, for example by better fitting parts, reducing clearances, introducing washers and insulating mats. Usually these are the cheapest methods, but they are not always effective enough, and in some types of equipment, such as those involved in material handling, they are impossible to use. In this case, one can try to isolate the noise source from people or people from the noise source. In this case, sound insulating or sound absorbing barriers or screens are used. However, this method in its classical formulation has a number of limitations, most often associated with decreasing effectiveness for low frequencies, which are often dominant in machinery and equipment noise and road

DOI: 10.1201/9781003273806-1

noise. Improved effectiveness is achieved by increasing the thickness and density of barrier materials, which in turn is unacceptable in many cases, because it involves increased size and weight, and impedes the dissipation of heat from the equipment insulated in this way. A special case of isolation is personal hearing protection in the form of earmuffs or earplugs, for example, when other methods are not applicable, or the user needs to operate processes that generate excessive noise. The above solutions are passive in nature, where there are not employed any components that require the use of any energy.

With the development of microelectronics and signal processing techniques in the last 30 years, active noise reduction methods are gaining popularity. Their hallmark is the introduction of external energy into the process. There are different physical justifications associated with different active control methods. The most common justification is based on the Young's principle of destructive wave interference – the superposition of two acoustic waves with the same amplitudes but opposite phases causes their mutual elimination. Starting from these premises, therefore, it is necessary to generate an additional acoustic wave, which will destructively interfere with the acoustic wave associated with noise. Such a wave should be generated using an additional secondary sound source, such as a loudspeaker or other vibrating structure, based on measurements made in the acoustic field, or by estimating appropriate quantities using mathematical models. Unfortunately, while the general form of the method seems to be very simple, its effective application in practice encounters many limitations. First of all, the acoustic field is complex in nature, and even if one could ensure that the above amplitude-phase condition could be satisfied at a single point in space, it is much more difficult to satisfy it over a larger area. In practice, this can produce so-called local quiet zones, the size of which depends largely on the frequencies dominating the noise. Enlarging these sizes requires suitably favourable geometrical conditions or the use of many or very many secondary sound sources. This involves considerable complexity of the algorithm, of the entire system, including the electronics, requiring multi-channel high-speed processing, and thus considerable and usually unacceptable cost. This complexity increases, in the case when the reduction of the sound pressure level is not the only goal, but it is also necessary to provide a rapid response to changes in the parameters of the sound field caused, for example, by non-stationary of noise, changes in the room itself, movement of users, temperature changes, etc.

Active noise reduction is also explained by the absorption of noise by secondary sources or by the control of the acoustic impedance of the medium by secondary sources. The use of such phenomena gave rise to the concept called active structural acoustic control, in which the vibration of the barrier separating two media is controlled in such a way as to reduce the transmission of sound between them. Many studies are known in the literature, especially concerning the vibration control of single or double-panel barriers. If the noise is transmitted mainly through these barriers and not through other paths, this approach allows a global noise reduction in the entire listening space.

The structural approach can be modified by dispensing with the need to inject additional energy into the sound field through semi-active controls. In the classical approach, this involves using a control system to change the parameters of piezoelectric elements bonded to an acoustic barrier in such a way as to alter its vibrational and sound transmission properties. Additional energy is therefore needed here only to change these parameters, not to excite vibrations of the structure or to generate secondary sound. The energy is therefore much less than for active control, and the complexity of the system is usually significant.

1.3 NOISE-CONTROLLING CASING

A common way to reduce the noise of many types of equipment is to enclose it in a sound-absorbing casing and thus isolate it from its surroundings. As a result, many casings are very heavy, thick, and have sound-absorbing materials, but this consumes a lot of raw materials and also makes heat dissipation difficult. Furthermore, such passive methods are not very effective for low frequencies. Classical active noise reduction methods, based on the principle of destructive interference of acoustic waves, require the use of a large number of sound sources and control and measurement apparatus, and yet allow only local zones of quiet, while beyond them the sound is reinforced.

The authors propose another, much more effective approach to reducing the noise of equipment—the control of vibrations of its casing. Thus, it is assumed that the device is enclosed in a thin-walled casing made of resilient material or the device can be placed in an additional casing with such properties. The above concept is an extension of the known method of controlling vibrations of a single wall or a system of two walls separating adjacent acoustic media—one in which noise is generated from the other in which noise reduction is desired. Casing vibration control introduces significant complications resulting from the interactions of the controlled walls and the multidimensionality of the overall problem. All three approaches described in the previous section can be applied to the casing control problem and many innovative solutions can be introduced—passive control, active control, semi-active control, and combinations thereof. Casing control aims at changes in its response that increase its acoustic isolation for specific frequency ranges associated with the generated noise. Properly designed, it can provide global noise reduction outside the casing without the need for additional and cumbersome secondary sources located in the environment.

1.4 OBJECTIVES AND SCOPE OF THE BOOK

The objective of this monograph is to present and convince the scientific community as well as practitioners and equipment manufacturers of the concept of effective global noise reduction of devices by controlling the vibration of their casings. This goal is to be achieved by combining theory with simulation and laboratory studies. A detailed presentation of design issues is necessary for this. Therefore, the authors decided to elaborate the monograph in the form of case studies.

At the very beginning, in **Chapter 2**, concrete experimental types of casings are presented, and thus their geometrical and material parameters are given. For this reason, concrete modelling and control results are presented, giving the reader the opportunity for direct reference. The authors also try to generalize the conclusions.

The first type of casing considered is a casing with a rigid structure consisting of a heavy frame to which walls constructed of one or two panels connected in parallel are attached. This design makes it possible to reduce the number of phenomena occurring by rigidly fixing the boundary conditions of the walls, thereby eliminating the influence of the vibrations of each wall on the boundary conditions of adjacent walls. It also facilitates modification of the space between the panels.

The degree of complication is increased in the second type of casing—a casing with flexible structure, made of folded and properly jointed walls, without the mediation of an additional frame. Then, under the influence of vibrations forced by noise or control, the casing undergoes temporary deformations of shape, increasing the research challenge.

The third type of casing—the casing of a real device—is definitely more challenging. A top-loading washing machine purchased from a supermarket was used as an example. Such a casing is characterized by a complex shape, different types of materials and connections, openings, and a multitude of phenomena occurring and interacting. Therefore, it requires a more generalized modelling approach and more compensation in control algorithms.

Each type of control considered in this monograph requires an appropriate object model with its associated components. **Chapter 3** is devoted to this subject and shows how to build models of a single wall, single-panel or double-panel one, as well as an entire casing. In this general case, a number of interacting subsystems are specified, as control of one panel causes vibrations of the medium between the panels and thus of the other panel in a two-panel arrangement, of the medium filling the casing and thus of the incident panels of the other walls, vibrations of a panel in a flexible casing cause changes in its boundary conditions and thus changes in the boundary conditions of the adjacent panels. The relationship between vibration and acoustic emission, the effect of the medium outside the casing on the outer panels, and the interactions can also be considered.

Various modelling approaches have been used, including a modal approach and a state-space description. The developed theoretical models have been verified by simulation using multiphysics packages and experimentally in laboratory tests using the casing designs described above. Experimental measurements have also been used to fine tune the parameters of the mathematical models. Principles are given and properties of observability and controllability of such objects are also analysed.

Chapter 4 deals with passive control. In the traditional view, passive control involves the use of a sound-absorbing material whose parameters are selected according to frequency. In general, suitably thick and thus heavy materials are required for low frequencies. The authors have shown that by attaching appropriately optimized point masses and ribs to the casing walls, it is possible to fully shape the frequency response of the structure and thus provide high isolation for noise dominant frequencies. Examples of different quality functions are used to show how different functionalities can be achieved with this method. A particular benefit of this approach is that significant sound reduction effects can be achieved at minimal, even near-zero cost, since the installation of these additional elements can be carried out at the manufacturing stage of the device casing.

An extension of this concept is the use of composite materials with variable properties along their cross section, which are known as functionally graded materials. Such an appropriately designed composite has more favourable sound insulation properties.

A particular type of passive reduction is the use of shunt circuits that collect energy from the surface of a vibrating plate, convert it into electrical energy and then dissipate it. The design of such a circuit is shown in detail.

Chapter 5 presents a semi-active approach to control. In brief, it consists of controlling changes in the parameters of an object, in particular switching, in such a way that energy is absorbed and dissipated accordingly. Energy is thus only needed to control these changes and is not introduced into the vibroacoustic system. The concept of shunt circuits was developed by moving it from a passive to a semi-active level. Shunt circuits known from literature are based on piezoelectric elements glued to a vibrating plate. The authors have additionally proposed a completely different approach. One of them, appropriate for two-panel walls, consists in the use of controlled links placed at selected locations between the panels, in such a way that, depending on the frequency ranges dominating the noise, the frequency response of the wall is altered so as to produce anti-resonant valleys in those ranges, thus increasing its acoustic isolation. Another approach involves a single-panel wall. It is based on using a concentrated mass and moving it on a mandrel attached to the panel, thereby changing the frequency response. Both of the proposed approaches have proven to be a very effective method of shaping the vibrating response of the structure.

Each time, the entire design process, including the hardware circuit and the algorithm, is given.

Chapter 6 deals with active control—the most complex in terms of hardware and algorithms, and thus involving the greatest implementation cost, but giving the greatest flexibility and the possibility of achieving higher levels of sound reduction. In the case of active control, proper selection of measurement and actuation elements, as well as their mounting locations, is crucial. Their presence affects the properties of the object due to the non-zero mass, especially in the case of vibration exciters, but most importantly the ability to observe and control subsequent modes. Therefore, the first part of this chapter presents a method of optimizing the placement of measuring and actuating elements based on a formal approach known from control theory. The method of construction of the electronic system and data processing is also described. The control structures—feedforward and feedback—are presented. Attention is paid to the internal acoustic feedback. The method of control with virtual microphones is presented, which enables the reduction of estimated sound pressure based on other measurements. Control algorithms based on Least Mean Squares family are discussed. The problems of stability and effective implementation are pointed out. Many design guidelines and specific algorithmic solutions for multidimensional control systems are given.

Chapter 7 provides a summary. It presents the main results and conclusions of the monograph. It also indicates the applicability of the proposed methods and the directions of further development of the proposed concept of control of device casings.

References lists literature related to the topics discussed.

The **Appendices** supplement the material presented in the main part of the monograph, giving the reader an opportunity to obtain additional detailed information.

2 Considered types of casings

2.1 INTRODUCTION

The presented concept of noise-reducing casing is inherently entangled with the vibroacoustical system it is based on. Hence, to gradually scale-up the complexity of the considered structures, both the actual research performed by the authors and introduction to considered types of casings given in this chapter have been divided into three stages. Each of them is characterized by different features affecting the adopted noise-controlling solutions.

In the first research stage, the rigid casing was considered, since the employment a heavy frame limits vibrational couplings between walls, facilitating initial attempts to develop a mathematical model and to control casing vibrations. Moreover, such structure can be easily reconfigurable, enabling experiments with different panel materials, thicknesses, and single- or double-panel walls.

In the second stage, the light-weight casing was used, which in contrast to previously employed structure, is made without an explicit frame, resulting in greater vibrational couplings. The casing is made of metal plates bolted directly together, therefore the structure is not reconfigurable. However, it resembles more real casings commonly used in the industry.

In the third and final stage, a real device casing was used as the subject of both modelling and noise control. A common washing machine was utilized as an example.

Each of aforementioned casings is described in this chapter, starting with details of a mechanical structure. Then, selection and arrangement of applied actuators and sensors are discussed. Finally, vibroacoustical analysis of the structures based on secondary paths and frequency response functions is presented.

2.2 RIGID CASING

The rigid casing discussed in this section is presented in Fig. 2.1, where dimensions, cross-sections and the method of mounting of casing walls are visualized. A photograph of the casing is given in Fig. 2.2. The casing has a heavy cubic frame made of 3 mm thick welded steel profiles. The high rigidity of the frame results in its resonance frequencies to be far above frequencies of the noise considered. The bottom of the casing is vibrationally and acoustically insulated. All walls of the casing are made of single or double panels. Each panel is attached to the structure by 20 screws embedded in the frame, and clamped with an additional steel square frame. Fully-clamped boundary conditions can be then assumed for the panels, achieving satisfactory modelling accuracy. For double panels the distance between them is 50 mm.

DOI: 10.1201/9781003273806-2

The panel closer to the casing interior is called the incident panel, and the outer panel is referred to as the radiating panel.

2.2.1 ACTUATORS AND SENSORS

In this stage of research, a loudspeaker placed on the casing floor was used as the primary noise source. It allowed for creating an environment more suitable for the

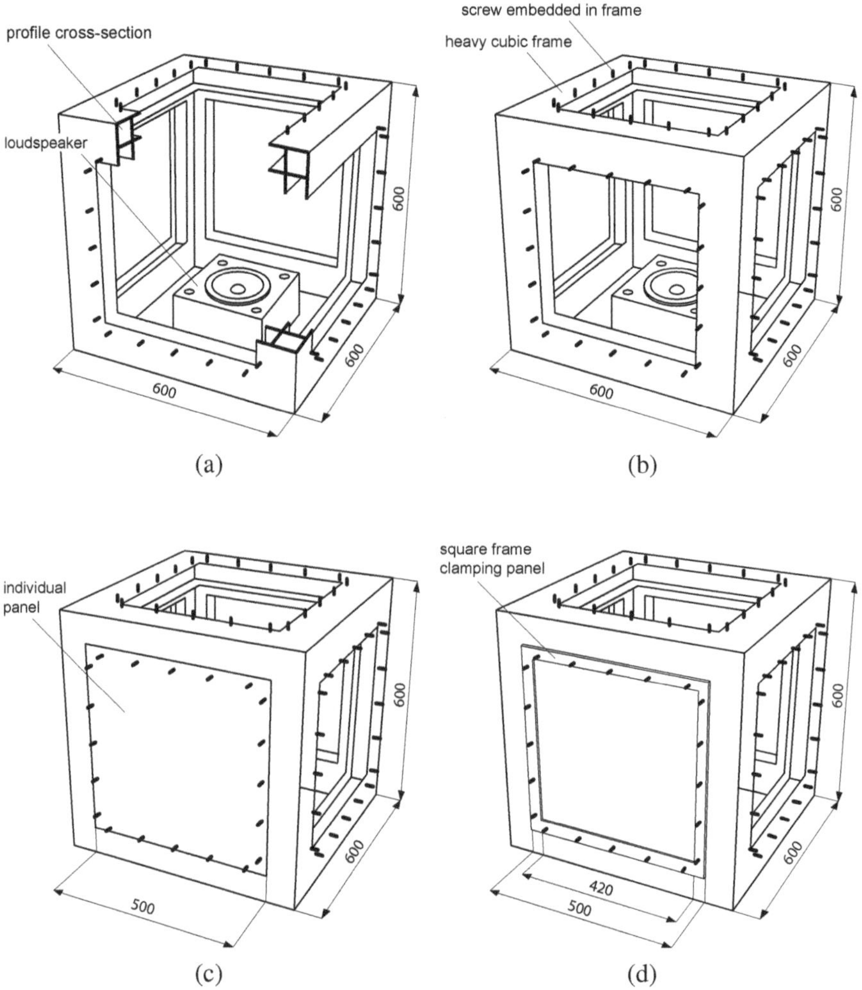

Figure 2.1 A schematic representation of the rigid active casing. All dimensions are given in [mm]. (a) A view of the enclosure interior with a loudspeaker. The cross-section of profiles is also visible. (b) A view of the whole frame. (c) A view with a panel mounted to the frame. (d) A view with attached clamping frame [164]. Reprinted from "Modelling and control of device casing vibrations for active reduction of acoustic noise", PhD thesis, Silesian University of Technology, S. Wrona.

Figure 2.2 A photograph of the rigid active casing.

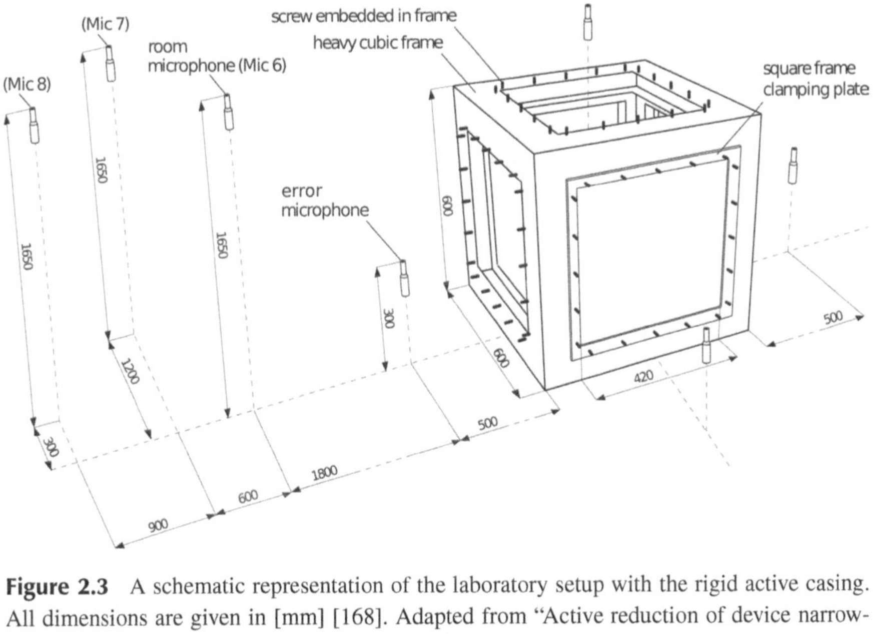

Figure 2.3 A schematic representation of the laboratory setup with the rigid active casing. All dimensions are given in [mm] [168]. Adapted from "Active reduction of device narrow-band noise by controlling vibration of its casing based on structural sensors", S. Wrona and M. Pawelczyk, 22nd International Congress on Sound and Vibration, Florence, Italy, 2015.

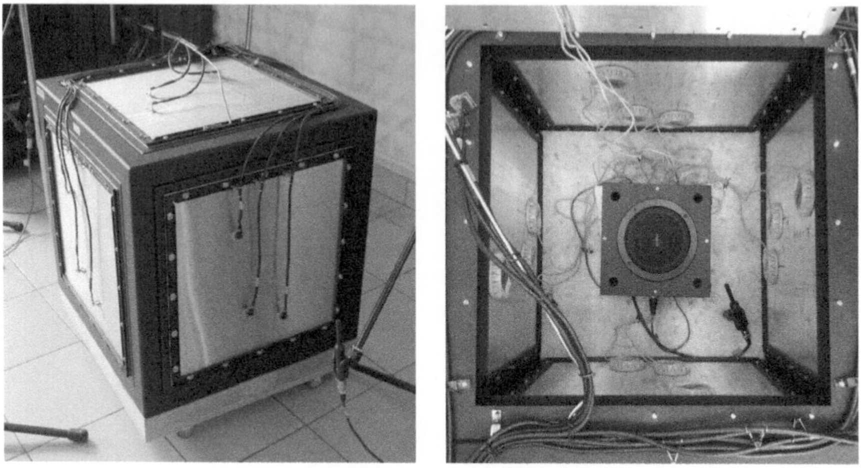

Figure 2.4 Photographs of the rigid active casing with mounted sensors and actuators. (a) A photograph from the outside. (b) A photograph from the inside [168]. Adapted from "Active reduction of device narrowband noise by controlling vibration of its casing based on structural sensors", S. Wrona and M. Pawelczyk, 22nd International Congress on Sound and Vibration, Florence, Italy, 2015.

research than a real operating device, which has been used in following stages. For feedforward control system implementations, the reference signal was obtained by a microphone placed next to the loudspeaker inside the casing enclosure (referred to further as the reference microphone). In front of each casing wall, a microphone is placed in the distance of 500 mm (referred to as the error microphone). These microphones are used mainly for control-related purposes. Additionally, to evaluate the noise reduction efficiency, multiple microphones were placed at several larger distances from the casing, corresponding to potential locations of the user (referred to as the room microphones). The laboratory setup is presented schematically in Fig. 2.3. Photographs of the rigid casing with mounted sensors and actuators are given in Fig. 2.4.

To control vibrations of the casing walls, inertial exciters NXT EX-1 were used. They weight 115 g and they are of small dimensions (70 mm), comparing to the size of the casing. In the performed control experiments, they are mounted on the incident plates from the inner side, three actuators per panel. Their placement has been optimized using a method that maximizes a measure of the controllability of the system. The impact of the mass of the actuators is included in the optimization procedure, as it is comparable with the mass of the casing walls and substantially affects their dynamical behaviour. The method of actuators positioning is described in details in Subsection 6.2.1.

As sensors for control purposes, microphones (Beyerdynamic MM-1) or accelerometers (Analog Devices ADXL203) were used, depending on the chosen control configuration. Accelerometers locations were calculated according to the method

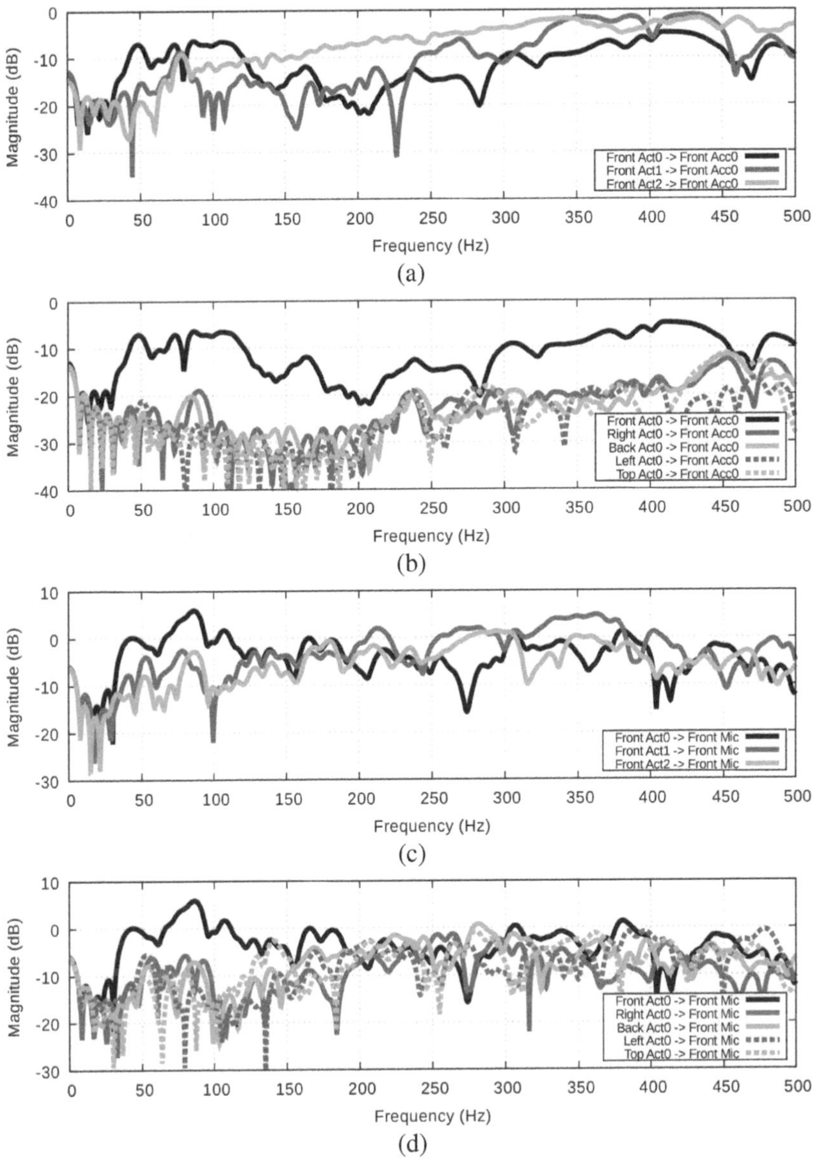

Figure 2.5 Exemplary amplitude responses of secondary paths for the rigid casing. (a) Direct paths between actuators no. 0–2 and the accelerometer no. 0 mounted on the front wall. (b) Cross paths between actuators no. 0 mounted on different walls and the accelerometer no. 0 mounted on the front wall. (c) Direct paths between actuators no. 0–2 and the outer microphone assigned to the front wall. (d) Cross paths between actuators no. 0 mounted on different walls and the outer microphone assigned to the front wall [164]. Reprinted from "Modelling and control of device casing vibrations for active reduction of acoustic noise", PhD thesis, Silesian University of Technology, S. Wrona.

that maximizes a measure of the observability of the system [167]. The placement of actuators and sensors was identical for each wall. In contrast to the inertial exciters, the employed accelerometers are light-weight (5 g) comparing to the mass of the casing walls, and therefore they have a marginal loading effect and their mass can be safely neglected in the mathematical modelling. However, if heavier sensors were used or an application would require a highest modelling accuracy, their mass could be modelled analogously as in the case of inertial exciters.

An additional type of the sensors used was a laser vibrometer Polytec PDV-100. As a specialized laboratory equipment, it was not used for active control experiments as a signal source (it would be infeasible in commercial applications), but for highly precise non-contact measurement of vibrations in laboratory environment, strictly for the research purposes (the obtained data have been used mainly to validate the mathematical modelling accuracy). Utilizing that the measurement does not affect anyhow the dynamical behaviour of the vibrating panel, all of experimentally measured mode shapes of casing walls presented are obtained with the laser vibrometer.

2.2.2 SECONDARY PATHS ANALYSIS

To present the vibroacoustical properties of the structure, a set of exemplary amplitude responses of secondary paths obtained for the rigid casing with single panels is shown in Figs. 2.5(a)–(d). It follows from the analysis that the direct paths between actuators and accelerometers mounted on the same wall are of similar magnitude in whole frequency range considered (see Fig. 2.5(a)). In turn, the magnitudes of cross paths between actuators mounted on one wall and accelerometers mounted on the other wall are many times weaker, comparing to magnitudes of direct paths within the same wall (see Fig. 2.5(b)). This is due to the heavy and rigid frame of the casing, isolating vibrationally individual walls. Hence, the interference with each other is mainly through the acoustic field. Therefore, since such separation has been noticed, for the mathematical modelling and control purposes, it is justified to consider each of the walls separately.

Analogous behaviour can be observed for the paths between actuators and outer microphones, but only for low frequencies up to approximately 250 Hz. Above this frequency, the cross paths between actuators mounted to one wall and an outer microphone placed in front of another wall become of similar magnitude, as the direct paths between actuators and an outer microphone assigned for the same wall (see Figs. 2.5(c) and (d)).

2.3 LIGHTWEIGHT CASING

2.3.1 ACTUATORS AND SENSORS

The light-weight device casing used in the second stage of research is presented in Fig. 2.6. In contrast to the rigid casing used in the previous stage, the light-weight casing was made without an explicit frame. It was made of steel plates (of 1.5 mm thick) bolted together, forming a closed cuboid of dimensions 500 mm × 630 mm ×

800 mm. Such structure results in greater vibrational couplings between individual walls, in addition to couplings through the acoustic field inside and, to a lesser extends, outside the casing. Moreover, due to the absence of the rigid frame, the walls were connected directly to each other, what results in boundary conditions which no longer behave as fully clamped—boundary conditions elastically restrained against both rotation and translation are more appropriate. Identification of spring constants representing elastic boundary conditions of light-weight device casing walls are performed using experimental data and a memetic algorithm. The procedure is described in details in [169].

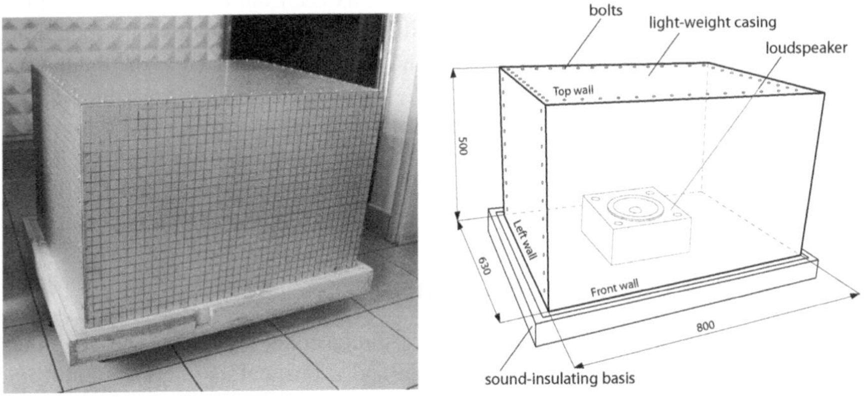

Figure 2.6 The light-weight active casing—a photograph and a schematic representation. All dimensions are given in [mm] [170]. Adapted from "Optimal placement of actuators for active structural acoustic control of a light-weight device casing", S. Wrona and M. Pawelczyk, 23rd International Congress on Sound and Vibration, Athens, Greece, 2016.

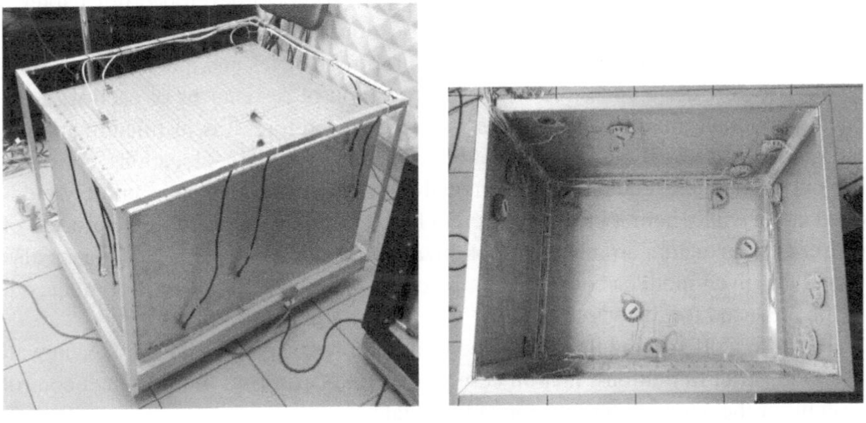

Figure 2.7 Photographs of the light-weight active casing with mounted sensors and actuators and a wire holder.

Similarly as in the previous stage, a loudspeaker placed on the sound-insulating basis is used as the primary noise source. A reference microphone placed next to the loudspeaker is used to obtain the reference signal. In front of each casing wall, a microphone was placed in the distance of 500 mm (also referred to as the error microphone). The room microphones are placed in similar configuration as in case of the rigid casing. Photographs of the light-weight active casing with mounted sensors and actuators are given in Fig. 2.7.

Inertial exciters NXT EX-1 are used to control vibrations of the casing walls. They are mounted on the walls from the inner side. The number of actuators depends on the particular wall—four actuators are mounted to front, right, back and left wall, and five actuators to the top wall. Their placement has been optimized using a method that maximizes a measure of the controllability of the system. Boundary conditions elastically restrained against both rotation and translation are used in the mathematical modelling [169]. As shown in Fig. 2.7, light-weight frames are added to the laboratory stand to support the actuators cables (inside the casing) and sensors cables (outside the casing). These frames are made in such a manner that they do not affect the vibrational or acoustical behaviour of the casing. The accelerometers used for control purposes (Analog Devices ADXL203) are collocated with the actuators.

2.3.2 FREQUENCY RESPONSE FUNCTIONS

The described light-weight casing is a three-dimensional structure. The couplings between individual walls, of both vibrational and acoustical nature, are significant. However, noteworthy is an analysis of spatially averaged (over the area of each wall) frequency responses of three walls of the casing, presented in Fig. 2.8 (left, front and top walls; right and back walls are omitted as they are symmetrical to the left and front walls, respectively). Resonances that originate on one wall, are clearly visible on the others. However, all of resonances visible as peaks in the given frequency range can be assigned to one of the walls where they originate. Such assignment of the eigenmodes is consistent with the mathematical model developed in Chapter 3 and validated for the light-weight casing in Section 3.5.

It leads to a conclusion that observed natural frequencies and mode shapes of the whole structure are a consequence of superposition of resonances of each wall excited individually (but as a part of the structure). Therefore, it is justified to analyse the walls separately for the purpose of optimization of actuators locations, considering only eigenmodes due to the given wall—if the resonance is appropriately modelled at the wall where it originates, it will be accordingly included in the process of sensors and actuators arrangement optimization for the whole casing. It means also that if the given mode is controlled with actuators at the wall where it originates, it will be reduced for the whole casing.

It is also worth noting that for the control purposes, separate control of each wall has more strict limitations, than in case of the rigid casing—a multichannel control system for the whole casing is more appropriate.

2.3.3 SECONDARY PATHS ANALYSIS

It follows from the analysis of secondary paths given in Fig. 2.9 that magnitudes of cross paths between actuators mounted on one wall and accelerometers mounted on the other wall are weaker than magnitudes of direct paths within the same wall only up to frequency of approximately 250 Hz (see Figs. 2.9(a) and (b)). Above this frequency, magnitudes of cross paths and direct paths are similar. It confirms, as expected, higher vibrational couplings between individual walls than in case of the rigid casing (where cross paths are many times weaker than direct paths in whole frequency range considered).

On the other hand, for the paths between actuators and outer microphones, cross paths and direct paths represent a similar magnitude in the whole frequency range

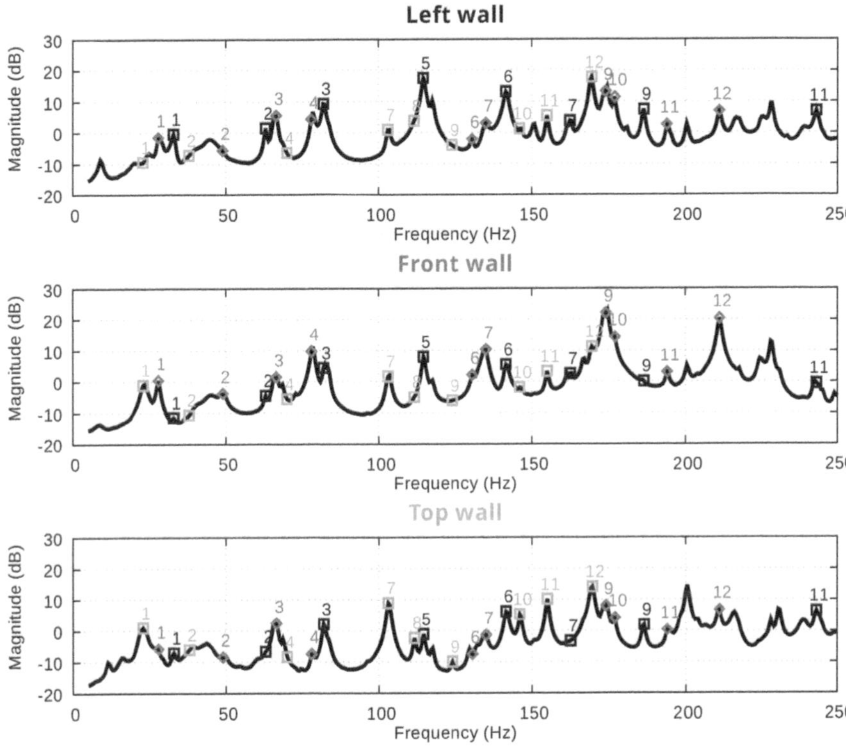

Figure 2.8 Spatially averaged frequency responses of casing walls. Two pairs of walls are symmetrical (left and right, front and back), hence only one of each pair is presented in the figure. Initial 12 eigenmodes originating at each wall are marked: eigenmodes originating at left wall are marked with circle, at front wall with diamond, at top wall with square [170]. Adapted from "Optimal placement of actuators for active structural acoustic control of a light-weight device casing", S. Wrona and M. Pawelczyk, 23rd International Congress on Sound and Vibration, Athens, Greece, 2016.

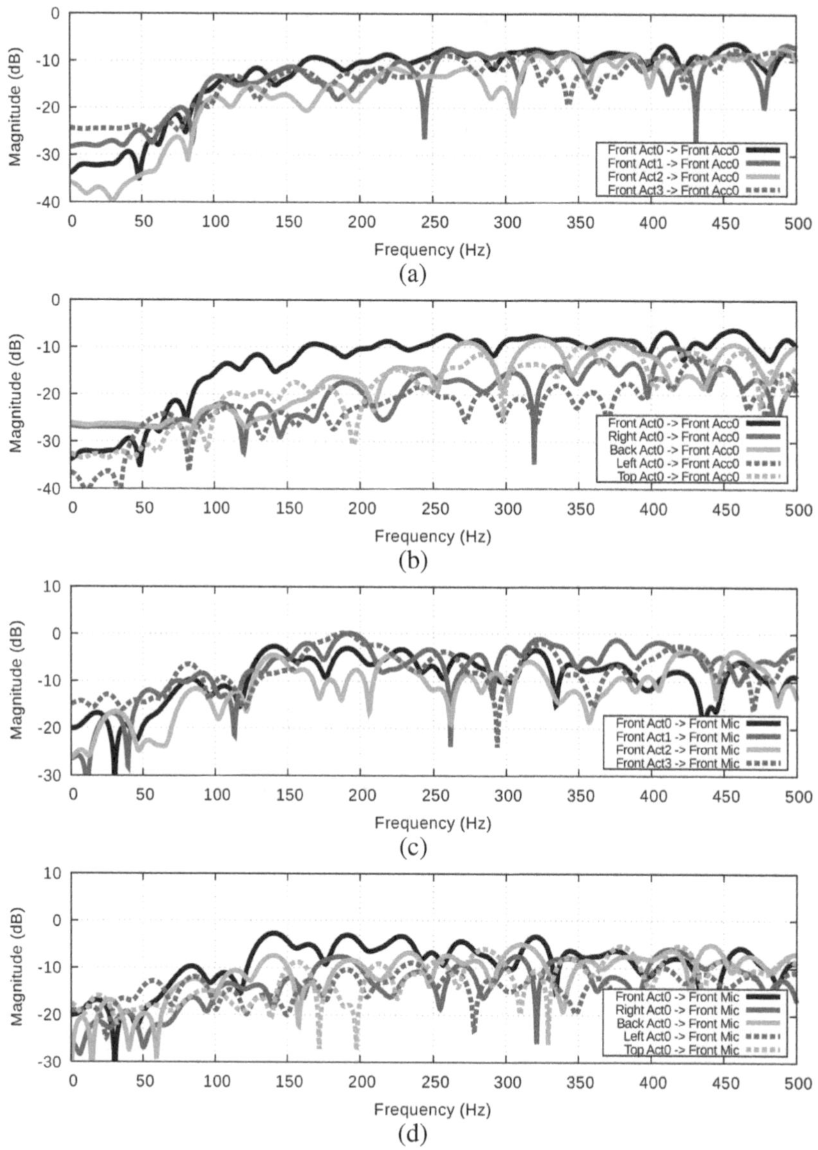

Figure 2.9 Amplitude responses of secondary paths for the light-weight casing. (a) Direct paths between actuators no. 0–3 and the accelerometer no. 0 mounted on the front wall. (b) Cross paths between act. no. 0 mounted on different walls and the acc. no. 0 mounted on the front wall. (c) Direct paths between act. no. 0–3 and the outer microphone assigned to the front wall. (d) Cross paths between act. no. 0 mounted on different walls and the outer mic. assigned to the front wall [164]. Reprinted from "Modelling and control of device casing vibrations for active reduction of acoustic noise", PhD thesis, Silesian University of Technology, S. Wrona.

(in contrast to the case of the rigid casing, where for low frequencies direct paths dominated over cross paths). Direct paths between actuators no. 0–3 and the outer microphone assigned to the front wall are shown in Fig. 2.9(c). Cross paths between actuators no. 0 mounted on different walls and the outer microphone assigned to the front wall are presented in Fig. 2.9(d). Such strong couplings generally make the synthesis of active noise control systems more difficult.

2.4 REAL CASING

The third type of the considered casings is a real device casing—ready-made and mass-produced. As an example, a washing machine has been chosen. The considered structure is presented in Fig. 2.10. In contrast to the laboratory casings described in the previous sections, the real device casing is often very irregular and inhomogeneous. Each of the walls represent different features, i.e. bendings, embossments, etc., what makes the task of fitting the mathematical model of each wall significantly more difficult.

The casing is a three-dimensional structure with strong couplings between walls (similarly to light-weight casing), however, different character of each wall is clearly noticeable, along with factual structural separation. Therefore, to simplify the analysis, each wall is considered separately for the purpose of actuators positioning, considering only eigenmodes due to the given wall (according to the concept that if the resonance is controlled with actuators at the wall where it originates, it will be reduced for whole casing).

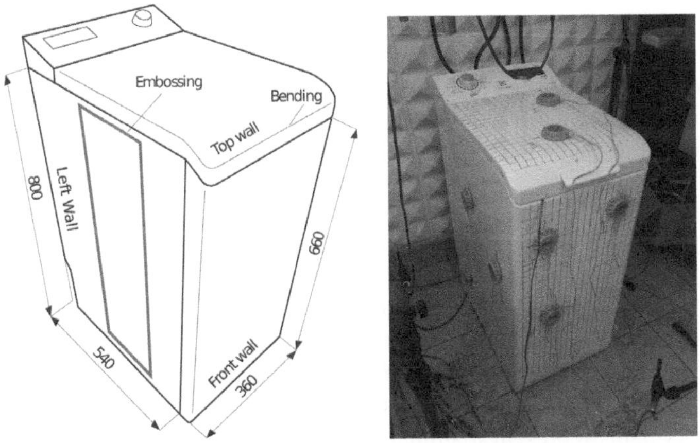

Figure 2.10 A schematic representation and a photograph of the real washing machine casing with bonded actuators. All dimensions are given in [mm] [166]. Adapted from "Optimal placement of actuators for active control of a washing machine casing", S. Wrona, K. Mazur, M. Pawelczyk, and J. Klamka, 13th Conference on Active Noise and Vibration Control Methods, Kazimierz Dolny, Poland, 2017.

To control vibrations of the device casing, inertial actuators are selected to be used. Front, right, left and top wall are considered for the active control. The back and the bottom of the casing are not controlled due to the fact, that the washing machine is most often placed at a wall. The impact of the mass of the actuators is included in the mathematical model of individual casing wall.

For the sake of brevity, the frequency response functions and secondary paths analysis are omitted for the real casing—they highly resemble ones obtained for the light-weight casing. The main difference is that parameters required for the mathematical model are more difficult to define. In fact, they constitute an approximation of the panels dynamical behaviour rather than direct representation of mechanical properties.

2.5 SUMMARY

The laboratory setups for control experiments consisting of several different casings have been presented and discussed in this chapter. The discussion includes practical application-related aspects of the laboratory stand assembly, vibroacoustic analysis of the introduced structures, set of assumptions setting a reference point for the mathematical models developed in the following chapter, and selection of sensors and actuators for the control systems.

Initially the rigid casing has been examined. Following conclusions are formulated basing on the analysis of secondary paths:

- the direct paths between actuators and accelerometers mounted on the same wall significantly dominate the cross paths between actuators mounted on one wall and accelerometers mounted on the other wall, what confirms high vibrational isolation between walls;
- for the paths between actuators and outer microphones, direct paths dominate the cross paths only for low frequencies up to approximately 250 Hz (above this frequency, they become of similar magnitude).

Afterwards, the light-weight casing has been considered. Studying its secondary paths and frequency response functions, following conclusions are formulated:

- direct vibrational paths dominate the cross paths only up to approximately 250 Hz. (above this frequency, magnitudes of cross paths and direct paths are similar, what confirms higher vibrational couplings);
- for the paths between actuators and outer microphones, cross paths and direct paths represent a similar magnitude in the whole frequency range considered;
- observed natural frequencies and mode shapes of the whole structure are a consequence of superposition of resonances of each wall excited individually (but as a part of the structure)—therefore, it is justified to analyse the walls separately for the purpose of optimization of actuators locations, considering only eigenmodes due to the given wall.

Finally, the real device casing has been considered. Analysis of its secondary paths and frequency response functions leads to the following conclusions:

- in contrast to the previously considered laboratory casings, the real device casing is very irregular and inhomogeneous, making the mathematical modelling significantly more difficult;
- the strong couplings between walls, similar to the light-weight casing, require employment of multichannel control system for the whole casing in order to be successful, however, for the purpose of optimization of actuators locations it is admissible to analyse the walls separately.

3 Modelling of casings

3.1 INTRODUCTION

Formulation of a mathematical model of the considered device casings brings a set of benefits. First, if the model is properly validated, it can be used for simulation and analysis of the vibrating casing dynamical behaviour what facilitates a better understanding of the structure. It can be also used for preliminary evaluations of control strategies. Moreover, the model can be used for optimization of the structure—to shape its frequency response to improve its passive properties or to maximize the susceptibility of the plant to the active or semi-active control. Also, it can be used to efficiently apply actuators and sensors, so the maximum advantage could be taken of them. All of the aforementioned benefits constitutes a significant value, what justify the undertaken effort to formulate the appropriate model.

In general, the difficulty of problems treated in science and engineering typically originates from the complexity of the systems under consideration. A system is understood as an object with distinguished input and output signals, whose properties/behaviour is to be studied. The general strategy used to understand behaviour of complex systems is to apply appropriate simplified description of the system, i.e. adequate model. More precisely, a mathematical model may be interpreted as an abstract, simplified, mathematical construct related to a part of reality and created for a particular purpose [8], e.g. to compute the time-space evolution of a physical system, assess how the system performs, make predictions about the system behaviour and quantify it. Generally, models provide adequate tools to make tractable problems related to complex systems. That is why, the adequate models place an important role in the analysis of complex systems. All mathematical models are composed of variables and a mathematical representation of the relationships between them. There are many different types of mathematical models that are used to address different aspects of systems. Mathematical models can be classified in many different ways depending on model operators, model parameters in relation to time, parameters types, parameters dimensionality, modelling goal, type of predicted outcome and many other aspects/criteria/factors.

In general, mathematical models can be classified into the following categories:

Empirical (input-output models) or physical (process based models)

Model is referred to as a physical, if it is based on a priori information about system it describes. In other words, physical model is based on mathematical description of phenomenon or process that is to be represented by the model. Physical models usually involve *physically interpretable parameters*, that is, parameters which represent real properties of the system what is one of their most important advantages. As an essential prerequisite, physical

DOI: 10.1201/9781003273806-3

models need a priori knowledge of the system, which is its characteristic feature. If nothing is known about the system, then empirical models have to be applied. An empirical model relies on observation rather than theory. Empirical models are focused on describing the data with the specification of very few assumptions about the data being analysed. In case of an empirical model the structure of the model is determined by the *observed* relationships among experimental data. An important advantage of empirical models is that they can be used in black box situations and that they typically require much less time and resources compared to physical models. Empirical models are universally applicable, easy to set up, but limited in scope. Physical models, in contrast, allow usually deeper insights into system performance and better predictions, but they require a priori knowledge about the system and often need more time and resources to be synthesized.

Pragmatic considerations should decide which type of model is used in practice.

Deterministic or stochastic

Division between deterministic and stochastic models is based on the type of the outcome they can predict. Deterministic models ignore random variation, and so always predict the same outcome from a given starting point. In other words, in deterministic models, the output of the model is fully determined by the parameters values and the initial conditions. Consequently, deterministic models perform the same way for a given set of initial conditions. With a deterministic model, the uncertain factors are external to the model. On the other hand, the model may be statistical in nature and so may predict the distribution of possible outcomes. Such models are said to be stochastic. The uncertain factors are built into the stochastic model. Stochastic models possess some inherent randomness. Consequently, the same set of parameter values and initial conditions will lead to an ensemble of different outputs. There are also used combined models, which are based on a deterministic basis and some uncertain behaviour is included in a stochastic part.

Discrete or continuous

In a continuous model, events can take place at every point in time. In a discrete model, events are categorized within time intervals and can take place only at discrete instants of time. Continuous models involve quantities that change continuously with time. Discrete models, on the other hand, involve quantities that change at discrete instants of time only.

In continuous models, the independent variables may assume arbitrary (typically real) values within some interval. In contrast, in discrete models the independent variables may assume some discrete values only.

Static or dynamic

A dynamic model accounts for time-dependent changes in the state of the system, while a static (or steady-state) model describes the system in equilibrium, and thus is time-invariant.

Unlike dynamic models, static models contain no time-dependent variables. It means that the static model describes the system in its state at a specific time instant. A static model is one which contains no internal history of either input values previously applied, values of internal variables, or output values. In static models relationships between outputs and inputs are typically described by algebraic equations. In this type of model, each output is dependent on some function of the inputs. With a static model, the value of output is determined only by the current value of input/inputs and reacts *instantaneously*. It is also possible for the output variables to be dependent on other output variables, although this results in a model which may not have closed-form solutions.

A dynamic model accounts for time-dependent changes in the state of the system. With dynamic model it takes some time the output to react to changes on the inputs. Dynamic models typically are described by difference equations or differential equations depending on whether they are discrete or continuous in time, respectively. If we are only interested in the equilibrium states of a dynamic system in which a given system may be, then we can restrict ourselves to a static model for such a dynamic system. In this context, static model can be considered as a steady-state of a dynamic model.

Stationary or non-stationary (instationary)

In non-stationary models, at least one of its parameters is a function of time (depends on time). Otherwise, if all model parameters are constant over time (independent of time) it is referred to as a stationary model. Non-stationary models are used to describe systems whose characteristics changes significantly in time. There are more flexible but harder to solve. Models with switched parameters can also be distinguished in this category.

Linear or non-linear

Mathematical models are usually composed by variables, which are abstractions of quantities of interest in the described systems, and operators that act on these variables, which can be algebraic operators, functions, differential operators, etc. If all the operators in a mathematical model exhibit linearity, the resulting mathematical model is referred to as a linear. Otherwise, a model is considered to be nonlinear. Non-linear models are the models where the operator uses non-linear dependencies between outputs and inputs. They are more difficult for analysis and do not have the superposition property when compared to linear models. From the analysis point of view it is important to distinguish two types of linearities, i.e.: model linear in its inputs and model linear in its parameters. Nonlinear models are typically more accurate but harder to solve.

Lumped (space-independent) or distributed (space-dependent)

A lumped model is one in which the dependent variables that are used to compose the model are a function of time alone. In general, this means that mathematical representation of the relationships between variables take the form of ordinary differential equations. A distributed model is one in which at least one of dependent variables that are used to compose the model of one or more spatial variables. This means that mathematical representation of the relationships between variables take the form of partial differential equations. There exist also so called of *spatiotemporal* distributed models in which dependent variables describing model depends on both spatial variables and time.

Macroscopic or Microscopic

Macroscopic models concern with the macroscopic behaviour of the system. In that case the system is characterized by average properties such as pressure, volume, temperature, etc. Microscopic models consider the individual behaviour of small constituents of the system at molecular or individual particle level.

1D, 2D, or 3D models

Depending on their spatial dimensionality, physical systems can be described using 1, 2 or 3 space variables. The number of space variables used to describe a physical system is called its dimension (frequently denoted 1D, 2D or 3D).

Direct or inverse

With direct model assuming given inputs and model parameters the output can be determined. In contrast, if for given inputs and outputs we want to determine model parameters or assuming given model parameters and outputs we want to determine the input the inverse model should be used. If we ask for parameter values, the resulting problem is also called a parameter identification problem. If we ask for input, the resulting problem is also called a control problem, since in this case the problem is to control the input in a way that generates some desired output.

Remark 3.1 (Dichotomy of models categories)
It is important to answer the question about the dichotomy of the presented models categories. It has to be pointed out that some of the above categorizations of mathematical models overlap. For example, physical models can be lumped or distributed, stationary or non-stationary, linear or nonlinear and so on. Consequently, presented categorization is not dichotomic (disjoint).

The process of developing a mathematical model is referred to as mathematical modelling. In general, in mathematical modelling of vibroacoustic systems under discussion we follow the philosophy presented by G. Box [21]:

"All models are approximations. Assumptions, whether implied or clearly stated, are never exactly true. ***All models are wrong, but some models are useful.*** *So the question you need to ask is not "Is the model true?" (it never is) but* ***"Is the model good enough for this particular application?"***

Consequently, what we expect from developed models is that they can supply a useful approximation to reality related to the modelled vibroacoustic systems.

Developed models are used to explain vibroacoustic systems they describe, to study the effects of different components, and to make predictions about their behaviour.

Remark 3.2 (The categories of models derived)
In the context of short introduction related to mathematical models presented in this section models that are synthesized in consecutive subsections of this chapter are physical, continuous in time and space, distributed in space (space-dependent), dynamic, stationary, linear, deterministic, 3D or 2D and macroscopic.

3.2 ANALYTICAL MODELLING OF INDIVIDUAL CASING WALLS

The successful application of the mathematical model depends on its accuracy, as it should appropriately reflect real behaviour of the plant. On the other hand, to facilitate practical applicability, the model should not be complicated more than it is necessary. Therefore, usually the model constitutes a trade-off between accuracy and complexity. In the presented modelling of the device casing, there are several simplifying assumptions adopted to obtain such balance.

Most importantly, basing on the vibroacoustical analysis given in the previous chapter, each wall of the casing can be considered separately. It is an intuitive approach for the rigid casing where the separation of the walls is clearly visible. However, same approach can also be adopted in case of the light-weight or real casings, as its frequency response can be decomposed into responses of individual walls. When modelling each of the walls separately, such model can provide valuable overview of the behaviour of the casing, however, certain walls interactions are missing. If the available resources allows for a more advanced model, the casing can be modelled as a whole what is also considered in details in this chapter. Other assumptions are related to the employed theories of individual vibrating panels—the assumption of small deflections of the panels, homogeneity of the panel material, etc. They are described in more details in the following sections.

A plate is a structural element with a small thickness compared to the planar dimensions. In most cases, the thickness is no greater than one-tenth of the smaller in-plane dimension [147]. The theory of plates is an approximation of the three-dimensional elasticity theory to two dimensions, which assumes that a mid-surface plane can be used to represent deformation of every point of the plate. The aim of the plate theory is to study the deformation and stresses in plate structures subjected to loads.

In this section, fundamental issues of two-dimensional mathematical models of plates are recalled to set a reference for further reading, where the final form of a model used for analysis of the casing walls is presented. Detailed derivations of presented models can be found, e.g. in [156, 147, 145].

3.2.1 KIRCHHOFF-LOVE PLATE THEORY

The Kirchhoff-Love theory is a most elementary mathematical model of plates considered in this monograph. This theory is an extension of Euler-Bernoulli beam theory and was developed in 1888 by Love using the Kirchhoff hypothesis [94], which consists of the following assumptions:

- straight lines perpendicular to the mid-surface (i.e. transverse normals) before deformation remain straight after deformation;
- the transverse normals rotate such that they remain perpendicular to the mid-surface after deformation;
- the transverse normals do not experience elongation (i.e. the thickness of the plate does not change during a deformation).

Such assumptions are eligible for thin plates with small deflections, where shear energy is negligible.

Due to the application of plate models for modelling of casing walls, hereinafter the plate theories are considered and presented for rectangular plates (as a most common shape of device casing walls). Hence, all following mathematical models are given in Cartesian coordinates system. However, presented theories are valid for plates of any shape and they can be easily expressed in different form, more suitable for other plate shapes.

Isotropic Kirchhoff plate

For an isotropic and homogeneous plate, which occupies the $X-Y$ plane in the reference stress-free state, free vibrations are governed by a differential system:

$$D\left(\frac{\partial^4 w}{\partial x^4}+2\frac{\partial^4 w}{\partial x^2 \partial y^2}+\frac{\partial^4 w}{\partial y^4}\right)+\rho_p h\frac{\partial^2 w}{\partial t^2}=0, \tag{3.1}$$

for

$$x\in(0,a), \quad y\in(0,b), \quad t>t_0>0, \tag{3.2}$$

where

$$D=\frac{Eh^3}{12(1-\nu^2)}. \tag{3.3}$$

Initial conditions are defined by:

$$w(x,y,t_0)=0, \qquad \left.\frac{\partial w(x,y,t)}{\partial t}\right|_{t=t_0}=0. \tag{3.4}$$

In Eq. (3.1)–(3.4) the function $w(x,y,t)$ denotes the displacement of the plate from the reference state to the z-direction at time $t>0$ and position (x,y); the lengths of rectangular plate sides are assumed to be equal to a and b, respectively; D is the flexural rigidity; E is the Young's modulus; ν is the Poisson's ratio; ρ_p is the mass density of the plate material; and h is the plate thickness.

The definitions of domains of spatial variables and time given in Eq. (3.2), for the sake of brevity, are not repeated in following considerations in this section. However, they are the same for all presented plate models.

Considering only the transverse motion and neglecting the effect of rotary inertia, the kinetic and strain energies of the plate, T_p and U_p, can be written as:

$$T_p = \frac{\rho_p h}{2} \iint\limits_{S_p} \left(\frac{\partial w}{\partial t}\right)^2 dx dy, \tag{3.5a}$$

$$U_p = \frac{D}{2} \iint\limits_{S_p} \left\{ \left(\frac{\partial^2 w}{\partial x^2}\right)^2 + \left(\frac{\partial^2 w}{\partial y^2}\right)^2 + 2\nu \frac{\partial^2 w}{\partial x^2}\frac{\partial^2 w}{\partial y^2} + 2(1-\nu)\left(\frac{\partial^2 w}{\partial x \partial y}\right)^2 \right\} dx dy, \tag{3.5b}$$

where S_p is the surface of the plate. The definition of the kinetic and strain energies of the plate is particularly important in this dissertation, as the Rayleigh-Ritz method is used to find an approximate solution of the differential system (the method is based on the definition of an energy functional).

Orthotropic Kirchhoff plate

Equation (3.1) can be extended to a case of orthotropic plates, which have material properties that differ along two orthogonal axes (in contrast to isotropic materials that have the same properties in every direction). It is an important subset of anisotropic plates. An example is a sheet metal formed by squeezing thick sections of metal between heavy rollers. Its properties differ between the direction it was rolled in and each of the two transverse directions. Sheet metal is commonly used for device casings, which are of particular interest in this dissertation.

Assuming that the orthotropic plate material have axes of symmetry along x and y directions, free vibrations of such plate are governed by the equation [145]:

$$D_x \frac{\partial^4 w}{\partial x^4} + 2\left(D_x \nu_y + 2D_{xy}\right) \frac{\partial^4 w}{\partial x^2 \partial y^2} + D_y \frac{\partial^4 w}{\partial y^4} + \rho_p h \frac{\partial^2 w}{\partial t^2} = 0, \tag{3.6}$$

where

$$D_x = \frac{E_x h^3}{12(1-\nu_x \nu_y)}, \quad D_y = \frac{E_y h^3}{12(1-\nu_x \nu_y)}, \quad D_{xy} = \frac{G h^3}{12}. \tag{3.7}$$

In (3.6)–(3.7) D_x, D_y and D_{xy} are orthotropic rigidities of the plate; E_x and E_y are the Young moduli along the x and y directions, respectively; G is the shear modulus; ν_x and ν_y are the Poisson's ratios corresponding to x and y direction, respectively. Four elastic constants E_x, E_y, G, ν_x are independent (while the isotropic plate model needs only two elastic constants to be defined). The coefficient ν_y can be determined according to the following relation:

$$\frac{\nu_x}{E_x}=\frac{\nu_y}{E_y}. \tag{3.8}$$

It can be easily shown that if $E_x = E_y$ and $\nu_x = \nu_y$, then Eq. (3.6) simplifies to (3.1). It is also helpful to express the shear modulus G in terms of Young modulus E and Poisson's ratio ν as:

$$G=\frac{E}{2(1+\nu)}. \tag{3.9}$$

The kinetic energy of the orthotropic plate can be calculated according to Eq. (3.5a), as in case of an isotropic plate. However, the strain energy is expressed as:

$$U_p=\frac{1}{2}\iint_{S_p}\left\{D_x\left(\frac{\partial^2 w}{\partial x^2}\right)^2+D_y\left(\frac{\partial^2 w}{\partial y^2}\right)^2+2D_x\nu_y\frac{\partial^2 w}{\partial x^2}\frac{\partial^2 w}{\partial y^2}+4D_{xy}\left(\frac{\partial^2 w}{\partial x\partial y}\right)^2\right\}dx\,dy. \tag{3.10}$$

3.2.2 MINDLIN-REISSNER PLATE THEORY

The Mindlin-Reissner theory is an extension of Kirchhoff-Love theory that takes into account rotary inertia and shear deformations of a plate [123]. It is obtained by relaxing the Kirchhoff's normality restriction and allowing for arbitrary but constant rotation of transverse normals after deformation (it is an analogy of extension of Euler-Bernoulli beam theory to Timoshenko beam theory). The Mindlin-Reissner theory is especially suitable for thick plates (i.e. with side to thickness ratios of the order of 20 or less), since Kirchhoff-Love theory tends to underpredict deflections and overpredict natural frequencies of such plates.

Isotropic Mindlin plate

Assuming that the functions $\Theta_x(x,y,t)$ and $\Theta_y(x,y,t)$ denote the rotations of a transverse normal at position (x,y) in the x and y directions, respectively, free vibrations the isotropic Mindlin plate are governed by the equations:.

$$\frac{\rho_p h^3}{12}\frac{\partial^2\Theta_x}{\partial t^2}-\frac{D}{2}\left[(1-\nu)\left(\frac{\partial^2\Theta_x}{\partial x^2}+\frac{\partial^2\Theta_x}{\partial y^2}\right)+(1+\nu)\left(\frac{\partial^2\Theta_x}{\partial x^2}+\frac{\partial^2\Theta_y}{\partial x\partial y}\right)\right]+\kappa hG\left(\frac{\partial w}{\partial x}+\Theta_x\right)=0, \tag{3.11a}$$

$$\frac{\rho_p h^3}{12}\frac{\partial^2\Theta_y}{\partial t^2}-\frac{D}{2}\left[(1-\nu)\left(\frac{\partial^2\Theta_y}{\partial x^2}+\frac{\partial^2\Theta_y}{\partial y^2}\right)+(1+\nu)\left(\frac{\partial^2\Theta_y}{\partial y^2}+\frac{\partial^2\Theta_x}{\partial x\partial y}\right)\right]+\kappa hG\left(\frac{\partial w}{\partial y}+\Theta_y\right)=0, \tag{3.11b}$$

$$\rho_p h \frac{\partial^2 w}{\partial t^2} - \kappa h G \left(\frac{\partial^2 w}{\partial x^2} + \frac{\partial^2 w}{\partial y^2} + \frac{\partial \Theta_x}{\partial x} + \frac{\partial \Theta_y}{\partial y} \right) = 0, \tag{3.11c}$$

where κ is the shear coefficient. Initial conditions are defined by:

$$w(x,y,t_0) = 0, \qquad \left. \frac{\partial w(x,y,t)}{\partial t} \right|_{t=t_0} = 0, \tag{3.12a}$$

$$\Theta_x(x,y,t_0) = 0, \qquad \left. \frac{\partial \Theta_x(x,y,t)}{\partial t} \right|_{t=t_0} = 0, \tag{3.12b}$$

$$\Theta_y(x,y,t_0) = 0, \qquad \left. \frac{\partial \Theta_y(x,y,t)}{\partial t} \right|_{t=t_0} = 0. \tag{3.12c}$$

Kinetic and strain energies of the plate, T_p and U_p, can be calculated as:

$$T_p = \frac{\rho_p h}{2} \iint_{S_p} \left\{ \frac{h^2}{12} \left[\left(\frac{\partial \Theta_x}{\partial t} \right)^2 + \left(\frac{\partial \Theta_y}{\partial t} \right)^2 \right] + \left(\frac{\partial w}{\partial t} \right)^2 \right\} dx\,dy, \tag{3.13a}$$

$$U_p = \frac{1}{2} \iint_{S_p} \left\{ D \left[\left(\frac{\partial \Theta_x}{\partial x} \right)^2 + \left(\frac{\partial \Theta_y}{\partial y} \right)^2 + 2\nu \frac{\partial \Theta_x}{\partial x} \frac{\partial \Theta_y}{\partial y} + \frac{1}{2}(1-\nu) \left(\frac{\partial \Theta_x}{\partial y} + \frac{\partial \Theta_y}{\partial x} \right)^2 \right] \right.$$
$$\left. + \kappa h G \left[\left(\frac{\partial w}{\partial x} + \Theta_x \right)^2 + \left(\frac{\partial w}{\partial y} + \Theta_y \right)^2 \right] \right\} dx\,dy. \tag{3.13b}$$

It is noteworthy that by setting $\Theta_x = -\frac{\partial w}{\partial x}$, $\Theta_y = -\frac{\partial w}{\partial y}$, and neglecting the effect of rotary inertia in (3.13a) and (3.13b), the energy definitions (3.5a) and (3.5b) for Kirchhoff plate can be obtained.

Orthotropic Mindlin plate

Assuming that the orthotropic Mindlin plate material has axes of symmetry along x and y directions (analogously like in case of an orthotropic Kirchhoff plate), free vibrations of such plate are governed by the equations:

$$\frac{\rho_p h^3}{12} \frac{\partial^2 \Theta_x}{\partial t^2} - D_x \left(\frac{\partial^2 \Theta_x}{\partial x^2} + \nu_y \frac{\partial^2 \Theta_y}{\partial x \partial y} \right) - D_{xy} \left(\frac{\partial^2 \Theta_x}{\partial y^2} + \frac{\partial^2 \Theta_y}{\partial x \partial y} \right) + \kappa_x h G_{xz} \left(\frac{\partial w}{\partial x} + \Theta_x \right) = 0, \tag{3.14a}$$

$$\frac{\rho_p h^3}{12} \frac{\partial^2 \Theta_y}{\partial t^2} - D_y \left(\frac{\partial^2 \Theta_y}{\partial y^2} + \nu_x \frac{\partial^2 \Theta_x}{\partial x \partial y} \right) - D_{xy} \left(\frac{\partial^2 \Theta_y}{\partial x^2} + \frac{\partial^2 \Theta_x}{\partial x \partial y} \right) + \kappa_y h G_{yz} \left(\frac{\partial w}{\partial y} + \Theta_y \right) = 0, \tag{3.14b}$$

$$\rho_p h \frac{\partial^2 w}{\partial t^2} - \kappa_x h G_{xz} \left(\frac{\partial^2 w}{\partial x^2} + \frac{\partial \Theta_x}{\partial x} \right) - \kappa_y h G_{yz} \left(\frac{\partial^2 w}{\partial y^2} + \frac{\partial \Theta_y}{\partial y} \right) = 0, \tag{3.14c}$$

where

$$D_x = \frac{E_x h^3}{12(1-\nu_x\nu_y)}, \quad D_y = \frac{E_y h^3}{12(1-\nu_x\nu_y)}, \quad D_{xy} = \frac{G_{xy} h^3}{12}. \tag{3.15}$$

In Eqs. (3.14a)–(3.15), G_{xy}, G_{xz} and G_{yz} are the shear moduli in $X-Y, X-Z$ and $Y-Z$ planes, respectively; κ_x and κ_y are the shear coefficients in the x and y directions, respectively. It can be easily shown, similarly as in case of Kirchhoff plate, that if given elastic constants corresponding to different directions become equal, then Eqs. (3.14) simplifies to (3.11).

The kinetic energy is calculated in the same manner as for isotropic Mindlin plate, given in Eq. (3.13a). However, the strain energy U_p is expressed by:

$$U_p = \frac{1}{2}\iint\limits_{S_p} \left\{ D_x \left(\frac{\partial \Theta_x}{\partial x}\right)^2 + D_y \left(\frac{\partial \Theta_y}{\partial y}\right)^2 + (\nu_y D_x + \nu_x D_y) \frac{\partial \Theta_x}{\partial x}\frac{\partial \Theta_y}{\partial y} + D_{xy}\left(\frac{\partial \Theta_x}{\partial y} + \frac{\partial \Theta_y}{\partial x}\right)^2 \right.$$
$$\left. + \kappa_x h G_{xz}\left(\frac{\partial w}{\partial x} + \Theta_x\right)^2 + \kappa_y h G_{yz}\left(\frac{\partial w}{\partial y} + \Theta_y\right)^2 \right\} dx dy. \tag{3.16}$$

As the most general of the considered models, the orthotropic Mindlin plate model is used in the following considerations and analysis. Therefore, it is helpful to express the kinetic and strain energies of the plate, given in Eqs. (3.13a) and (3.16), in non-dimensional coordinates, which facilitate the solution of the differential system. Introducing non-dimensional coordinates $\xi = \frac{x}{a}$, $\eta = \frac{y}{b}$, and a non-dimensional parameter $\alpha_p = \frac{a}{b}$, the kinetic and strain energies of the plate can be calculated as:

$$T_p = \frac{\rho_p h a b}{2} \int_0^1\int_0^1 \left\{ \frac{h^2}{12}\left[\left(\frac{\partial \Theta_x}{\partial t}\right)^2 + \left(\frac{\partial \Theta_y}{\partial t}\right)^2\right] + \left(\frac{\partial w}{\partial t}\right)^2 \right\} d\xi d\eta\,, \tag{3.17a}$$

$$U_p = \frac{1}{2\alpha_p} \int_0^1\int_0^1 \left\{ D_x \left(\frac{\partial \Theta_x}{\partial \xi}\right)^2 + \alpha_p^2 D_y \left(\frac{\partial \Theta_y}{\partial \eta}\right)^2 + \alpha_p (\nu_y D_x + \nu_x D_y) \frac{\partial \Theta_x}{\partial \xi}\frac{\partial \Theta_y}{\partial \eta} \right.$$
$$\left. + D_{xy}\left(\alpha_p \frac{\partial \Theta_x}{\partial \eta} + \frac{\partial \Theta_y}{\partial \xi}\right)^2 + \kappa_x h G_{xz}\left(\frac{\partial w}{\partial \xi} + a\Theta_x\right)^2 + \alpha_p^2 \kappa_y h G_{yz}\left(\frac{\partial w}{\partial \eta} + b\Theta_y\right)^2 \right\} d\xi d\eta. \tag{3.17b}$$

BOUNDARY CONDITIONS

Type of mounting of the plate edges affects strongly its behaviour (both natural frequencies and mode shapes). Therefore, the boundary conditions should be always carefully considered and appropriately modelled. In the literature, classical boundary conditions have been thoroughly evaluated—simply-supported, fully-clamped or free edges (e.g. in [86]). However, the actual boundary conditions of a real system

Table 3.1

Summary of different boundary conditions, depending on spring constants, at the exemplary plate edge $x = 0$. The dashed line represents the reference state of the plate [164]. Reprinted from "Modelling and control of device casing vibrations for active reduction of acoustic noise", PhD thesis, Silesian University of Technology, S. Wrona.

<table>
<tr><th></th><th>$k_{tx0} = 0$</th><th>$0 < k_{tx0} < \infty$</th><th>$k_{tx0} = \infty$</th></tr>
<tr><td>$k_{rx0} = 0$</td>
<td>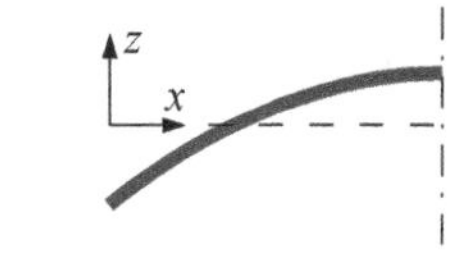
free edge
$\frac{\partial^2 \Theta_x}{\partial x^2} = 0,$
$\frac{\partial \Theta_x}{\partial x} = 0.$</td>
<td>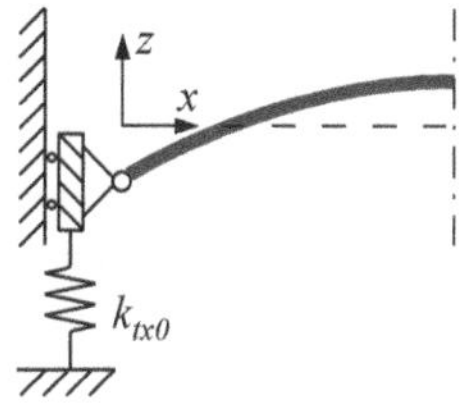
$k_{tx0} w = -D_x \frac{\partial^2 \Theta_x}{\partial x^2},$
$\frac{\partial \Theta_x}{\partial x} = 0.$</td>
<td>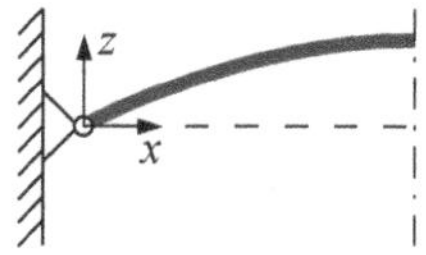
simply-supported edge
$w = 0,$
$\frac{\partial \Theta_x}{\partial x} = 0.$</td></tr>
<tr><td>$0 < k_{rx0} < \infty$</td>
<td>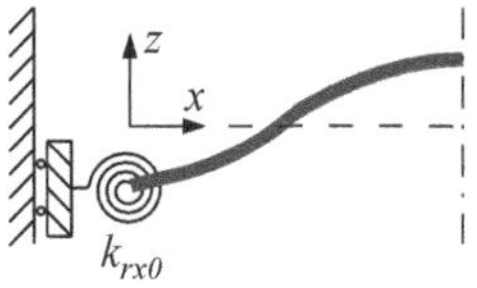
$\frac{\partial^2 \Theta_x}{\partial x^2} = 0,$
$k_{rx0} \Theta_x = D_x \frac{\partial \Theta_x}{\partial x}.$</td>
<td>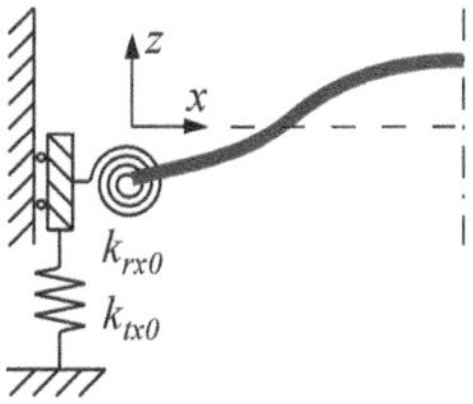
$k_{tx0} w = -D_x \frac{\partial^2 \Theta_x}{\partial x^2},$
$k_{rx0} \Theta_x = D_x \frac{\partial \Theta_x}{\partial x}.$</td>
<td>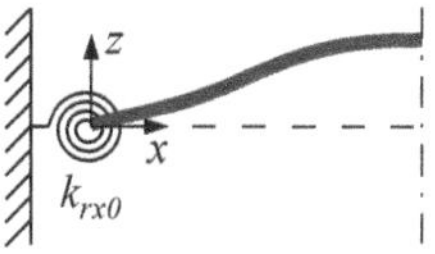
$w = 0,$
$k_{rx0} \Theta_x = D_x \frac{\partial \Theta_x}{\partial x}.$</td></tr>
<tr><td>$k_{rx0} = \infty$</td>
<td>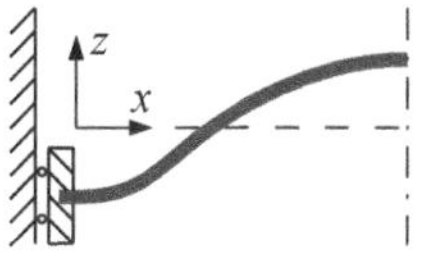
$\frac{\partial^2 \Theta_x}{\partial x^2} = 0,$
$\Theta_x = 0.$</td>
<td>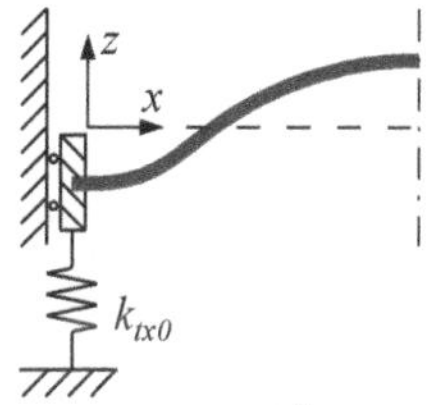
$k_{tx0} w = -D_x \frac{\partial^2 \Theta_x}{\partial x^2},$
$\Theta_x = 0.$</td>
<td>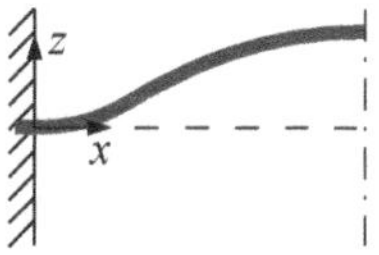
fully clamped edge
$w = 0,$
$\Theta_x = 0.$</td></tr>
</table>

are mostly not classical, but somewhere in between these conditions. In this dissertation, a general formulation of boundary conditions is employed—plate edges are elastically restrained against both translation and rotation. Such mounting is represented by translational and rotational springs distributed linearly along plate edges and defined by spring constants (it corresponds to a majority of practical application of plates, including casing walls). It can be easily showed that the classical boundary conditions of the plate can be obtained as limiting cases when the spring constants approach their natural limits of zero or infinity (it is schematically presented in Table 3.1). Detailed considerations and derivation of elastic boundary conditions can be found, e.g. in [145].

For the sake of brevity, necessary definitions of the boundary conditions are presented strictly for the purpose of application of the Rayleigh-Ritz method. In this method, appropriate admissible functions, which satisfy the geometric boundary conditions have to be chosen. These functions, in fact, determine, which geometry of the plate is considered, and what boundary conditions are adopted (together with an appropriate energy functional). To obtain the admissible functions, in this dissertation products of characteristic orthogonal polynomials having the properties of Timoshenko beam functions are used (a rectangular plate is considered, hence there are two beams assumed, corresponding to x and y directions). Therefore, equations representing the boundary conditions are presented as for the Timoshenko beam, but in fact, an edge of the beam represents homogeneous boundary conditions of a corresponding edge of the plate.

To describe in such manner boundary conditions of four plate edges, eight spring constants needs to be defined, as shown in Fig. 3.1. Translational spring constants in the x direction at $x = 0$ and $x = a$, and in the y direction at $y = 0$ and $y = b$ are designated as k_{tx0}, k_{tx1}, k_{ty0} and k_{ty1}, respectively. Rotational spring constants for plate edges given in the same manner as above, are noted as k_{rx0}, k_{rx1}, k_{ry0} and k_{ry1}, respectively. With such notation, the boundary conditions for elastically restrained edges are:

$$k_{tx0} w = -D_x \frac{\partial^2 \Theta_x}{\partial x^2}, \qquad k_{rx0} \Theta_x = D_x \frac{\partial \Theta_x}{\partial x} \qquad \text{at} \quad x = 0, \tag{3.18a}$$

$$k_{tx1} w = D_x \frac{\partial^2 \Theta_x}{\partial x^2}, \qquad k_{rx1} \Theta_x = -D_x \frac{\partial \Theta_x}{\partial x} \qquad \text{at} \quad x = a, \tag{3.18b}$$

$$k_{ty0} w = -D_y \frac{\partial^2 \Theta_y}{\partial y^2}, \qquad k_{ry0} \Theta_y = D_y \frac{\partial \Theta_y}{\partial y} \qquad \text{at} \quad y = 0, \tag{3.18c}$$

$$k_{ty1} w = D_y \frac{\partial^2 \Theta_y}{\partial y^2}, \qquad k_{ry1} \Theta_y = -D_y \frac{\partial \Theta_y}{\partial y} \qquad \text{at} \quad y = b. \tag{3.18d}$$

A summary of the equations representing boundary conditions depending on spring constants (at the exemplary plate edge $x = 0$) are given in Table 3.1.

As needed in the Rayleigh-Ritz method, the strain energy stored in translational and rotational springs, U_b, utilizing the non-dimensional coordinates (ξ, η), is given by:

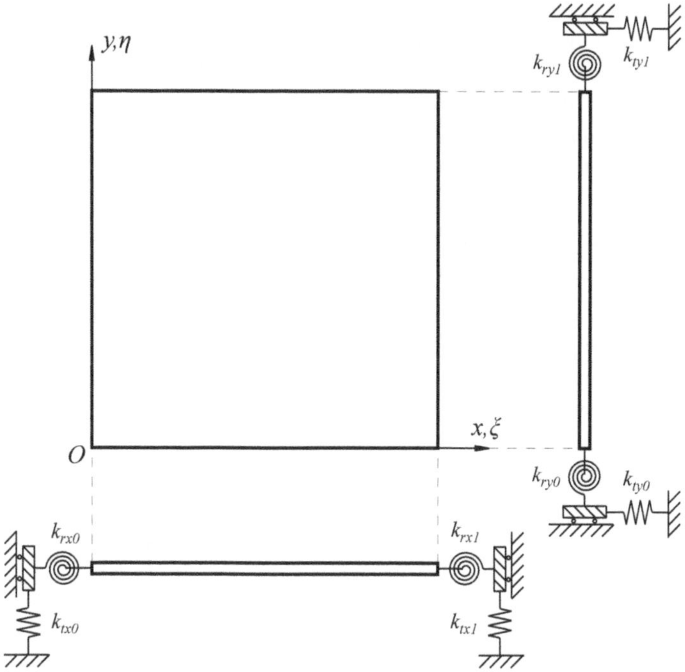

Figure 3.1 A multiview orthographic projection of the rectangular plate with boundary conditions represented as rotational and translational springs [164]. Reprinted from "Modelling and control of device casing vibrations for active reduction of acoustic noise", PhD thesis, Silesian University of Technology, S. Wrona.

$$U_b = \frac{1}{2\alpha_p}\left[\int_0^1 a\left\{\left(k_{tx0}w^2 + k_{rx0}\Theta_x^2\right)\Big|_{\xi=0} + \left(k_{tx1}w^2 + k_{rx1}\Theta_x^2\right)\Big|_{\xi=1}\right\}d\eta \right.$$
$$\left. + \alpha_p^2 \int_0^1 b\left\{\left(k_{ty0}w^2 + k_{ry0}\Theta_y^2\right)\Big|_{\eta=0} + \left(k_{ty1}w^2 + k_{ry1}\Theta_y^2\right)\Big|_{\eta=1}\right\}d\xi\right]. \tag{3.19}$$

ADDITIONAL ELEMENTS

The casing walls considered in this dissertation are a subject of both passive and active control. For this purpose, different kinds of element are bonded to the walls surface—actuators, sensors, additional masses and ribs (stiffeners). They have strong impact on the plate dynamical behaviour and have to be included in the mathematical modelling, so the model remains valid after the elements are attached to the plate surface.

In this section, mathematical modelling of additional masses and ribs bonded to plate surface is considered. Presence of sensors and actuators is also modelled as additional masses. Modelling of ribs includes the Engesser theory associated with the consideration of torsion [46]. The idea is presented schematically in Fig. 3.2.

For the sake of brevity, mathematical modelling of additional elements is presented apart from differential system of the vibrating plate, defining only the kinetic and potential energy related to the elements (as they are most important for the Rayleigh-Ritz method used to solve the differential system).

Energy related to additional masses

Elements represented by additional masses in this dissertation (actuators, sensors and passive masses), can be for simplification (not limiting the analysis) considered of small size, compared to the dimensions of the plate. Therefore, an influence of strain caused by these elements bonded to the plate surface is neglected. Moreover, assuming perfect bonding, the total energy introduced to the system by the additional masses is considered to be the kinetic energy, expressed in the non-dimensional coordinates (ξ, η) as:

$$\begin{aligned} T_m = & \sum_{i=0}^{N_a} \frac{1}{2} \left\{ m_{a,i} \left(\frac{\partial w}{\partial t} \right)^2 + I_{ax,i} \left(\frac{\partial \Theta_x}{\partial t} \right)^2 + I_{ay,i} \left(\frac{\partial \Theta_y}{\partial t} \right)^2 \right\} \Bigg|_{\substack{\xi=\xi_{a,i} \\ \eta=\eta_{a,i}}} \\ & + \sum_{i=0}^{N_s} \frac{1}{2} \left\{ m_{s,i} \left(\frac{\partial w}{\partial t} \right)^2 + I_{sx,i} \left(\frac{\partial \Theta_x}{\partial t} \right)^2 + I_{sy,i} \left(\frac{\partial \Theta_y}{\partial t} \right)^2 \right\} \Bigg|_{\substack{\xi=\xi_{s,i} \\ \eta=\eta_{s,i}}} \\ & + \sum_{i=0}^{N_m} \frac{1}{2} \left\{ m_{m,i} \left(\frac{\partial w}{\partial t} \right)^2 + I_{mx,i} \left(\frac{\partial \Theta_x}{\partial t} \right)^2 + I_{my,i} \left(\frac{\partial \Theta_y}{\partial t} \right)^2 \right\} \Bigg|_{\substack{\xi=\xi_{m,i} \\ \eta=\eta_{m,i}}}, \end{aligned} \tag{3.20}$$

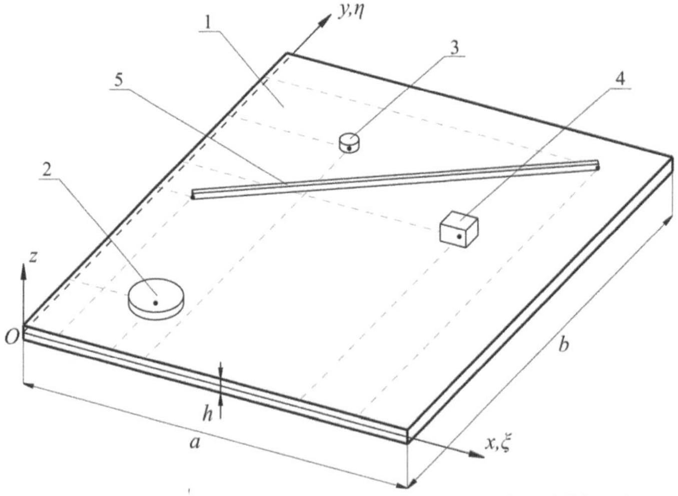

Figure 3.2 Rectangular plate (1) with actuators (2), sensors (3), additional masses (4) and ribs (5) bonded to its surface—a visualization in an isometric projection [164]. Adapted from "Modelling and control of device casing vibrations for active reduction of acoustic noise", PhD thesis, Silesian University of Technology, S. Wrona.

where N_Γ, $m_{\Gamma,i}$, $I_{\Gamma x,i}$, $I_{\Gamma y,i}$, $\xi_{\Gamma,i}$ and $\eta_{\Gamma,i}$ (depending of the kind of elements—Γ stand for: a for actuators, s for sensors and m for passive masses) are the number of elements bonded to the plate surface, mass of the ith element, moments of inertia of the ith element with respect to x and y directions, and coordinates of the ith element, respectively.

Energy related to ribs

Assuming N_r stiffening ribs mounted on the plate surface, the total energy functionals related to the ribs can be expressed as:

$$T_r = \sum_{i=0}^{N_r} T_{r,i}, \tag{3.21a}$$

$$U_r = \sum_{i=0}^{N_r} U_{r,i}, \tag{3.21b}$$

where $T_{r,i}$ and $U_{r,i}$ are kinetic and potential energy of the ith rib, respectively.

In the present considerations, the ribs can be mounted on the plate surface with any orientation, defined for the ith rib by angle α_i. Therefore, a local coordinates $(\tilde{x}_{r,i}, \tilde{y}_{r,i})$ system has to be introduced (shown in Fig. 3.3), in which the energy func-

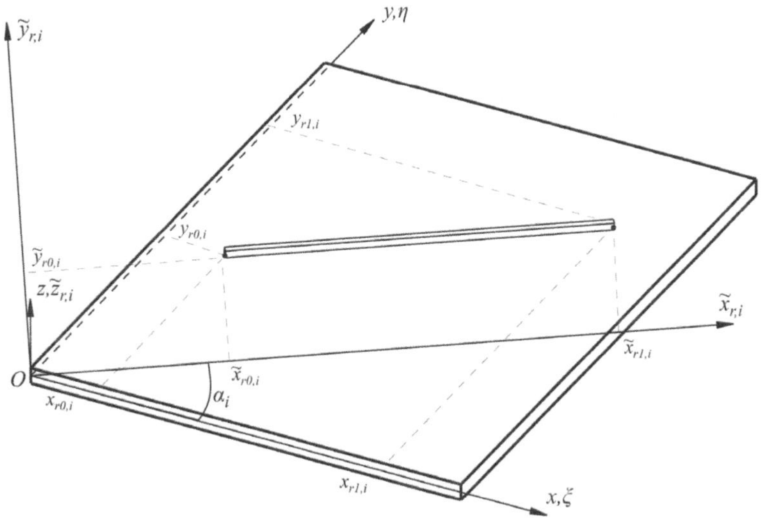

Figure 3.3 Rectangular plate with the ith rib bonded to the plate surface, utilizing global and local coordinates systems [164]. Adapted from "Modelling and control of device casing vibrations for active reduction of acoustic noise", PhD thesis, Silesian University of Technology, S. Wrona.

tionals of the ith rib are expressed as:

$$T_{r,i}=\frac{1}{2}A_{r,i}\rho_{r,i}\int_{\tilde{x}_{r0,i}}^{\tilde{x}_{r1,i}}\left\{\left(\frac{\partial w}{\partial t}\right)^2+k_{r,i}\left[\left(\frac{\partial\Theta_{\tilde{x}_{r,i}}}{\partial t}\right)^2+\left(\frac{\partial\Theta_{\tilde{y}_{r,i}}}{\partial t}\right)^2\right]\right\}\Bigg|_{\tilde{y}_{r,i}=\tilde{y}_{r0,i}}d\tilde{x}_{r,i}, \qquad (3.22a)$$

$$U_{r,i}=\frac{1}{2}\int_{\tilde{x}_{r0,i}}^{\tilde{x}_{r1,i}}\left\{E_{r,i}I_{r,i}\left(\frac{\partial\Theta_{\tilde{x}_{r,i}}}{\partial\tilde{x}_{r,i}}\right)^2+\frac{G_{r,i}A_{r,i}}{\beta_{r,i}}\left(\Theta_{\tilde{x}_{r,i}}+\frac{\partial w}{\partial\tilde{x}_{r,i}}\right)+G_{r,i}J_{r,i}\left(\frac{\partial\Theta_{\tilde{y}_{r,i}}}{\partial\tilde{x}_{r,i}}\right)^2\right\}\Bigg|_{\tilde{y}_{r,i}=\tilde{y}_{r0,i}}d\tilde{x}_{r,i}, \qquad (3.22b)$$

where $\tilde{x}_{r0,i}$ and $\tilde{x}_{r1,i}$ are the coordinates of the start and end points of the ith rib, assuming that the ribs is mounted along the $\tilde{x}_{r,i}$ direction; $\tilde{y}_{r0,i}$ is the location of the ith rib along the $\tilde{y}_{r,i}$ direction; the functions $\Theta_{\tilde{x}_{r,i}}$ and $\Theta_{\tilde{y}_{r,i}}$ denote the cross-sectional rotations of the plate mid-plane in the $\tilde{x}_{r,i}$ and $\tilde{y}_{r,i}$ directions, respectively; $E_{r,i}$ is a Young modulus; $I_{r,i}$ is a second moment of inertia about the plate mid-plane; $G_{r,i}$ is a shear modulus; $A_{r,i}$ is a cross-sectional area; $\beta_{r,i}$ is a shape factor; $J_{r,i}$ is a torsional constant; $\rho_{r,i}$ is a mass density; and $k_{r,i}$ is a radius of gyration of the ith rib.

The relation between the global (x,y) and the local $(\tilde{x}_{r,i},\tilde{y}_{r,i})$ coordinate systems can be expressed as:

$$x=\tilde{x}_{r,i}\cos\alpha_i-\tilde{y}_{r,i}\sin\alpha_i, \qquad (3.23a)$$

$$y=\tilde{x}_{r,i}\sin\alpha_i+\tilde{y}_{r,i}\cos\alpha_i. \qquad (3.23b)$$

The partial derivatives are related by:

$$\frac{\partial}{\partial\tilde{x}_{r,i}}=\cos\alpha_i\frac{\partial}{\partial x}+\sin\alpha_i\frac{\partial}{\partial y}. \qquad (3.24)$$

The relationship between the rotations of the plate mid-plane in global coordinates and in local coordinates may be derived as:

$$\Theta_{\tilde{x}_{r,i}}=\Theta_x\cos\alpha_i+\Theta_y\sin\alpha_i, \qquad (3.25a)$$

$$\Theta_{\tilde{y}_{r,i}}=-\Theta_x\sin\alpha_i+\Theta_y\cos\alpha_i. \qquad (3.25b)$$

Considering (3.23)–(3.25), the energy functionals (3.22) for the ith rib can be transformed from the local coordinate system into the global coordinate system as below:

$$T_{r,i}=\frac{1}{2\cos\alpha_i}A_{r,i}\rho_{r,i}\int_{x_{r0,i}}^{x_{r1,i}}\left\{\left(\frac{\partial w}{\partial t}\right)^2+k_{r,i}\left[\left(\frac{\partial\Theta_x}{\partial t}\right)^2+\left(\frac{\partial\Theta_y}{\partial t}\right)^2\right]\right\}\Bigg|_{y=f_{r,i}(x)}dx, \qquad (3.26a)$$

$$U_{r,i} = \frac{1}{2\cos\alpha_i} \int_{x_{r0,i}}^{x_{r1,i}} \Bigg\{ E_{r,i} I_{r,i} \left[\cos^2\alpha_i \frac{\partial \Theta_x}{\partial x} + \sin\alpha_i \cos\alpha_i \left(\frac{\partial \Theta_x}{\partial y} + \frac{\partial \Theta_y}{\partial x} \right) + \sin^2\alpha_i \frac{\partial \Theta_y}{\partial y} \right]^2$$
$$+ \frac{G_{r,i} A_{r,i}}{\beta_{r,i}} \left[\cos\alpha_i \left(\Theta_x + \frac{\partial w}{\partial x} \right) + \sin\alpha_i \left(\Theta_y + \frac{\partial w}{\partial y} \right) \right]^2$$
$$+ G_{r,i} J_{r,i} \left[\cos^2\alpha_i \frac{\partial \Theta_y}{\partial x} - \sin\alpha_i \cos\alpha_i \left(\frac{\partial \Theta_x}{\partial x} - \frac{\partial \Theta_y}{\partial y} \right) - \sin^2\alpha_i \frac{\partial \Theta_x}{\partial y} \right]^2 \Bigg\} \Bigg|_{y=f_{r,i}(x)} dx. \tag{3.26a}$$

Then, utilizing the non-dimensional coordinates (ξ, η), (3.26) can be expressed as:

$$T_{r,i} = \frac{a}{2\cos\alpha_i} A_{r,i} \rho_{r,i} \int_{\xi_{r0,i}}^{\xi_{r1,i}} \left\{ \left(\frac{\partial w}{\partial t} \right)^2 + k_{r,i} \left[\left(\frac{\partial \Theta_x}{\partial t} \right)^2 + \left(\frac{\partial \Theta_y}{\partial t} \right)^2 \right] \right\} \Bigg|_{\eta=g_{r,i}(\xi)} d\xi, \tag{3.27a}$$

$$U_{r,i} = \frac{a}{2\cos\alpha_i} \int_{\xi_{r0,i}}^{\xi_{r1,i}} \Bigg\{ E_{r,i} I_{r,i} \left[\frac{\cos^2\alpha_i}{a} \frac{\partial \Theta_x}{\partial \xi} + \sin\alpha_i \cos\alpha_i \left(\frac{1}{b} \frac{\partial \Theta_x}{\partial \eta} + \frac{1}{a} \frac{\partial \Theta_y}{\partial \xi} \right) + \frac{\sin^2\alpha_i}{b} \frac{\partial \Theta_y}{\partial \eta} \right]^2$$
$$+ \frac{G_{r,i} A_{r,i}}{\beta_{r,i}} \left[\cos\alpha_i \left(\Theta_x + \frac{1}{a} \frac{\partial w}{\partial \xi} \right) + \sin\alpha_i \left(\Theta_y + \frac{1}{b} \frac{\partial w}{\partial \eta} \right) \right]^2$$
$$+ G_{r,i} J_{r,i} \left[\frac{\cos^2\alpha_i}{a} \frac{\partial \Theta_y}{\partial \xi} - \sin\alpha_i \cos\alpha_i \left(\frac{1}{a} \frac{\partial \Theta_x}{\partial \xi} - \frac{1}{b} \frac{\partial \Theta_y}{\partial \eta} \right) - \frac{\sin^2\alpha_i}{b} \frac{\partial \Theta_x}{\partial \eta} \right]^2 \Bigg\} \Bigg|_{\eta=g_{r,i}(\xi)} d\xi. \tag{3.27b}$$

TOTAL ENERGY FUNCTIONAL

In the previous three sections of this chapter, the mathematical model of the vibrating plate, elastic boundary conditions and additional elements bonded to the plate surface is presented. To employ the Rayleigh-Ritz method, the total energy functional has to be defined. Summarizing, the maximum strain energy of the system, U, can be expressed as:

$$U = U_p + U_b + U_r, \tag{3.28}$$

where U_p, U_b and U_r are potential energies corresponding to plate strain, boundary restrains and ribs, respectively.

The maximum kinetic energy of the system, T, can be defined as:

$$T = T_p + T_m + T_r, \tag{3.29}$$

where T_p, T_m and T_r are kinetic energies of the plate, additional masses and ribs, respectively.

THE RAYLEIGH-RITZ METHOD

The Rayleigh-Ritz method is used to calculate approximate solution of the presented differential system, obtaining natural frequencies and mode shapes of the system. To utilize this method, the total energy of the system needs to be defined (which is derived in the previous sections) and appropriate admissible functions, which satisfy the geometric boundary conditions have to be chosen. More detailed information regarding the Rayleigh-Ritz method is provided in [180].

For free vibration of the plate, the solution of w, Θ_x and Θ_y can be expressed in a required form using a set of appropriate trial functions:

$$w(\xi,\eta,t) = \sum_{i=1}^{N} \phi_i(\xi,\eta) q_i(t) = \phi^{\mathrm{T}}\mathbf{q}, \tag{3.30a}$$

$$\Theta_x(\xi,\eta,t) = \sum_{i=1}^{N} \psi_{x,i}(\xi,\eta) p_{x,i}(t) = \psi_x^{\mathrm{T}}\mathbf{p}_x, \tag{3.30b}$$

$$\Theta_y(\xi,\eta,t) = \sum_{i=1}^{N} \psi_{y,i}(\xi,\eta) p_{y,i}(t) = \psi_y^{\mathrm{T}}\mathbf{p}_y, \tag{3.30c}$$

where $\mathbf{q}$, $\mathbf{p}_x$ and $\mathbf{p}_y$ are a generalized plate displacement vector, generalized plate rotations vectors in the x and y directions, respectively; ϕ, ψ_x and ψ_y are vectors, which represent a set of time-invariant trial functions—in this dissertation, characteristic orthogonal polynomials having the property of Timoshenko beam functions are used; the superscript T denotes the transpose. All of mentioned vectors are of dimension $(N \times 1)$. The procedure for forming orthogonal polynomial trial functions for Mindlin plates is described in details in [70].

Energy definition

Utilizing (3.30), the total potential and kinetic energies defined in (3.28) and (3.29) can also be written as functions of generalized plate displacement and rotations vectors $\mathbf{q}$, $\mathbf{p}_x$, $\mathbf{p}_y$, mass matrix $\mathbf{M}$ of dimension $(3N \times 3N)$ and stiffness matrix $\mathbf{K}$ of dimension $(3N \times 3N)$:

$$U = \frac{1}{2} \begin{bmatrix} \mathbf{q} \\ \mathbf{p}_x \\ \mathbf{p}_y \end{bmatrix}^{\mathrm{T}} \mathbf{K} \begin{bmatrix} \mathbf{q} \\ \mathbf{p}_x \\ \mathbf{p}_y \end{bmatrix}, \tag{3.31}$$

$$T = \frac{1}{2} \begin{bmatrix} \dot{\mathbf{q}} \\ \dot{\mathbf{p}}_x \\ \dot{\mathbf{p}}_y \end{bmatrix}^{\mathrm{T}} \mathbf{M} \begin{bmatrix} \dot{\mathbf{q}} \\ \dot{\mathbf{p}}_x \\ \dot{\mathbf{p}}_y \end{bmatrix}, \tag{3.32}$$

where matrices $\mathbf{K}$ and $\mathbf{M}$ are defined in following subsections.

Stiffness matrix

The overall stiffness matrix $\mathbf{K}$ is calculated as a sum of stiffness matrices related to different energy components:

$$\mathbf{K} = \mathbf{K}_p + \mathbf{K}_b + \mathbf{K}_r, \tag{3.33}$$

where $\mathbf{K}_p$, $\mathbf{K}_b$ and $\mathbf{K}_r$ correspond to strain energy of the plate, potential energy stored in boundary restrains and potential energy related to ribs mounted on the plate surface, respectively.

Stiffness matrices introduced in (3.33) are defined as:

$$\mathbf{K}_p = \begin{bmatrix} \mathbf{K}_{pcc} & \mathbf{K}_{pcd} & \mathbf{K}_{pce} \\ & \mathbf{K}_{pdd} & \mathbf{K}_{pde} \\ symm. & & \mathbf{K}_{pee} \end{bmatrix}, \quad \mathbf{K}_b = \begin{bmatrix} \mathbf{K}_{bcc} & \mathbf{0}_{N,N} & \mathbf{0}_{N,N} \\ & \mathbf{K}_{bdd} & \mathbf{0}_{N,N} \\ symm. & & \mathbf{K}_{bee} \end{bmatrix}, \quad \mathbf{K}_r = \begin{bmatrix} \mathbf{K}_{rcc} & \mathbf{K}_{rcd} & \mathbf{K}_{rce} \\ & \mathbf{K}_{rdd} & \mathbf{K}_{rde} \\ symm. & & \mathbf{K}_{ree} \end{bmatrix}, \tag{3.34}$$

where $\mathbf{0}_{N,N}$ denotes a zero matrix of dimension $(N \times N)$. Detailed definitions of submatrices used in (3.34) are given in Appendix A for convenience of the reader. They, in fact, can be obtained for the considered problem based on the literature [90], but the derivation is nontrivial and combines more components.

Mass matrix

The overall mass matrix $\mathbf{M}$ is calculated similarly to the stiffness matrix—as a sum of matrices related to different energy components:

$$\mathbf{M} = \mathbf{M}_p + \mathbf{M}_m + \mathbf{M}_r, \tag{3.35}$$

where $\mathbf{M}_p$, $\mathbf{M}_m$ and $\mathbf{M}_r$ correspond to kinetic energy of the plate, kinetic energy of additional masses and kinetic energy related to ribs, respectively.

Mass matrices introduced in (3.35) are defined as:

$$\mathbf{M}_p = \begin{bmatrix} \mathbf{M}_{pcc} & \mathbf{0}_{N,N} & \mathbf{0}_{N,N} \\ & \mathbf{M}_{pdd} & \mathbf{0}_{N,N} \\ symm. & & \mathbf{M}_{pee} \end{bmatrix}, \quad \mathbf{M}_m = \begin{bmatrix} \mathbf{M}_{mcc} & \mathbf{0}_{N,N} & \mathbf{0}_{N,N} \\ & \mathbf{M}_{mdd} & \mathbf{0}_{N,N} \\ symm. & & \mathbf{M}_{mee} \end{bmatrix}, \quad \mathbf{M}_r = \begin{bmatrix} \mathbf{M}_{rcc} & \mathbf{0}_{N,N} & \mathbf{0}_{N,N} \\ & \mathbf{M}_{rdd} & \mathbf{0}_{N,N} \\ symm. & & \mathbf{M}_{ree} \end{bmatrix}. \tag{3.36}$$

Detailed definitions of submatrices used in (3.36) are given in Appendix B. Similarly, as in case of Appendix A, they could be obtained using [131], but presented definitions are given in a convenient form consistent with previous derivations.

Equation of the vibrating structure and a harmonic solution

When the stiffness and mass matrices are defined, by using the Lagrange equation of the second kind, the equation of a vibrating structure can be obtained as:

$$\mathbf{M} \begin{bmatrix} \ddot{\mathbf{q}} \\ \ddot{\mathbf{p}}_x \\ \ddot{\mathbf{p}}_y \end{bmatrix} + \mathbf{K} \begin{bmatrix} \mathbf{q} \\ \mathbf{p}_x \\ \mathbf{p}_y \end{bmatrix} = \mathbf{Q}, \tag{3.37}$$

where $\mathbf{Q}$ is the vector of generalized forces of dimension $(3N \times 1)$. In this dissertation, inertial actuators are considered. Hence, for the purpose of positioning, their action can be simplified and taken into account as a force acting on a point. Therefore, the control vector $\mathbf{u}$ of dimension $(N_a \times 1)$ can be defined as:

$$\mathbf{u} = \left[F_1, F_2, ..., F_{N_a}\right]^{\mathrm{T}}, \tag{3.38}$$

where F_i is a force generated by a ith actuator. Then, the vector of generalized forces can be expressed as:

$$\mathbf{Q} = \begin{bmatrix} \begin{bmatrix} \phi\Big|_{\substack{\xi=\xi_{a,1}\\ \eta=\eta_{a,1}}}, \phi\Big|_{\substack{\xi=\xi_{a,2}\\ \eta=\eta_{a,2}}}, ..., \phi\Big|_{\substack{\xi=\xi_{a,Na}\\ \eta=\eta_{a,Na}}} \end{bmatrix} \mathbf{u} \\ \mathbf{0}_{2N,1} \end{bmatrix}, \tag{3.39}$$

The harmonic solution to (3.37) gives the eigenvector matrix $\mathbf{\Phi}$ of dimension $(3N \times 3N)$ and $(3N)$ eigenfrequencies ω_i. Replacing $\left[\ \mathbf{q},\ \mathbf{p}_x,\ \mathbf{p}_y\ \right]^{\mathrm{T}}$ by $\mathbf{\Phi v}$, and multiplying (3.37) on the left by $\mathbf{\Phi}^{\mathrm{T}}$, it gives:

$$\mathbf{\Phi}^{\mathrm{T}}\mathbf{M}\mathbf{\Phi}\ddot{\mathbf{v}} + \mathbf{\Phi}^{\mathrm{T}}\mathbf{K}\mathbf{\Phi}\mathbf{v} = \mathbf{\Phi}^{\mathrm{T}}\mathbf{Q}, \tag{3.40}$$

where $\mathbf{v}$ denotes a modal displacement vector of dimension $(3N \times 1)$:

$$\mathbf{v} = \left[v_1, v_2, ..., v_{(3N)}\right]^{\mathrm{T}}. \tag{3.41}$$

Taking advantage of the orthonormality of eigenvectors in matrix $\mathbf{\Phi}$, the modal mass matrix becomes a unit matrix $\mathbf{I}_{(3N)}$ of dimension $(3N \times 3N)$ and the corresponding modal stiffness matrix becomes a diagonal matrix Ω of $(3N)$ eigenvalues ω_i^2 [34]:

$$\mathbf{\Phi}^{\mathrm{T}}\mathbf{M}\mathbf{\Phi} = \mathbf{I}_{(3N)}, \tag{3.42}$$

$$\mathbf{\Phi}^{\mathrm{T}}\mathbf{K}\mathbf{\Phi} = \Omega = \left[\mathrm{diag}(\omega_1^2, \omega_2^2, ..., \omega_{(3N)}^2)\right]. \tag{3.43}$$

Then, by substituting (3.42) and (3.43) to (3.40), it gives:

$$\ddot{\mathbf{v}} + \Omega\mathbf{v} = \mathbf{\Phi}^{\mathrm{T}}\mathbf{Q}. \tag{3.44}$$

For a better reference to a real system behaviour this equation is extended to the following one:

$$\ddot{\mathbf{v}} + \Xi\dot{\mathbf{v}} + \Omega\mathbf{v} = \mathbf{\Phi}^{\mathrm{T}}\mathbf{Q}, \tag{3.45}$$

where $\Xi\dot{\mathbf{v}}$ is a term introduced to include the damping in the system, and Ξ is a diagonal matrix of dimension $(3N \times 3N)$ defined as:

$$\Xi = \left[\mathrm{diag}(2\xi_{d,1}\omega_1, 2\xi_{d,2}\omega_2, ..., 2\xi_{d,(3N)}\omega_{(3N)})\right]. \tag{3.46}$$

In (3.46) the damping ratios, $0 < \xi_{d,i} < 1$, are calculated with use of the thermoelastic damping model for elastic plates described in details in [132]. The damping mechanism could be also included at the beginning of the modelling in the form of complex bending rigidities. However, it would substantially complicate the derivation. Introducing it at this point preserves the brevity of derivation and leads to equivalent solution. Such approach was also used in [88].

STATE SPACE MODEL

Equation (3.45) can be written in the usual state-space form:

$$\dot{\mathbf{x}} = \mathbf{A}\mathbf{x} + \mathbf{B}\mathbf{u} \tag{3.47a}$$

$$\mathbf{y} = \mathbf{C}\mathbf{x} + \mathbf{D}\mathbf{u} \tag{3.47b}$$

with the output vector $\mathbf{y}$ of dimension $(N_s \times 1)$:

$$\mathbf{y} = \left[y_1, y_2, ..., y_{N_s}\right]^{\mathrm{T}}, \tag{3.48}$$

and the state vector $\mathbf{x}$ of dimension $(6N \times 1)$:

$$\mathbf{x} = \left[\dot{v}_1, \omega_1 v_1, \dot{v}_2, \omega_2 v_2, ..., \dot{v}_{3N}, \omega_{3N} v_{3N}\right]^{\mathrm{T}}. \tag{3.49}$$

The state matrix $\mathbf{A} = \left[\mathrm{diag}(\mathbf{A}_1, \mathbf{A}_2, ..., \mathbf{A}_{3N})\right]$ of dimension $(6N \times 6N)$, is defined by:

$$\mathbf{A}_i = \begin{bmatrix} -2\xi_{d,i}\omega_i & -\omega_i \\ \omega_i & 0 \end{bmatrix}, \qquad i = 1, 2, ..., 3N. \tag{3.50}$$

Matrix $\mathbf{B}$ of dimension $(6N \times N_a)$ can be expressed as:

$$\mathbf{B} = \left[\mathrm{diag}(\mathbf{b}_1, \mathbf{b}_2, ..., \mathbf{b}_{3N})\right] \mathbf{\Phi}^{\mathrm{T}} \begin{bmatrix} \left[\phi\Big|_{\substack{\xi=\xi_{a,1} \\ \eta=\eta_{a,1}}}, \phi\Big|_{\substack{\xi=\xi_{a,2} \\ \eta=\eta_{a,2}}}, ..., \phi\Big|_{\substack{\xi=\xi_{a,N_a} \\ \eta=\eta_{a,N_a}}} \right] \\ \mathbf{0}_{2N,N_a} \end{bmatrix}, \tag{3.51}$$

where $\mathbf{b}_i = \left[1\,0\right]^{\mathrm{T}}$.

Considering accelerometers as the sensors, the output matrix $\mathbf{C}$ of dimension $(N_s \times 6N)$ is defined by:

$$\mathbf{C} = \left[\begin{matrix} \phi^{\mathrm{T}}\Big|_{\substack{\xi=\xi_{s,1} \\ \eta=\eta_{s,1}}} \\ \phi^{\mathrm{T}}\Big|_{\substack{\xi=\xi_{s,2} \\ \eta=\eta_{s,2}}} \\ \vdots \\ \phi^{\mathrm{T}}\Big|_{\substack{\xi=\xi_{s,N_s} \\ \eta=\eta_{s,N_s}}} \end{matrix}, \mathbf{0}_{N_s,2N}\right] \mathbf{\Phi}\left[\mathrm{diag}(\mathbf{c}_1, \mathbf{c}_2, ..., \mathbf{c}_{3N})\right], \tag{3.52}$$

where $\mathbf{c}_i = \left[-2\xi_{d,i}\omega_i, \; -\omega_i\right]$. Matrix $\mathbf{D}$ of dimension $(N_s \times N_a)$ can be calculated as:

$$\mathbf{D} = \begin{bmatrix} \phi^{\mathrm{T}}\Big|_{\substack{\xi=\xi_{s,1} \\ \eta=\eta_{s,1}}} \\ \phi^{\mathrm{T}}\Big|_{\substack{\xi=\xi_{s,2} \\ \eta=\eta_{s,2}}} \\ \vdots \\ \phi^{\mathrm{T}}\Big|_{\substack{\xi=\xi_{s,N_s} \\ \eta=\eta_{s,N_s}}} \end{bmatrix} \left[\phi\Big|_{\substack{\xi=\xi_{a,1} \\ \eta=\eta_{a,1}}}, \phi\Big|_{\substack{\xi=\xi_{a,2} \\ \eta=\eta_{a,2}}}, ..., \phi\Big|_{\substack{\xi=\xi_{a,N_a} \\ \eta=\eta_{a,N_a}}} \right]. \tag{3.53}$$

ACOUSTIC RADIATION

An estimate of the radiated acoustic power corresponding to the i-th vibration mode of the plate under consideration is derived in this subsection. In the scenario considered, it is assumed that the plate is placed in an infinite rigid baffle. Adopting an appropriate Green's function, derived in [146], the modal sound pressure amplitude $p_i(x,y,z)$ can be calculated as

$$p_i(x,y,z) = \frac{k_e a\, b}{4\pi^2}\rho_e c \iint\limits_{-\infty}^{+\infty} \exp\left[\imath\,(\xi x + \eta y + \gamma z)\right] M_i(\xi,\eta)\frac{\mathrm{d}\xi\,\mathrm{d}\eta}{\gamma}\,, \tag{3.54}$$

for

$$z > 0, \tag{3.55}$$

where

$$M_i(\xi,\eta) = \frac{-2\imath\,\omega_i}{ab}\iint\limits_{S_p} \Phi_i^{\mathrm{T}}\phi\exp\left[-\imath\,(\xi x + \eta y)\right]\mathrm{d}x\mathrm{d}y. \tag{3.56}$$

In Eqs. (3.54)–(3.56) the symbol $k_e = \omega_i/c$ is the acoustic wavenumber; ξ, η and γ are the components of the acoustic wavevector; ρ_e and c are the density of air and the sound speed in air, respectively; $\imath$ is the imaginary number $\sqrt{-1}$; Φ_i is the ith eigenvector (the ith column in the eigenvector matrix Φ).

To determine an estimate of the modal acoustic power P_i, the squared modal sound pressure $p_i(x,y,z)$ can be integrated over a surface S_e, which encloses the vibrating plate. Hence P_i can be expressed as

$$P_i = \iint\limits_{S_e} \frac{|p_{RMS,i}(x,y,z)|^2}{\rho_e c}\mathrm{d}S_e, \tag{3.57}$$

where $p_{RMS,i}(x,y,z) = \frac{1}{\sqrt{2}}p_i(x,y,z)$ is the root mean square of $p_i(x,y,z)$. In theoretical analyses the surface S_e is often taken to be a surface directly adjacent to the plate. However, to allow for experimental validation of the model, the surface S_e will be assumed to be a limited plane parallel to the plate and at a distance greater than zero. This may slightly modify the overall estimate of the modal acoustic power, but this effect is negligible for the objective of frequency response shaping.

CONTROLLABILITY AND OBSERVABILITY

Controllability and observability similarly as stability are fundamental concepts in mathematical control theory [7, 71, 72, 75]. These are qualitative properties of dynamical control systems and are of particular importance in control theory. Systematic study of controllability and observability was started at the beginning of sixties in the last century, when the theory of controllability and observability based mainly on the description in the form of state space for both time-invariant and time-varying different types of linear control and nonlinear control systems was worked out.

Roughly speaking, controllability generally means, that it is possible to steer dynamical control system from an arbitrary initial state to an arbitrary final state using the set of admissible controls. On the other side, observability means, that it is possible to compute uniquely state of the control system knowing initial control and output signal in a given period of time.

It should be mentioned, that in the literature there are many different definitions of controllability and observability, which strongly depend on one hand on a class of dynamical control systems and on the other hand on the form of admissible controls (see, e.g. [71, 72, 75, 153] for more details).

Controllability and observability problems for different types of dynamical systems require the application of numerous mathematical concepts and methods taken directly from differential geometry, functional analysis, topology, matrix analysis and theory of ordinary and partial differential equations and theory of difference equations. Mainly state-space models of dynamical systems are used, which provide a robust and universal method for studying controllability and observability of various classes of systems. There are various important relationships between controllability, observability, stability and stabilizability of linear both finite-dimensional and infinite-dimensional control systems. It should be pointed out, that in finite dimensional case controllability and observability are also strongly related with the theory of realization, minimal realization and canonical forms. For linear, finite dimensional, time-invariant control systems we have many canonical forms such as for example Kalman canonical form, Jordan canonical form or Luenberger canonical form. It should be mentioned, that for many mainly linear dynamical systems there exist a formal duality between the concepts of controllability and observability. Moreover, controllability is strongly connected with minimum energy control problem for many classes of linear finite dimensional, infinite dimensional dynamical systems, and delayed systems both deterministic and stochastic [71, 72, 75]. For controllable dynamical system there exist generally many different admissible controls, which steer the system from a given initial state to the final desired state at given final time. Therefore, we may look for the admissible control, which is an optimal in the sense of the quadratic performance index representing energy of control. The admissible control, which minimizes the performance index is called the minimum energy control. For infinite-dimensional dynamical systems we may introduce two general kinds of controllability, i.e. approximate (weak) controllability and exact (strong) controllability [76, 77]. However, it should be mentioned, that in the case when the semigroup associated with the dynamical system is compact or the control operator is compact, then dynamical system is never exactly controllable in infinite-dimensional state space.

In real control systems admissible controls are always constrained. Therefore, it is very important to verify constrained controllability of dynamical systems. Constrained controllability of linear both continuous-time or discrete-time control systems was considered in the series of papers [65, 66, 73, 74]. Moreover, constrained controllability of nonlinear or semilinear both continuous-time or discrete-time control systems was considered in the papers [73, 74, 152].

Controllability and observability concepts have many important applications not only in control theory and systems theory, but also in many engineering areas including vibration and acoustic control.

Controllability and observability measures for the plate

Taking advantage of the fact that the model is expressed in the state-space form, classical methods can be used to describe the controllability and observability of the system [75]. The method presented in this section represents the energy-based approach, and it is used later in the optimization process for active control scenarios.

The control energy required to reach the desired state $\mathbf{x}_{T_1}$ at time $t = T_1$, assuming the optimal solution, can be expressed as:

$$E_c = \int_0^{T_1} \mathbf{u}^{\mathrm{T}}(t)\mathbf{u}(t)\,dt = (e^{\mathbf{A}T_1}\mathbf{x}_0 - \mathbf{x}_{T_1})^{\mathrm{T}}\mathbf{W}^{-1}(T_1)(e^{\mathbf{A}T_1}\mathbf{x}_0 - \mathbf{x}_{T_1}), \tag{3.58}$$

where $\mathbf{W}(T_1)$ is the controllability Gramian matrix of dimension $(6N \times 6N)$. To minimize control required energy with respect to the actuators locations, a measure of the Gramian matrix should be maximized. It has been shown in the literature that instead of using $\mathbf{W}(T_1)$, a steady state controllability Gramian matrix, $\mathbf{W}_c$, can be used for stable systems, when time tends to infinity [6].

Analogously, the output energy received by the sensors, when the system starts in initial state $\mathbf{x}_0$ and is not controlled, can be written as:

$$E_o = \int_0^{\infty} \mathbf{y}^{\mathrm{T}}(t)\mathbf{y}(t)\,dt = \mathbf{x}_0^{\mathrm{T}}\left(\int_0^{\infty} e^{\mathbf{A}^{\mathrm{T}}t}\mathbf{C}^{\mathrm{T}}\mathbf{C}e^{\mathbf{A}^{\mathrm{T}}t}\,dt\right)\mathbf{x}_0 = \mathbf{x}_0^{\mathrm{T}}\mathbf{W}_o\mathbf{x}_0, \tag{3.59}$$

where $\mathbf{W}_o$ is the observability Gramian matrix of dimension $(6N \times 6N)$. To maximize output energy with respect to the sensors locations, again a measure of the Gramian matrix should be maximized.

Both controllability and observability Gramian matrices can be calculated by solving the Lyapunov equations:

$$\mathbf{A}\mathbf{W}_c + \mathbf{W}_c\mathbf{A}^T + \mathbf{B}\mathbf{B}^T = 0. \tag{3.60a}$$

$$\mathbf{A}^T\mathbf{W}_o + \mathbf{W}_o\mathbf{A} + \mathbf{C}^T\mathbf{C} = 0. \tag{3.60b}$$

The controllability and observability Gramian matrices are convenient to use, because if the $(2i)$th value at the diagonal of the matrix, corresponding to the ith eigenmode is small, the eigenmode is difficult to control (it can be regulated only if a large control energy is available) or is not observable well, respectively. Such information can be an important criterion in the optimization process of actuators and sensors placement.

Formally, controllability and observability are dichotomous properties, but "controllable" and "observable" does not say how high control effort is needed to reach the final state and how much energy is received by the sensors.

3.3 ANALYTICAL MODELLING OF THE WHOLE CASING

This section is devoted to the modelling study of the whole active casing. To construct the model of the casing an approach based on the decomposition of the system under discussion to the simpler subsystems is applied. Following this philosophy the process of model construction consists in combining the results of performing consecutive analysis and synthesis steps. The main idea behind first stage of model synthesis is to perform the analysis which is aimed at singling out such subsystems in the system under discussion that are simpler to analyse from the synthesis of mathematical model perspective. Then, in the next step the synthesis is performed to formulate mathematical models for all singled out in previous step subsystems. Next, once again the analysis is conducted to identify interactions between the subsystems. Finally, in the next synthesis step models of identified interactions are formulated. Modelled interactions describe cross-couplings that occur between the subsystems. Consequently, incorporating identified interactions, models of the subsystems are coupled resulting in formulation of a complete model of the whole active casing. In this context, to synthesize the model of the whole casing the following actions are performed:

- subsystems of the acoustic fields in appropriate regions of active casing are singled out and modelled,
- subsystems of vibrating flexible casing walls/panels are singled out and modelled and then,
- fluid-structure and structure-structure (in case of lightweight casing) interactions between singled out subsystems are identified and modelled.

In the synthesis of mathematical model of the whole casing the philosophy assuming that modelling aims at simplification, rather than construction of complex models of complex systems is applied [161]. This can be observed in Section 3.3.1 where many simplifying assumptions are formulated. In this context the best model is the simplest one which is still complex enough to help us understand a system and to solve the problem. Consecutive sections start with the short description of notation used and simplifying assumptions under which corresponding model is synthesized. Given assumptions determine the model usability range. *What is important and has to be pointed out, all assumptions applied to derive the models are chosen so as to relate the model to the realities of the utility.* For this reason, the dynamics of flexible casing walls is described using Kirchhoff–Love plate theory that allows for sufficient precision of the model and reduction of model complexity compared to Mindlin-Reissner plate theory. The resulting partial differential equations describing walls' panels motion are of fourth order in space variables and second order in time. The acoustic fields in appropriate regions of active casing are described by the acoustic wave equations. Consequently, derived models have a structure of a system of coupled higher order partial differential equations (PDEs). PDEs describing the model are accompanied by appropriate system of boundary conditions that reflects aspects related with mechanical construction of the casing. The model takes into account the dynamic behaviour of each casing wall/panel, coupled with the acoustic fields in the singled out regions of active casing. In this context, the model derived incorporates fluid-structure interactions of both vibroacoustic and acoustic-vibration character.

Taking into account the form of the models derived, active casing is treated as a sophisticated vibroacoustic system. The model synthesis is based on methods of continuum mechanics and acoustics, and the theory of partial differential equations.

3.3.1 SINGLE-PANEL RIGID CASING

As described in Section 2.2, the modelling begins with the casing which is assumed to be of a rectangular parallelepiped shape consisting of single-panel thin flexible walls mounted to a rigid frame (Fig. 2.1). The walls are assumed to be able to radiate sound when made to vibrate. Inside the casing there is an acoustic field due to enclosed certain source of noise (e.g. machine, device, another appliance, etc.).

For the purpose of further analysis an orthogonal system of coordinates is introduced with the origin at one of vertices of the casing (Fig. 3.4).

Axes directions are assumed to form a right-handed coordinate system. Lengths of casing sides are assumed to be equal to l_x, l_y, l_z, respectively. To make the exposition of results as clear as possible each casing wall is assigned a unique number. The wall lying in the Z–Y plane has number "1", and the parallel wall intersecting the X-axis at the point $x = l_x$ has number "2". In turn, the wall lying in the X–Z plane has number "3", and the parallel wall intersecting Y-axis at the point $y = l_y$ has number "4". Finally, the wall lying in the X–Y plane has number "5" and the parallel wall intersecting Z-axis at the point $z = l_z$ has number "6" (Fig. 3.4).

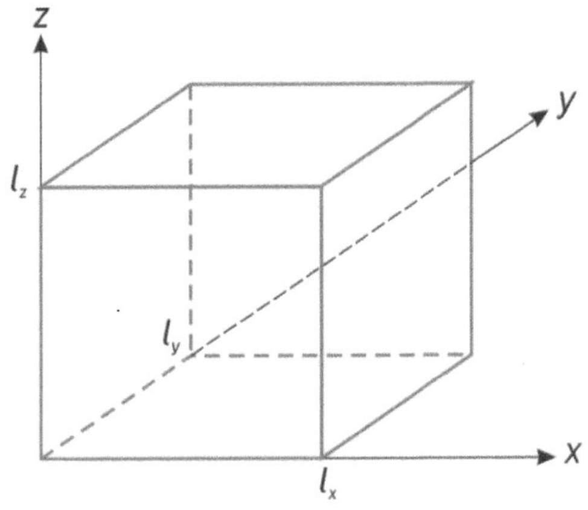

Figure 3.4 The casing structure with the system of coordinates [179]. Reprinted from Applied Mathematical Modelling, 50:219–236, J. Wyrwal, R. Zawiski, M. Pawelczyk, and J. Klamka, "Modelling of coupled vibroacoustic interactions in an active casing for the purpose of control", Copyright (2022), with permission from Elsevier (STM).

Let Ω be a bounded open domain in the Euclidian space $\boldsymbol{R}^3$ defined as:

$$\Omega := \{(x,y,z) \in \mathbf{R}^3 : 0 < x < l_x, 0 < y < l_y, 0 < z < l_z\} \tag{3.61}$$

Let Ω_i, $i = 1, 2, 3$ be bounded open domains in the Euclidian space $\boldsymbol{R}^2$ defined as:

$$\Omega_1 := \left\{(y,z) \in \mathbf{R}^2 : 0 < y < l_y, 0 < z < l_z\right\}, \tag{3.62}$$

$$\Omega_2 := \left\{(x,z) \in \mathbf{R}^2 : 0 < x < l_x, 0 < z < l_z\right\}, \tag{3.63}$$

$$\Omega_3 := \left\{(x,y) \in \mathbf{R}^2 : 0 < x < l_x, 0 < y < l_y\right\}. \tag{3.64}$$

The model is synthesized under the following most significant simplifying assumptions:

- A1. The thickness of the wall ($h_i \quad (i = 1,2,\ldots,6)$) is small compared to its lateral dimensions, i.e. $h_i \ll l_x \quad (i = 3,4,5,6)$, $h_i \ll l_y \quad (i = 1,2,5,6)$, $h_i \ll l_z \quad (i = 1,2,3,4)$. In practice the wall is considered to be thin, when the ratio of its thickness to the smaller lateral dimension is less than $\frac{1}{20}$.
- A2. Vibration-induced deformations are only in the direction of the wall normal vector.
- A3. The middle plane of the wall does not undergo an in-plane deformation.
- A4. The influence of a transverse shear deformation is neglected.
- A5. The transverse normal strain under transverse loading can be neglected.
- A6. Size and weight of actuators are negligible compared to wall size and weight.
- A7. Air density and ambient temperature are constant.
- A8. Normal displacements of the walls are positive in the direction of axes X, Y, Z, respectively.
- A9. Walls of the casing are fasten in a rigid frame.
- A10. Casing walls are assumed to interact with the acoustic field enclosed inside the casing only. The interactions with acoustic field outside the casing are neglected since taking into account active casing operation conditions the level of acoustic pressure inside the casing is much greater then outside.

To derive a model of the casing the vibroacoustic system under discussion is split into simpler subsystems. Then, mathematical models for all subsystems are formulated. Finally, by incorporating interactions between the subsystems, their models are interconnected to obtain a complete model of the overall active casing. In the adopted approach the following most important physical phenomena are taken into account:

- The acoustic field in the fluid enclosed within the casing due to primary noise source operation and actuator-forced vibration of casing walls.
- The dynamics of casing walls excited both acoustically by the enclosed fluid, and mechanically by the actuators.
- Fluid-structure interactions—the interaction between casing walls and fluid enclosed within the casing.

- The influence of fluid loading effect on casing walls vibration.

Model synthesis begins with the analysis of casing walls dynamics.

Dynamics of casing walls excited by the enclosed fluid and exciters

In general, the active casing consists of six light-weight single-panel thin flexible walls that are mounted in a frame which is assumed to be rigid, what justifies neglecting structural interactions between walls. Point-operating exciters are bonded to all walls, except the wall at the bottom of the casing (numbered as the 5th wall). Masses of the exciters and their size are neglected, what is justified as they are much smaller than masses and areas of corresponding walls.

Assuming classical plate theory for thin plates (rotary inertia and shear strain are then neglected), the equation of transverse motion of thin isotropic walls of active casing is given by [145]:

$$\begin{aligned} & D_i\left(\frac{\partial^4}{\partial\eta_{1i}^4}+2\frac{\partial^4}{\partial\eta_{1i}^2\partial\eta_{2i}^2}+\frac{\partial^4}{\partial\eta_{2i}^2}\right)w_i\left(\eta_{1i},\eta_{2i},t\right) \\ & +R_i\left(\frac{\partial^4}{\partial\eta_{1i}^4}+2\frac{\partial^4}{\partial\eta_{1i}^2\eta_{2i}^2}+\frac{\partial^4}{\partial\eta_{2i}^2}\right)\frac{\partial w_i\left(\eta_{1i},\eta_{2i},t\right)}{\partial t} \\ & \qquad +\rho_i h_i\frac{\partial^2 w_i\left(\eta_{1i},\eta_{2i},t\right)}{\partial t^2}=f\left(\eta_{1i},\eta_{2i},t\right), \\ & \qquad (\eta_{1i},\eta_{2i})\in\Omega_{[i/2]},t>0,\ i=1,2,\ldots,6 \end{aligned} \tag{3.65}$$

where:

$$(\eta_{1i},\eta_{2i})=\begin{cases}(y,z)\in\Omega_1 & \text{for } i=1,2,\\ (x,z)\in\Omega_2 & \text{for } i=3,4,\\ (x,y)\in\Omega_3 & \text{for } i=5,6,\end{cases}\quad \eta_{3i}=\begin{cases}x,\ i=1,2,\\ y,i=3,4,\\ z,i=5,6,\end{cases} \tag{3.66}$$

$$D_i=\frac{E_ih_i^3}{12\left(1-v_i^2\right)},\ i=1,2,\ldots,6 \tag{3.67}$$

$$f\left(\eta_{1i},\eta_{2i},t\right)=\begin{cases}-p\left(0,\eta_{1i},\eta_{2i},t\right)-u_1\left(\eta_{1i},\eta_{2i},t\right),i=1,\left(\eta_{11},\eta_{21}\right)=(y,z)\in\Omega_1\\ p\left(l_x,\eta_{1i},\eta_{2i},t\right)+u_2\left(\eta_{1i},\eta_{2i},t\right),i=2,\left(\eta_{12},\eta_{22}\right)=(y,z)\in\Omega_1\\ -p\left(\eta_{13},0,\eta_{23},t\right)-u_3\left(\eta_{13},\eta_{23},t\right),i=3,\left(\eta_{13},\eta_{23}\right)=(x,z)\in\Omega_2\\ p\left(\eta_{14},l_y,\eta_{24},t\right)+u_4\left(\eta_{14},\eta_{24},t\right),i=4,\left(\eta_{14},\eta_{24}\right)=(x,z)\in\Omega_2\\ -p\left(\eta_{15},\eta_{25},0,t\right),i=5,\left(\eta_{15},\eta_{25}\right)=(x,y)\in\Omega_3\\ p\left(\eta_{16},\eta_{26},l_z,t\right)+u_6\left(\eta_{16},\eta_{26},t\right),\ i=6,\left(\eta_{16},\eta_{26}\right)=(x,y)\in\Omega_3\end{cases} \tag{3.68}$$

$$u_i\left(\eta_{1i},\eta_{2i},t\right)=\sum_{m=1}^{K_i}B_{im}F_{im}(t)\delta(\eta_{1i}-\eta_{1im})\delta(\eta_{2i}-\eta_{2im}),\ i=1,2,\ldots,6, \tag{3.69}$$

$$\delta\left(\eta_{1i}-\eta_{1im}\right)\delta(\eta_{2i}-\eta_{2im})=\begin{cases}\delta\left(y-y_{im}\right)\delta\left(z-z_{im}\right),\ i=1,2,\\ \delta\left(x-x_{im}\right)\delta\left(z-z_{im}\right),i=3,4,\\ \delta\left(x-x_{im}\right)\delta\left(y-y_{im}\right),i=5,6,\end{cases} \tag{3.70}$$

$$\eta_{1im} = \begin{cases} y_{im},\ i=1,2, \\ x_{im}, i=3,4,5,6, \end{cases}, \eta_{2im} = \begin{cases} z_{im},\ i=1,2,3,4, \\ y_{im}, i=5,6, \end{cases} \tag{3.71}$$

$$w_i(\cdot,\cdot,t) = \begin{cases} w_i(y,z,t) \text{ for } t>0,\ (y,z)\in\Omega_1 \text{ and } i=1,2, \\ w_i(x,z,t) \text{ for } t>0,\ (x,z)\in\Omega_2 \text{ and } i=3,4, \\ w_i(x,y,t) \text{ for } t>0,\ (x,y)\in\Omega_3 \text{ and } i=5,6. \end{cases} \tag{3.72}$$

Equation (3.65) describes the transverse motion of the ith $(i=1,2,\ldots,6)$ elastic wall of the casing, which occupies the appropriate plane in the reference and stress-free state. Function $w_i(\cdot,\cdot,t)$ denotes the displacement of the wall from the reference state to the η_{3i}-direction at time $t>0$ at point $(\cdot,\cdot)\in\Omega_k, k=[i/2]$ ($[x]$ denotes the smallest integer number greater than or equal to x).

Since Eq. (3.65) is of fourth order with respect to spatial variables, four boundary conditions for each spatial variable have to be formulated. Boundary conditions for Eq. (3.65) take the following form:

$$w_i(\eta_{1i},\eta_{2i},t)=0 \text{ for } \begin{cases} \eta_{1i}=0,\ 0\le\eta_{2i}\le l_{1i}, t\ge 0 \\ \eta_{1i}=l_{2i},\ 0\le\eta_{2i}\le l_{1i}, t\ge 0 \\ \eta_{2i}=0,\ 0\le\eta_{1i}\le l_{2i}, t\ge 0 \\ \eta_{2i}=l_{1i},\ 0\le\eta_{1i}\le l_{2i}, t\ge 0 \end{cases},\ i=1,2,\ldots,6, \tag{3.73a}$$

$$\frac{\partial^2 w_i(\eta_{1i},\eta_{2i},t)}{\partial\eta_{1i}^2} = -\frac{\xi_i}{l_{2i}}\frac{w_i(\eta_{1i},\eta_{2i},t)}{\partial\eta_{1i}} \text{ for } \eta_{1i}=0,\ 0\le\eta_{2i}\le l_{1i}, t\ge 0,\ i=1,2,\ldots,6, \tag{3.73b}$$

$$\frac{\partial^2 w_i(\eta_{1i},\eta_{2i},t)}{\partial\eta_{1i}^2} = \frac{\xi_i}{l_{2i}}\frac{w_i(\eta_{1i},\eta_{2i},t)}{\partial\eta_{1i}} \text{ for } \eta_{1i}=l_{2i},\ 0\le\eta_{2i}\le l_{1i}, t\ge 0,\ i=1,2,\ldots,6, \tag{3.73c}$$

$$\frac{\partial^2 w_i(\eta_{1i},\eta_{2i},t)}{\partial\eta_{2i}^2} = -\frac{\xi_i}{l_{1i}}\frac{w_i(\eta_{1i},\eta_{2i},t)}{\partial\eta_{2i}} \text{ for } \eta_{2i}=0,\ 0\le\eta_{1i}\le l_{2i}, t\ge 0,\ i=1,2,\ldots,6, \tag{3.73d}$$

$$\frac{\partial^2 w_i(\eta_{1i},\eta_{2i},t)}{\partial\eta_{2i}^2} = \frac{\xi_i}{l_{1i}}\frac{w_i(\eta_{1i},\eta_{2i},t)}{\partial\eta_{2i}} \text{ for } \eta_{2i}=l_{1i},\ 0\le\eta_{1i}\le l_{2i}, t\ge 0,\ i=1,2,\ldots,6, \tag{3.73e}$$

where:

$$l_{1i} = \begin{cases} l_z,\ i=1,2,\ldots,4, \\ l_y, i=5,6 \end{cases},\ l_{2i} = \begin{cases} l_y,\ i=1,2, \\ l_x, i=3,4,5,6. \end{cases} \tag{3.74}$$

Remark 3.3 (The type of boundary conditions for walls)

The edges of the walls are considered to be rigidly supported and elastically restrained against rotation. The degree of restraint against rotation on the edges of the ith wall is determined by the value of the boundary restraint coefficients $\xi_i\in\langle 0,\infty)$, i=1,2,…,6. This kind of boundary conditions corresponds to an imperfect nature of fastening wall edges to the rigid frame. If $\xi_i=0$ then assumed boundary conditions coincide with hinged or simply supported edges when the deflection and normal bending moment resultant on the wall edges are equal to zero. On the other hand, if $\xi_i\to\infty$, the assumed boundary conditions simplify to fully clamped (fixed or built-in) edges. In this case the deflection and slope normal to the edges are equal to zero.

Assuming that the wall edges are perfectly fastened upon the rigid frame of the active casing the fully clamped boundary conditions are obtained. Considering elastically restrained against rotation boundary conditions makes it possible to track the behaviour of the system in case of both perfect and imperfect fastening of wall edges by changing the value of boundary restraint coefficients ξ_i, $i=1,2,\ldots,6$.

Since Eq. (3.65) is of second order with respect to time, two initial conditions are formulated:

$$w_i(\eta_{1i},\eta_{2i},0) = w_{0i}(\eta_{1i},\eta_{2i}),\ (\eta_{1i},\eta_{2i}) \in \Omega_{[i/2]},\ i = 1,2\ldots,6, \tag{3.75}$$

$$\left.\frac{\partial w_i(\eta_{1i},\eta_{2i},t)}{\partial t}\right|_{t=0} = w_{1i}(\eta_{1i},\eta_{2i}),\ (\eta_{1i},\eta_{2i}) \in \Omega_{[i/2]},\ i = 1,2\ldots,6, \tag{3.76}$$

where functions $w_{0i}(\eta_{1i},\eta_{2i})$ and $w_{1i}(\eta_{1i},\eta_{2i})$, $i = 1,2,\ldots,6$, are assumed to be given. Function $w_{0i}(\eta_{1i},\eta_{2i})$ describes the initial displacement of the walls from the reference and stress-free state, whereas $w_{1i}(\eta_{1i},\eta_{2i})$ represents the initial distribution of velocity on the surface of casing walls. Additionally, functions $w_{0i}(\eta_{1i},\eta_{2i})$ and $w_{0i}(\eta_{1i},\eta_{2i})$ are assumed to satisfy the consistency conditions:

$$\begin{array}{l} w_{0i}(\eta_{1i},\eta_{2i}) = 0 \\ w_{1i}(\eta_{1i},\eta_{2i}) = 0 \end{array} \ \text{for} \begin{cases} \eta_{1i} = 0,\ 0 \le \eta_{2i} \le l_{1i}, t \ge 0 \\ \eta_{1i} = l_{2i},\ 0 \le \eta_{2i} \le l_{1i}, t \ge 0 \\ \eta_{2i} = 0,\ 0 \le \eta_{1i} \le l_{2i}, t \ge 0 \\ \eta_{2i} = l_{1i},\ 0 \le \eta_{1i} \le l_{2i}, t \ge 0 \end{cases},\ i = 1,2,\ldots,6. \tag{3.77}$$

where l_{1i} and l_{2i}, $i = 1,2,\ldots,6$, are given by (3.74).

Acoustic field within the fluid enclosed in the casing

The behaviour of acoustic field inside the casing through a homogeneous, viscous, isotropic and compressible fluid is described by the acoustic wave equation [142, 48]:

$$\left(\frac{\partial^2}{\partial x^2} + \frac{\partial^2}{\partial y^2} + \frac{\partial^2}{\partial z^2}\right) p(x,y,z,t) + \mu_0 \left(\frac{\partial^2}{\partial x^2} + \frac{\partial^2}{\partial y^2} + \frac{\partial^2}{\partial z^2}\right) \frac{\partial p(x,y,z,t)}{\partial t} - \frac{1}{c_0^2}\frac{\partial^2 p(x,y,z,t)}{\partial t^2} = -\rho_0 \sum_{k=1}^{K_s} \frac{\partial q_k}{\partial t}, \tag{3.78}$$

$$(x,y,z) \in \Omega, t > 0,\ c_0 = \frac{\gamma P_0}{\rho_0}, \tag{3.79}$$

where c_0 is the frequency-independent speed of sound, P_0 is the mean fluid pressure, ρ_0 is the mean fluid density, γ is the adiabatic bulk modulus of the fluid, $p(x,y,z,t)$ is the sound pressure within the casing at point $(x,y,z) \in \Omega$, and time $t > 0$, q_k are the acoustic source strengths (the distribution of source volume velocity per unit volume) and K_s is the number of sound sources. The right-hand side of (3.78) represents a rate of change of mass flux per unit area. It is assumed that the compressible fluid contained within the active casing is excited by acoustic waves generated by K_s sources of secondary noise.

Since Eq. (3.78) is of second order with respect to time, two initial conditions are formulated:

$$p(x,y,z,0) = p_0(x,y,z), (x,y,z) \in \Omega, \tag{3.80}$$

$$\left.\frac{\partial p(x,y,z,t)}{\partial t}\right|_{t=0} = p_1(x,y,z), (x,y,z) \in \Omega, \tag{3.81}$$

where functions $p_0(x,y,z)$ and $p_1(x,y,z)$ are assumed to be given. Function $p_0(x,y,z)$ describes the initial distribution of acoustic pressure and $p_1(x,y,z)$ represents the initial distribution of the rate of change of acoustic pressure in the fluid enclosed within the casing.

Interactions between the casing walls and the fluid enclosed within the casing

There are many systems of practical interest, in which a structure is in contact with the fluid contained within a finite volume defined by the structure physical boundaries.

It should be emphasized that acoustic field propagating in the fluid contained within the casing stimulates the vibrations of the casing walls. Any solid structure exposed to the sound field in a contiguous fluid responds to some degree to the fluctuating pressure acting at the interface between the two media. The fluid pressure on the surface of casing walls is the agent by which the fluid influences the structural motion of the casing walls. This phenomenon was incorporated into the model by adding term $p(^\circ,^\circ,^\circ,t)$ in the right-hand side of Eq. (3.65) describing transverse motion of casing walls. It is described by the function $f(\eta_{1i},\eta_{2i},t)$ that appears in the right-hand side of Eq. (3.65) and is given by (3.68). The term $p(^\circ,^\circ,^\circ,t)$ represents a distributed transverse force per unit area acting on the walls surface and stimulating walls vibrations due to the acoustic field in the fluid enclosed within the casing while the terms $u_i(^\circ,^\circ,t), i = 1,2,\ldots,6$ represent the forces generated by point actuators bonded to wall surface.

On the other hand, casing walls that can be made to vibrate by actuators become sources of sound, and the sound emitted by each wall contributes to the acoustic field in the fluid enclosed within the casing. The normal surface accelerations of the casing walls are the agents by which the casing structure influence the enclosed fluid field. This phenomenon is incorporated into the model by the boundary conditions for acoustic wave equation (3.65).

In order to take into account how the casing structure influences the enclosed fluid field, the acoustic wave equation (3.78) should be solved subject to boundary conditions imposed by the vibrating surface of casing walls of the following form:

$$\frac{\partial p(x,y,z,t)}{\partial x} = \begin{cases} -\rho_0 \frac{\partial^2 w_1(y,z,t)}{\partial t^2}; 0 \le y \le l_y, 0 \le z \le l_z, x = 0, t \ge 0 \\ +\rho_0 \frac{\partial^2 w_2(y,z,t)}{\partial t^2}; 0 \le y \le l_y, 0 \le z \le l_z, x = l_x, t \ge 0 \end{cases} \tag{3.82a}$$

$$\frac{\partial p(x,y,z,t)}{\partial y} = \begin{cases} -\rho_0 \frac{\partial^2 w_3(x,z,t)}{\partial t^2}; 0 \le x \le l_x, 0 \le z \le l_z, y = 0, t \ge 0 \\ +\rho_0 \frac{\partial^2 w_4(x,z,t)}{\partial t^2}; 0 \le x \le l_x, 0 \le z \le l_z, y = l_y, t \ge 0 \end{cases} \tag{3.82b}$$

$$\frac{\partial p(x,y,z,t)}{\partial z} = \begin{cases} -\rho_0 \frac{\partial^2 w_5(x,y,t)}{\partial t^2}; 0 \le x \le l_x, 0 \le y \le l_y, z = 0, t \ge 0 \\ +\rho_0 \frac{\partial^2 w_6(x,y,t)}{\partial t^2}; 0 \le x \le l_x, 0 \le y \le l_y, z = l_z, t \ge 0 \end{cases} \tag{3.82c}$$

Remark 3.4 (Interactions in the model)
The model proposed, on the one hand, reflects the influence of casing structure on the enclosed fluid field. There are fluid-structure interactions of vibroacoustic nature that are modelled by the boundary conditions (3.82) *for the acoustic wave equation* (3.78). *On the other hand, model derived reflects the influence of the fluid field on the structural vibrations of the casing structure. There are fluid-structure interactions of acoustic-vibration character that are modelled by* $f(\eta_{1i}, \eta_{2i}, t)$ *given by* (3.68).

3.3.2 DOUBLE-PANEL RIGID CASING

A vibroacoustic system considered in this section can be treated as a direct extension of that analysed in Section 3.3.1. As previously, it is assumed to be a rectangular parallelepiped casing but now double-panel thin flexible walls mounted to a rigid frame are considered (Fig. 3.5).

In general, each panel can be made of a different material. Moreover, each cavity between the walls' panels can be filled with a different fluid. The panels of the walls are assumed to be able to radiate sound when forced to vibrate. Inside the casing there is an acoustic field due to enclosed certain source of noise (e.g. machine, device, another appliance, etc.).

For the purpose of further analysis an orthogonal system of coordinates is introduced with the origin at one of vertices of the casing (Fig. 3.5). Axes directions are assumed to form a right-handed coordinate system. Lengths of panels' sides are assumed to be equal to l_x, l_y, l_z along x, y and z axes, respectively. The distance between the panels in the ith wall is denoted by $d_i\,(i = 1,2,\ldots,6)$ (Fig. 3.6). To make the exposition of results as clear as possible each casing wall is assigned a unique number (Fig. 3.5). The wall lying in the Z–Y plane has number "1" and its panels denoted by numbers "1" and "2" intersect the X-axis at the points $x = 0$ and $x = d_1$, respectively. The parallel wall has number "2" and its panels denoted by numbers "2" and "1" intersect the X-axis at the points $x = l_x+2a - d_2$ and $x = l_x+2a$, respectively (Fig. 3.6). In turn, the wall lying in the X–Z plane has number "3" and its panels denoted by numbers "1" and "2" intersect the Y-axis at the points $y = 0$ and $y = d_3$, respectively. The parallel wall has number "4" and its panels denoted by numbers "2" and "1" intersect the Y-axis at the points $y = l_y+2a - d_4$ and $y = l_y+2a$, respectively (Fig. 3.6). Finally, the wall lying in the X–Y plane has number "5" and its panels denoted by numbers "1" and "2" intersect the Z-axis at the points $z = 0$ and $z = d_5$, respectively. The parallel wall has number "6" and its panels denoted by numbers "2" and "1" intersect the Y-axis at the points $z = l_z+2a - d_6$ and $z = l_z+2a$, respectively (Fig. 3.6).

Let Ω_i (i=0,1,...,6) be a bounded open domain in the Euclidian space $\boldsymbol{R}^3$ defined as:

$$\Omega_0 := \{(x,y,z) \in \mathbf{R}^3 : d_1 + h_{12} < x < l_x + 2a - d_2 - h_{22}, \\ d_3 + h_{32} < y < l_y + 2a - d_4 - h_{42}, \\ d_5 + h_{52} < z < l_z + 2a - d_6 - h_{62}\} \quad (3.83)$$

$$\Omega_1 := \{(x,y,z) \in \mathbf{R}^3 : 0 < x < d_1, a < y < l_y + a, a < z < l_z + a\} \quad (3.84)$$

$$\Omega_2 := \{(x,y,z) \in \mathbf{R}^3 : l_x + 2a - d_2 < x < l_x + 2a, a < y < l_y + a, a < z < l_z + a\} \quad (3.85)$$

$$\Omega_3 := \{(x,y,z) \in \mathbf{R}^3 : a < x < l_x + a, 0 < y < d_3, a < z < l_z + a\} \quad (3.86)$$

$$\Omega_4 := \{(x,y,z) \in \mathbf{R}^3 : a < x < l_x + a, l_y + 2a - d_4 < y < l_y + 2a, a < z < l_z + a\} \quad (3.87)$$

$$\Omega_5 := \{(x,y,z) \in \mathbf{R}^3 : a < x < l_x + a, a < y < l_y + a, 0 < z < d_5\} \quad (3.88)$$

$$\Omega_6 := \{(x,y,z) \in \mathbf{R}^3 : a < x < l_x + a, a < y < l_y + a, l_z + 2a - d_6 < z < l_z + 2a\} \quad (3.89)$$

The domain Ω_0 refers to the interior of the active casing while domain Ω_i, (i=1,2,...,6), refer to the interior of the cavity between panels of the ith wall.

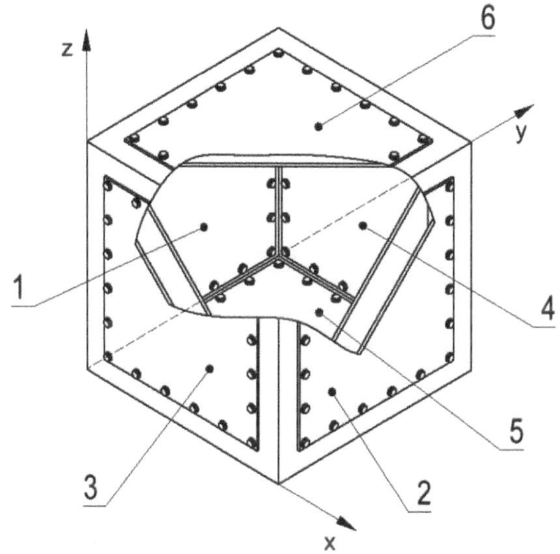

Figure 3.5 An active casing with numeration of double-panel walls and the system of coordinates [178]. Reprinted from Mechanical Systems and Signal Processing, 152:107371, J. Wyrwal, M. Pawelczyk, L. Liu, and Z. Rao, "Double-panel active noise reducing casing with noise source enclosed inside-modelling and simulation study", Copyright (2022), with permission from Elsevier.

Let Ω' and Ω'_0 be domains in the Euclidian space $\boldsymbol{R}^3$ defined as:

$$\Omega' := \{(x,y,z) \in \mathbf{R}^3 : -h_{11} < x < l_x + 2a + h_{21}, \\ -h_{31} < y < l_y + 2a + h_{41}, \\ -h_{51} < z < l_z + 2a + h_{61}\}, \tag{3.90}$$

$$\Omega'_0 = \mathbf{R}^3 \backslash \Omega' . \tag{3.91}$$

The unbounded domain Ω'_0 refers to the exterior of the active casing. Moreover, let Ω_{ij} be bounded open domains in the Euclidian space $\boldsymbol{R}^2$ defined as:

$$\Omega_{12} := \{(y,z) \in \mathbf{R}^2 : a < y < l_y + a, a < z < l_z + a\}, \tag{3.92}$$

$$\Omega_{34} := \{(x,z) \in \mathbf{R}^2 : a < x < l_x + a, a < z < l_z + a\}, \tag{3.93}$$

$$\Omega_{56} := \{(x,y) \in \mathbf{R}^2 : a < x < l_x + a, a < y < l_y + a\}. \tag{3.94}$$

Two-dimensional domains Ω_{ij} $(i = 1,3,5, \;\; j = i+1)$ refer to surfaces of panels in ith and jth walls.

The small deflection theory of thin plates (classical plate theory) is used to describe the motion of casing walls panels [86, 156]. The procedure of model synthesis takes into account following most significant simplifying assumptions:

A1. The thickness of the panel is small compared to its lateral dimensions, i.e. $h_{ij} \ll l_x \;\; (i = 3,4,5,6)$, $h_{ij} \ll l_y \;\; (i = 1,2,5,6)$, $h_{ij} \ll l_z$ $(i = 1,2,3,4)\,(j = 1,2)$. In practice the panel is considered to be thin, when the ratio of its thickness to the smaller lateral dimension is less than $\frac{1}{20}$.
A2. Vibration-induced deformations are only in the direction of the wall normal vector.
A3. The middle plane of the wall does not undergo an in-plane deformation.
A4. The influence of a transverse shear deformation is neglected.
A5. The transverse normal strain under transverse loading can be neglected.
A6. Size and weight of actuators and sensors are negligible compared to panels size and weight.
A7. Air density and ambient temperature are constant.
A8. Normal displacements of the walls panels are positive in the direction of axes X, Y, Z, respectively.
A9. Walls of the casing are fasten in a rigid frame.
A10. Casing walls panels are assumed to interact with the acoustic field in the fluid enclosed inside the casing, the acoustic fields in the fluids enclosed inside walls cavities and the acoustic field in the fluid surrounding the casing from the outside.

Remark 3.5 (Assumptions for model synthesis)
As can be seen, the assumptions used to derive the double-panel casing model are basically similar to those adopted for the single-panel casing, except for the assumption A10. It can be observed that in contrast to Section 3.3.1 where the model

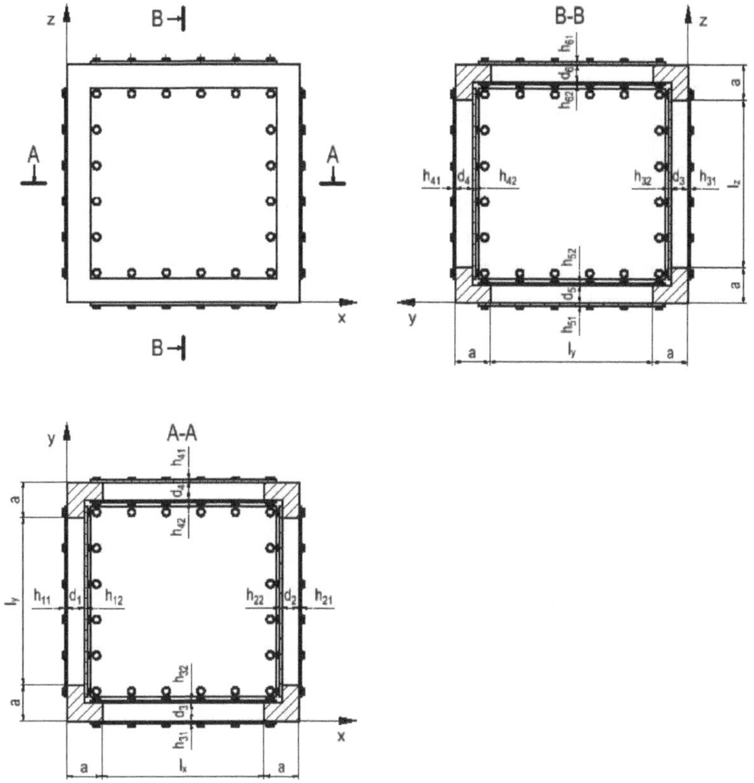

Figure 3.6 Side view and cross sections of the active casing with its dimensions [178]. Reprinted from Mechanical Systems and Signal Processing, 152:107371, J. Wyrwal, M. Pawelczyk, L. Liu, and Z. Rao, "Double-panel active noise reducing casing with noise source enclosed inside-modelling and simulation study", Copyright (2022), with permission from Elsevier.

for single-panel casing is derived in the current section the interactions with acoustic field in the fluid surrounding the casing from the outside are incorporated into the model (assumption A10). It complicates the model obviously but gives the opportunity to study the effect of sound insulation what is the essential purpose of the active casing. Additionally, by incorporating this phenomenon the model derived is expected to be more accurate.

In the adopted construction solution a double-panel casing is assumed. It means that each casing wall consists of two thin flexible panels: the internal one and the external one. The internal panel will be referred to as the *incident* panel (index j=2) since it accepts the incoming noise from the noise source located in the interior of the active casing. On the other hand, the external panel will be referred to as the *radiating* panel (index j=1) since it radiates the sound to the exterior of the active casing.

To derive a model of the casing the vibroacoustic system under discussion is split into simpler subsystems. Then, mathematical models for all subsystems are formulated. Finally, by incorporating interactions between the subsystems, their models are interconnected to obtain a complete model of the overall active casing. In the adopted approach the following most important physical phenomena are taken into account (Fig. 3.7):

- acoustic field in the fluid enclosed within the casing due to primary noise source operation and actuator-forced vibration of casing walls panels.
- dynamics of casing double-panel walls excited both acoustically by the enclosed fluids, and mechanically by the actuators.
- fluid-structure interactions including:
  - interactions between incident panels and the acoustic field in the fluid enclosed in the interior of the casing.
  - interactions between casing walls' panels and the acoustic field in the fluids enclosed in the cavities between panels of walls,
  - interactions between casing radiating panels and the acoustic field in fluid surrounding the casing from the outside.
- influence of the fluid loading effect on walls' panels vibration.
- influence of point-operating actuators on walls' panels vibration.

Model synthesis begins with the analysis of double-panel walls dynamics.

Dynamics of double panel casing walls excited by the acoustic fields and actuators

In general, the active casing consists of six *light-weight thin flexible double-panel* walls that are mounted in a rigid frame, what justifies neglecting structural interactions between walls. Point-operating actuators (exciters) are bonded to all walls' panels, except panels of the wall at the bottom of the casing (numbered as the 5th wall), i.e. $K_{5j} = 0$ $(j = 1,2)$. Masses of the actuators and sensors as well as their size are neglected, what is justified as they can be much smaller compared to the masses and size of corresponding walls panels. If heavier actuators are used, e.g. electrodynamic ones, it can be necessary to include their mass, what will complicate the model further.

Each wall of the active casing is considered as a composite system consisting of two light-weight thin flexible panels and the cavity between its panels that is filled with a homogeneous, viscous, isotropic and compressible fluid.

To describe the behaviour of each wall it is necessary to model the motion of its panels and the acoustic field in the fluid contained within the cavity between its panels. Moreover, the model has to incorporate the mutual interactions between the wall components as well as their interactions with acoustic field enclosed within the interior of the active casing. Therefore, behaviour of ith wall $(i = 1,2,\dots,6)$, is described by the following system of three coupled second order with respect to time partial differential equations:

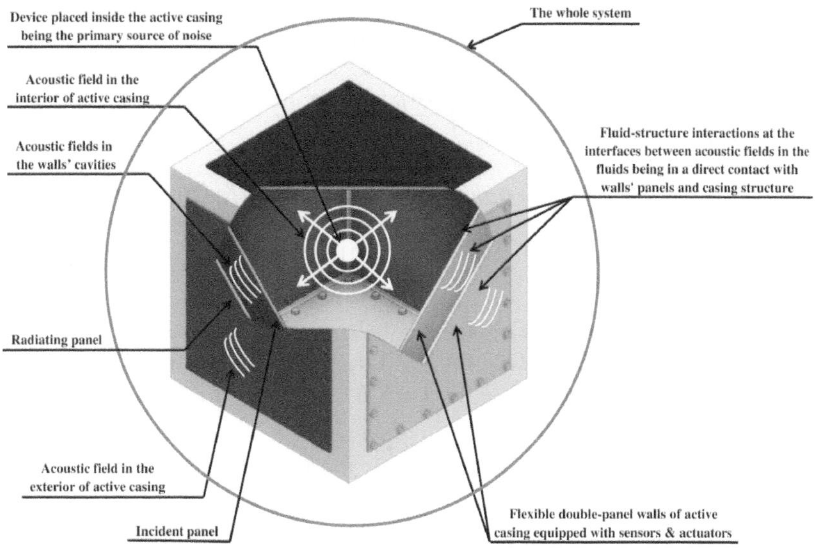

Figure 3.7 Active casing with fluid-structure interactions [178]. Reprinted from Mechanical Systems and Signal Processing, 152:107371, J. Wyrwal, M. Pawelczyk, L. Liu, and Z. Rao, "Double-panel active noise reducing casing with noise source enclosed inside-modelling and simulation study", Copyright (2022), with permission from Elsevier.

– radiating panel equation of motion with internal damping inside the panel material:

$$D_{i1}\left(\frac{\partial^4}{\partial\eta_{1i}^4}+2\frac{\partial^4}{\partial\eta_{1i}^2\partial\eta_{2i}^2}+\frac{\partial^4}{\partial\eta_{2i}^2}\right)w_{i1}(\eta_{1i},\eta_{2i},t)$$
$$+R_{i1}\left(\frac{\partial^4}{\partial\eta_{1i}^4}+2\frac{\partial^4}{\partial\eta_{1i}^2\partial\eta_{2i}^2}+\frac{\partial^4}{\partial\eta_{2i}^2}\right)\frac{\partial w_{i1}(\eta_{1i},\eta_{2i},t)}{\partial t}$$
$$+\rho_{i1}h_{i1}\frac{\partial^2 w_{i1}(\eta_{1i},\eta_{2i},t)}{\partial t^2}=f_{i1}(\eta_{1i},\eta_{2i},t),$$
$$(\eta_{1i},\eta_{2i})\in\Omega_{(2[i/2]-1)(2[i/2])},t>0,\ i=1,2,\ldots,6, \quad (3.95)$$

– damped acoustic wave equation describing the acoustic field in the fluid contained in the cavity between wall panels:

$$\left(\frac{\partial^2}{\partial x^2}+\frac{\partial^2}{\partial y^2}+\frac{\partial^2}{\partial z^2}\right)p_i(x,y,z,t)$$
$$+\mu_i\left(\frac{\partial^2}{\partial x^2}+\frac{\partial^2}{\partial y^2}+\frac{\partial^2}{\partial z^2}\right)\frac{\partial p_i(x,y,z,t)}{\partial t}-\frac{1}{c_i^2}\frac{\partial^2 p_i(x,y,z,t)}{\partial t^2}=0$$
$$(x,y,z)\in\Omega_i,\ t>0,\ i=1,2,\ldots,6, \quad (3.96)$$

– incident panel equation of motion with internal damping inside the panel material:

$$D_{i2}\left(\frac{\partial^4}{\partial\eta_{1i}^4}+2\frac{\partial^4}{\partial\eta_{1i}^2\partial\eta_{2i}^2}+\frac{\partial^4}{\partial\eta_{2i}^2}\right)w_{i2}(\eta_{1i},\eta_{2i},t)$$
$$+R_{i2}\left(\frac{\partial^4}{\partial\eta_{1i}^4}+2\frac{\partial^4}{\partial\eta_{1i}^2\partial\eta_{2i}^2}+\frac{\partial^4}{\partial\eta_{2i}^2}\right)\frac{\partial w_{i2}(\eta_{1i},\eta_{2i},t)}{\partial t}$$
$$+\rho_{i2}h_{i2}\frac{\partial^2 w_{i2}(\eta_{1i},\eta_{2i},t)}{\partial t^2}=f_{i2}(\eta_{1i},\eta_{2i},t),$$
$$(\eta_{1i},\eta_{2i})\in\Omega_{(2[i/2]-1)(2[i/2])},t>0,\ i=1,2,\ldots,6,\quad(3.97)$$

where:

$$(\eta_{1i},\eta_{2i})=\begin{cases}(y,z)\in\Omega_{12}\ \text{for}\ i=1,2,\\(x,z)\in\Omega_{34}\ \text{for}\ i=3,4,\\(x,y)\in\Omega_{56}\ \text{for}\ i=5,6,\end{cases}\quad \eta_{3i}=\begin{cases}x,\ i=1,2,\\y,i=3,4,\\z,i=5,6,\end{cases}\quad(3.98)$$

$$D_{ij}=\frac{E_{ij}h_{ij}^3}{12\left(1-\nu_{ij}^2\right)},c_i^2=\frac{\gamma_i\,\mathrm{P}_i}{\rho_i},\ i=1,2,\ldots,6,\ j=1,2,\quad(3.99)$$

$$f_{1i}(\eta_{1i},\eta_{2i},t)=\begin{cases}-p_1(0,\eta_{1i},\eta_{2i},t)+p_s(-h_{11},\eta_{1i},\eta_{2i},t)-u_{11}(\eta_{1i},\eta_{2i},t),\\ \quad i=1,(\eta_{11},\eta_{21})=(y,z)\in\Omega_{12}\\ p_2(l_x+2a,\eta_{1i},\eta_{2i},t)-p_s(l_x+2a+h_{21},\eta_{1i},\eta_{2i},t)+u_{21}(\eta_{1i},\eta_{2i},t),\\ \quad i=2,(\eta_{12},\eta_{22})=(y,z)\in\Omega_{12}\\ -p_3(\eta_{1i},0,\eta_{2i},t)+p_s(\eta_{1i},-h_{31},\eta_{2i},t)-u_{31}(\eta_{1i},\eta_{2i},t),\\ \quad i=3,(\eta_{13},\eta_{23})=(x,z)\in\Omega_{34}\\ p_4(\eta_{1i},l_y+2a,\eta_{2i},t)-p_s(\eta_{1i},l_y+2a+h_{41},\eta_{2i},t)+u_{41}(\eta_{1i},\eta_{2i},t),\\ \quad i=4,(\eta_{14},\eta_{24})=(x,z)\in\Omega_{34}\\ -p_5(\eta_{1i},\eta_{2i},0,t)+p_s(\eta_{1i},\eta_{2i},-h_{51},t),\\ \quad i=5,(\eta_{15},\eta_{25})=(x,y)\in\Omega_{56}\\ p_6(\eta_{1i},\eta_{2i},l_z+2a,t)-p_s(\eta_{1i},\eta_{2i},l_z+2a+h_6,t)+u_{61}(\eta_{1i},\eta_{2i},t),\\ \quad i=6,(\eta_{16},\eta_{26})=(x,y)\in\Omega_{56}\end{cases}\quad(3.100)$$

$$f_{2i}(\eta_{1i},\eta_{2i},t)=\begin{cases}p_1(d_1,\eta_{1i},\eta_{2i},t)-p(d_1+h_{12},\eta_{1i},\eta_{2i},t)-u_{12}(\eta_{1i},\eta_{2i},t),\\ \quad i=1,(\eta_{11},\eta_{21})=(y,z)\in\Omega_{12}\\ -p_2(l_x+2a-d_2,\eta_{1i},\eta_{2i},t)+p(l_x+2a-d_2-h_{22},\eta_{1i},\eta_{2i},t)+\\ \quad +u_{22}(\eta_{1i},\eta_{2i},t),\ i=2,(\eta_{12},\eta_{22})=(y,z)\in\Omega_{12}\\ p_3(\eta_{1i},d_3,\eta_{2i},t)-p(\eta_{1i},d_3+h_{32},\eta_{2i},t)-u_{32}(\eta_{1i},\eta_{2i},t),\\ \quad i=3,(\eta_{13},\eta_{23})=(x,z)\in\Omega_{34}\\ -p_4(\eta_{1i},l_y+2a-d_4,\eta_{2i},t)+p(\eta_{1i},l_y+2a-d_4-h_{42},\eta_{2i},t)+\\ \quad +u_{42}(\eta_{1i},\eta_{2i},t),i=4,(\eta_{14},\eta_{24})=(x,z)\in\Omega_{34}\\ p_5(\eta_{1i},\eta_{2i},d_5,t)-p(\eta_{1i},\eta_{2i},d_5+h_{52},t),\\ \quad i=5,(\eta_{15},\eta_{25})=(x,y)\in\Omega_{56}\\ -p_6(\eta_{1i},\eta_{2i},l_z+2a-d_6,t)+p(\eta_{1i},\eta_{2i},l_z+2a-d_6-h_{62},t)+\\ \quad +u_{62}(\eta_{1i},\eta_{2i},t),i=6,(\eta_{16},\eta_{26})=(x,y)\in\Omega_{56}\end{cases}\quad(3.101)$$

$$u_{ij}(\eta_{1i},\eta_{2i},t)=\sum_{m=1}^{K_{ij}}B_{\mathrm{ijm}}F_{\mathrm{ijm}}(t)\delta(\eta_{1i}-\eta_{1ijm}^a)\delta(\eta_{2i}-\eta_{2ijm}^a),\ i=1,2,\ldots,6,j=1,2\quad(3.102)$$

$$a < \eta^a_{1ijm} < a + l_{1i} \text{ and } a < \eta^a_{2ijm} < a + l_{2i},\ i = 1,2,\ldots 4,6,\ j = 1,2, m = 1,2,\ldots,K_{ij} \tag{3.103}$$

$$l_{1i} = \begin{cases} l_y,\ i = 1,2, \\ l_x, i = 3,4,5,6, \end{cases},\ l_{2i} = \begin{cases} l_z,\ i = 1,2,3,4, \\ l_y, i = 5,6, \end{cases} \tag{3.104}$$

$$\eta^a_{1ijm} = \begin{cases} y^a_{\mathrm{ijm}},\ i = 1,2, j = 1,2, \\ x^a_{\mathrm{ijm}}, i = 3,4,5,6, j = 1,2, \end{cases},\ \eta^a_{2ijm} = \begin{cases} z^a_{\mathrm{ijm}},\ i = 1,2,3,4, j = 1,2, \\ y^a_{\mathrm{ijm}}, i = 5,6, j = 1,2, \end{cases} \tag{3.105}$$

$$\delta\left(\eta_{1i} - \eta^a_{1ijm}\right)\delta(\eta_{2i} - \eta^a_{2ijm}) = \begin{cases} \delta\left(y - y^a_{\mathrm{ijm}}\right)\delta\left(z - z^a_{\mathrm{ijm}}\right),\ i = 1,2, j = 1,2, \\ \delta\left(x - x^a_{\mathrm{ijm}}\right)\delta\left(z - z^a_{\mathrm{ijm}}\right), i = 3,4, j = 1,2, \\ \delta\left(x - x^a_{\mathrm{ijm}}\right)\delta\left(y - y^a_{\mathrm{ijm}}\right), i = 5,6, j = 1,2, \end{cases} \tag{3.106}$$

$$w_{ij}\left(\eta_{1i}, \eta_{2i}, t\right) = \begin{cases} w_{ij}\left(y,z,t\right) \text{ for } t > 0,\ (y,z) \in \Omega_{12},\ i = 1,2, j = 1,2, \\ w_{ij}\left(x,z,t\right) \text{ for } t > 0,\ (x,z) \in \Omega_{34},\ i = 3,4, j = 1,2, \\ w_{ij}\left(x,y,t\right) \text{ for } t > 0,\ (x,y) \in \Omega_{56},\ i = 5,6, j = 1,2. \end{cases} \tag{3.107}$$

Function $u_{ij}\left(\eta_{1i}, \eta_{2i}, t\right)$ $(i = 1,2,\ldots,6, j = 1,2)$ given by (3.102) describes control corresponding to actuators bonded to the jth panel in the ith wall.

Assuming classical plate theory for thin plates (rotary inertia and shear strain are then neglected), Eqs. (3.95) and (3.97) describe the transverse motion of the elastic thin isotropic panels of ith wall, which:

- for $i = 1,3,5$ occupy the $\eta_{1i} - \eta_{2i}$ plane and the plane that is parallel to $\eta_{1i} - \eta_{2i}$ plane intersecting η_{3i} axis at the point $\eta_{3i} = d_i$, respectively, in the reference and stress-free state,
- for $i = 2,4,6$ occupy the planes that are parallel to $\eta_{1i} - \eta_{2i}$ plane and intersect η_{3i} axis at the points $\eta_{3i} = l_{3i} + 2a - d_i$ and $\eta_{3i} = l_{3i} + 2a$, respectively, in the reference and stress-free state.

Functions $w_{i1}\left(\eta_{1i}, \eta_{2i}, t\right)$ and $w_{i2}\left(\eta_{1i}, \eta_{2i}, t\right)$ denote the displacement of the corresponding panel “1” and “2” from the reference state to the η_{3i}-direction at time $t > 0$ at point $\left(\eta_{1i}, \eta_{2i}\right) \in \Omega_{(2[i/2]-1)(2[i/2])}$ (Fig. 3.8). The homogeneous acoustic wave Eq. (3.96) describes the behaviour of acoustic field inside the cavity that is located between panels of the ith wall filled with a homogeneous, viscous, isotropic and compressible fluid [142, 48].

The term $\left(\left(-1\right)^i p_i\left(\cdot,\cdot,\cdot,t\right) + \left(-1\right)^{i+1} p_s\left(\cdot,\cdot,\cdot,t\right)\right)$ in the right hand side of Eq. (3.95) given by (3.100) describes the influence of the acoustic field in the cavity of the ith wall as well as the acoustic field in the exterior of the active casing on the motion of (radiating) panel “1” in this wall. In turn, the term $\left(\left(-1\right)^{i+1} p_i\left(\cdot,\cdot,\cdot,t\right) + \left(-1\right)^i p\left(\cdot,\cdot,\cdot,t\right)\right)$ in the right hand side of Eq. (3.97) given by (3.101) describes the influence of the acoustic field in the cavity of the ith wall as well as the acoustic field in the interior of the active casing on the motion of (incident) panel “2” in this wall.

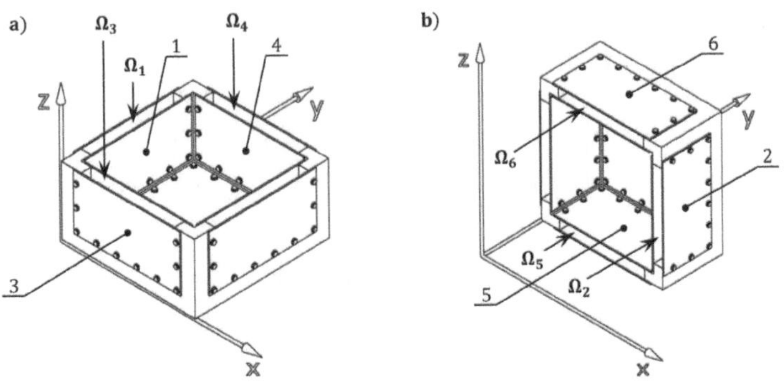

Figure 3.8 The double-panel walls of active casing and domains Ω_i corresponding to walls cavities of the active casing: (a) A 2-D cross-sectional view parallel to the *X*–*Y* plane, (b) A 2-D cross-sectional view parallel to the *X*–*Z* plane [178]. Reprinted from Mechanical Systems and Signal Processing, 152:107371, J. Wyrwal, M. Pawelczyk, L. Liu, and Z. Rao, "Double-panel active noise reducing casing with noise source enclosed inside-modelling and simulation study", Copyright (2022), with permission from Elsevier.

Obviously, partial differential Eqs. (3.95) and (3.97) have to be complemented by the appropriate boundary conditions that reflect fastening of panels edges to the rigid frame and influence of the panels motion on the acoustic field inside the wall cavity. Boundary conditions for the *j*th panel of the *i*th wall described by Eqs. (3.95) (*j*=1) and (3.97) (*j*=2), take the following form:

$$w_{ij}(\eta_{1i},\eta_{2i},t)=0 \quad \text{for} \begin{cases} \eta_{1i}=a,\ a\leq\eta_{2i}\leq a+l_{2i}, t\geq 0 \\ \eta_{1i}=a+l_{1i},\ a\leq\eta_{2i}\leq a+l_{2i}, t\geq 0 \\ \eta_{2i}=a,\ a\leq\eta_{1i}\leq a+l_{1i}, t\geq 0 \\ \eta_{2i}=a+l_{2i},\ a\leq\eta_{1i}\leq a+l_{1i}, t\geq 0 \end{cases}, \tag{3.108}$$

$$\frac{\partial^2 w_{ij}(\eta_{1i},\eta_{2i},t)}{\partial\eta_{1i}^2}=-\frac{\xi_{ij}}{l_{1i}}\frac{\partial w_{ij}(\eta_{1i},\eta_{2i},t)}{\partial\eta_{1i}} \quad \text{for } \eta_{1i}=a,\ a\leq\eta_{2i}\leq a+l_{2i}, t\geq 0, \tag{3.109}$$

$$\frac{\partial^2 w_{ij}(\eta_{1i},\eta_{2i},t)}{\partial\eta_{1i}^2}=\frac{\xi_{ij}}{l_{1i}}\frac{\partial w_{ij}(\eta_{1i},\eta_{2i},t)}{\partial\eta_{1i}} \quad \text{for } \eta_{1i}=a+l_{1i},\ a\leq\eta_{2i}\leq a+l_{2i}, t\geq 0, \tag{3.110}$$

$$\frac{\partial^2 w_{ij}(\eta_{1i},\eta_{2i},t)}{\partial\eta_{2i}^2}=-\frac{\xi_{ij}}{l_{2i}}\frac{\partial w_{ij}(\eta_{1i},\eta_{2i},t)}{\partial\eta_{2i}} \quad \text{for } \eta_{2i}=a,\ a\leq\eta_{1i}\leq a+l_{1i}, t\geq 0, \tag{3.111}$$

$$\frac{\partial^2 w_{ij}(\eta_{1i},\eta_{2i},t)}{\partial\eta_{2i}^2}=\frac{\xi_{ij}}{l_{2i}}\frac{\partial w_{ij}(\eta_{1i},\eta_{2i},t)}{\partial\eta_{2i}} \quad \text{for } \eta_{2i}=a+l_{2i},\ a\leq\eta_{1i}\leq a+l_{1i}, t\geq 0, \tag{3.112}$$

all for $i = 1,2,\ldots,6, j = 1,2$, where: η_{1i}, η_{2i} are given by (3.98) and l_{1i}, l_{2i} by (3.104), respectively.

Remark 3.6 (The type of boundary conditions for walls panels)
Once again, the edges of the walls panels are considered to be rigidly supported and elastically restrained against rotation. The degree of restraint against rotation on the edges of the jth panel of the ith wall is determined by the value of the boundary restraint coefficients $\xi_{ij}\langle 0,\infty)$, i=1,2,...,6, j=1,2. See Remark 3.3 for more comments and details related to the explanation of the choice of the boundary conditions type. Generally speaking, considering elastically restrained against rotation boundary conditions makes it possible to track the behaviour of the system in case of both perfect ($\xi_{ij} \rightarrow \infty$, fully clamped boundary conditions) and imperfect (ξ_{ij}—nonzero finite) fastening of panels edges by changing the value of boundary restraint coefficients $\xi_{ij}\langle 0,\infty)$, i=1,2,...,6, j=1,2.

Boundary conditions for Eq. (3.96) so for the acoustic field inside the cavity between incident and radiating panels of the ith wall describe interactions in system under discussion and they will be discussed in detail later in the section devoted to fluid-structure interactions.

Since Eqs. (3.95) and (3.97) are second order with respect to time, two initial conditions are formulated for each panel:

$$w_{ij}(\eta_{1i},\eta_{2i},0) = w_{ij}^0(\eta_{1i},\eta_{2i}),\ (\eta_{1i},\eta_{2i}) \in \Omega_{(2[i/2]-1)(2[i/2])}, i = 1,2\ldots,6, j = 1,2, \tag{3.113}$$

$$\left.\frac{\partial w_{ij}(\eta_{1i},\eta_{2i},t)}{\partial t}\right|_{t=0} = w_{ij}^1(\eta_{1i},\eta_{2i}),\ (\eta_{1i},\eta_{2i}) \in \Omega_{(2[i/2]-1)(2[i/2])},$$
$$i = 1,2\ldots,6, j = 1,2, \tag{3.114}$$

where functions $w_{ij}^0(\eta_{1i},\eta_{2i})$ and $w_{ij}^1(\eta_{1i},\eta_{2i})$ $(i = 1,2\ldots,6, j = 1,2)$, are assumed to be given. Function $w_{ij}^0(\eta_{1i},\eta_{2i})$ describes the initial displacement of the jth panel of the ith wall from the reference and stress-free state, whereas $w_{ij}^1(\eta_{1i},\eta_{2i})$ represents the initial distribution of velocity on the surface of the jth panel of the ith wall. Additionally, to assure that the problem is well-posed functions $w_{ij}^0(\eta_{1i},\eta_{2i})$ and $w_{ij}^1(\eta_{1i},\eta_{2i})$ $(i = 1,2\ldots,6, j = 1,2)$, are assumed to satisfy the consistency conditions:

$$\begin{matrix} w_{ij}^0(\eta_{1i},\eta_{2i}) = 0 \\ w_{ij}^1(\eta_{1i},\eta_{2i}) = 0 \end{matrix} \quad \text{for} \begin{cases} \eta_{1i} = a,\ a \le \eta_{2i} \le a + l_{2i}, t \ge 0 \\ \eta_{1i} = a + l_{1i},\ a \le \eta_{2i} \le a + l_{2i}, t \ge 0 \\ \eta_{2i} = a,\ a \le \eta_{1i} \le a + l_{1i}, t \ge 0 \\ \eta_{2i} = a + l_{2i},\ a \le \eta_{1i} \le a + l_{1i}, t \ge 0 \end{cases}, i = 1,2,\ldots,6, j = 1,2. \tag{3.115}$$

Since Eq. (3.96) is second order with respect to time, two initial conditions are formulated for the acoustic wave equation:

$$p_i(x,y,z,0) = p_i^0(x,y,z), (x,y,z) \in \Omega_i, i = 1,2\ldots,6, \tag{3.116}$$

$$\left.\frac{\partial p_i(x,y,z,t)}{\partial t}\right|_{t=0} = p_i^1(x,y,z),(x,y,z) \in \Omega_i, i = 1,2\ldots,6, \tag{3.117}$$

where functions $p_i^0(x,y,z)$ and $p_i^1(x,y,z)$ are assumed to be given. Function $p_i^0(x,y,z)$ describes the initial distribution of acoustic pressure and $p_i^1(x,y,z)$ represents the initial distribution of the rate of change of acoustic pressure in the fluid enclosed within the cavity between panels of the *i*th wall.

Acoustic field in the fluid enclosed in the casing

The behaviour of acoustic field inside the casing through a homogeneous, viscous, isotropic and compressible fluid is described by the nonhomogeneous acoustic wave equation [142, 48]:

$$\begin{aligned}
&\left(\frac{\partial^2}{\partial x^2}+\frac{\partial^2}{\partial y^2}+\frac{\partial^2}{\partial z^2}\right)p(x,y,z,t)\\
&+\mu_0\left(\frac{\partial^2}{\partial x^2}+\frac{\partial^2}{\partial y^2}+\frac{\partial^2}{\partial z^2}\right)\frac{\partial p(x,y,z,t)}{\partial t}-\frac{1}{c_0^2}\frac{\partial^2 p(x,y,z,t)}{\partial t^2}=-\rho_0\sum_{k=1}^{K_s}\frac{\partial q_k}{\partial t},\\
&(x,y,z)\in\Omega_0\ ,\ t>0,\ c_0^2=\frac{\gamma\,\mathrm{P}_0}{\rho_0}.
\end{aligned} \tag{3.118}$$

The right-hand side of Eq. (3.118) represents a rate of change of mass flux per unit area. It is assumed that the compressible fluid contained within the active casing is excited by acoustic waves generated by K_s sources of primary noise of strengths q_k $(k = 1,2,\ldots,K_s)$.

Since Eq. (3.118) is second order with respect to time, two initial conditions are formulated:

$$p(x,y,z,0) = p^0(x,y,z),(x,y,z) \in \Omega_0 \tag{3.119}$$

$$\left.\frac{\partial p(x,y,z,t)}{\partial t}\right|_{t=0} = p^1(x,y,z),(x,y,z) \in \Omega_0, \tag{3.120}$$

where functions $p_0(x,y,z)$ and $p_1(x,y,z)$ are assumed to be given. Function $p_0(x,y,z)$ describes the initial distribution of acoustic pressure and $p_1(x,y,z)$ represents the initial distribution of the rate of change of acoustic pressure in the fluid enclosed within the casing.

Acoustic field in the fluid surrounding the casing

The behaviour of acoustic field outside the casing through a homogeneous, viscous, isotropic and compressible fluid is described by the homogeneous acoustic wave equation [142, 48]:

$$\left(\frac{\partial^2}{\partial x^2}+\frac{\partial^2}{\partial y^2}+\frac{\partial^2}{\partial z^2}\right)p_s(x,y,z,t)$$
$$+\mu_s\left(\frac{\partial^2}{\partial x^2}+\frac{\partial^2}{\partial y^2}+\frac{\partial^2}{\partial z^2}\right)\frac{\partial p_s(x,y,z,t)}{\partial t}-\frac{1}{c_s^2}\frac{\partial^2 p_s(x,y,z,t)}{\partial t^2}=0,$$
$$(x,y,z)\in\Omega_0^{'},\ \mathrm{t}>0, c_s^2=\frac{\gamma_s P_s}{\rho_s}, \tag{3.121}$$

where c_s is the frequency-independent speed of sound, P_s is the mean fluid pressure, ρ_s is the mean fluid density, γ_s is the adiabatic bulk modulus of the fluid, $p_s(x,y,z,t)$ is the sound pressure outside the casing at point $(x,y,z)\in\Omega_0^{'}$, and time $t>0$.

Since Eq. (3.20) is second order with respect to time, two initial conditions are formulated:

$$p_s(x,y,z,0)=p_s^0(x,y,z);(x,y,z)\in\Omega_0^{'} \tag{3.122}$$
$$\left.\frac{p_s(x,y,z,t)}{t}\right|_{t=0}=p_s^1(x,y,z);(x,y,z)\in\Omega_0^{'}, \tag{3.123}$$

where functions $p_s^0(x,y,z)$ and $p_s^1(x,y,z)$ are assumed to be given. Function $p_s^0(x,y,z)$ describes the initial distribution of acoustic pressure and $p_s^0(x,y,z)$ represents the initial distribution of the rate of change of acoustic pressure in the fluid enclosed within the casing.

Fluid-structure interactions in the system

It should be emphasized that acoustic fields propagating in the fluids contained within the walls cavities and casing interior as well as in the fluid in the casing exterior stimulate the vibrations of the casing walls panels. Any solid structure exposed to the sound field in a continuous fluid responds to some degree to the fluctuating pressure acting at the interface between the two media. The fluids pressures on the surface of casing panels are the agents by which the fluids influence the structural motion of the casing panels. This type of fluid-structure interaction is referred to as an interaction of acoustic-vibration character. This phenomenon was incorporated into the model by adding term depending on $p(^\circ,^\circ,^\circ,t)$, $p_i(^\circ,^\circ,^\circ,t)$ $(i=1,2,\ldots,6)$ and $p_s(^\circ,^\circ,^\circ,t)$, in the right-hand side of Eqs. (3.95) and (3.97) describing transverse motion of casing panels. They are described by the functions $f_{1i}(\eta_{1i},\eta_{2i},t)$ and $f_{2i}(\eta_{1i},\eta_{2i},t)$, given by (3.100) and (3.101) being the stimuli (forcing terms) for Eqs. (3.95) and (3.97). The terms $p_i(^\circ,^\circ,^\circ,t)$ $(i=1,2,\ldots,6)$ $p_s(^\circ,^\circ,^\circ,t)$ and $p(^\circ,^\circ,^\circ,t)$ represent a distributed transverse force per unit area acting on the corresponding panels surface and stimulating panels vibrations due to the acoustic fields in the fluids enclosed within the walls cavities and interior of the active casing as well as in the fluid in the casing exterior, respectively. The sign of particular component in the corresponding term depends on direction of the field force that a fluid exerts on the surface of panels being the result of positive increase in pressure of appropriate fluid.

On the other hand, casing walls panels that can be made to vibrate by actuators become sources of sound, and the sound emitted by each panel contributes to the acoustic fields in all fluids being in direct contact with active casing, i.e.:

- acoustic field in the fluid enclosed in the interior of the active casing,
- acoustic fields in the fluids enclosed in the interiors of walls cavities of the active casing,
- acoustic field in the fluid surrounding the active casing from the exterior.

This type of fluid-structure interaction is referred to as an interaction of vibro-acoustic character. The normal surface accelerations of the casing panels are the agents by which the casing structure influence the acoustic fields in the fluids being in contact with the casing structure. The influence of walls panels motion on the acoustic fields in the fluids enclosed within the walls cavities is incorporated into the model by the boundary conditions for acoustic wave Eqs. (3.96) (i = 1,2,...,6) that take the following form:

$$\frac{\partial p_i(x,y,z,t)}{\partial \eta_{3i}} = \begin{cases} -\rho_i \frac{\partial^2 w_{i1}(_{1i},_{2i},t)}{\partial t^2}; a \le \eta_{1i} \le a + l_{1i}, a \le \eta_{2i} \le a + l_{2i},_{3i} = 0, t \ge 0 \\ \rho_i \frac{\partial^2 w_{i2}(_{1i},_{2i},t)}{\partial t^2}; a \le \eta_{1i} \le a + l_{1i}, a \le \eta_{2i} \le a + l_{2i},_{3i} = d_i, t \ge 0 \end{cases} \tag{3.124a}$$

$$\frac{\partial p_i(x,y,z,t)}{\partial \eta_{1i}} = \begin{cases} 0; 0 \le \eta_{3i} \le d_i, a \le \eta_{2i} \le a + l_{2i}, \eta_{1i} = a, t \ge 0, i = 1,3,5, \\ 0; 0 \le \eta_{3i} \le d_i, a \le \eta_{2i} \le a + l_{2i}, \eta_{1i} = l_{1i} + a, t \ge 0, i = 1,3,5, \end{cases} \tag{3.124b}$$

$$\frac{\partial p_i(x,y,z,t)}{\partial \eta_{2i}} = \begin{cases} 0; 0 \le \eta_{3i} \le d_i, a \le \eta_{1i} \le a + l_{1i}, \eta_{2i} = a, t \ge 0, i = 1,3,5, \\ 0; 0 \le \eta_{3i} \le d_i, a \le \eta_{1i} \le a + l_{1i}, \eta_{2i} = l_{2i} + a, t \ge 0, i = 1,3,5, \end{cases} \tag{3.124c}$$

all for $i = 1,3,5$ and:

$$\frac{\partial p_i(x,y,z,t)}{\partial \eta_{3i}} = \begin{cases} \rho_i \frac{\partial^2 w_{i1}(_{1i},_{2i},t)}{\partial t^2}; a \le \eta_{1i} \le a + l_{1i}, a \le \eta_{2i} \le a + l_{2i}, \eta_{3i} = l_{3i} + 2a, \\ -\rho_i \frac{\partial^2 w_{i2}(_{1i},_{2i},t)}{\partial t^2}; a \le \eta_{1i} \le a + l_{1i}, a \le \eta_{2i} \le a + l_{2i}, \eta_{3i} = l_{3i} + 2a - d_i, \end{cases} \tag{3.125a}$$

$$\frac{\partial p_i(x,y,z,t)}{\partial \eta_{1i}} = \begin{cases} 0; l_{3i} + 2a - d_i \le \eta_{3i} \le l_{3i} + 2a, a \le \eta_{2i} \le a + l_{2i}, \eta_{1i} = a, \\ 0; l_{3i} + 2a - d_i \le \eta_{3i} \le l_{3i} + 2a, a \le \eta_{2i} \le a + l_{2i}, \eta_{1i} = l_{1i} + a, \end{cases} \tag{3.125b}$$

$$\frac{\partial p_i(x,y,z,t)}{\partial \eta_{2i}} = \begin{cases} 0; l_{3i} + 2a - d_i \le \eta_{3i} \le l_{3i} + 2a, a \le \eta_{1i} \le a + l_{1i}, \eta_{2i} = a, \\ 0; l_{3i} + 2a - d_i \le \eta_{3i} \le l_{3i} + 2a, a \le \eta_{1i} \le a + l_{1i}, \eta_{2i} = l_{2i} + a, \end{cases} \tag{3.125c}$$

all for $i = 2,4,6$, $t \ge 0$, where: $\eta_{1i}, \eta_{2i}, \eta_{3i}$ are given by (3.98) and:

$$l_{3i} = \begin{cases} l_x, \ i = 2, \\ l_y, i = 4, \\ l_z, i = 6. \end{cases} \tag{3.126}$$

Terms appearing in the right hand side of Eqs. (3.124a) and (3.125a) describe the influence of radiating and incident panels motion on the acoustic field in the fluid inside the cavity of ith wall. Under the assumption that the frame is rigid boundary conditions (3.124b), (3.125b), (3.124c) and (3.125c) that describe behaviour of acoustic fields at the boundaries of the walls cavities that are in contact with the frame are homogeneous (Fig. 3.8).

In turn, the influence of incident panels motion on the acoustic field in the fluid enclosed in the interior of the active casing is incorporated into the model by the nonhomogeneous Neumann boundary conditions for the acoustic wave Eq. (3.118):

$$\frac{\partial p(x,y,z,t)}{\partial x}=\begin{cases}-\rho_0\frac{\partial^2 w_{12}(y,z,t)}{\partial t^2};a\leq y\leq l_y+a,a\leq z\leq l_z+a,x=d_1+h_{12},t\geq 0\\ \rho_0\frac{\partial^2 w_{22}(y,z,t)}{\partial t^2};a\leq y\leq l_y+a,a\leq z\leq l_z+a,x=l_x+2a-d_2-h_{22},t\geq 0\end{cases} \tag{3.127a}$$

$$\frac{\partial p(x,y,z,t)}{\partial y}=\begin{cases}-\rho_0\frac{\partial^2 w_{32}(x,z,t)}{\partial t^2};a\leq x\leq l_x+a,a\leq z\leq l_z+a,y=d_3+h_{32},t\geq 0\\ \rho_0\frac{\partial^2 w_{42}(x,z,t)}{\partial t^2};a\leq x\leq l_x+a,a\leq z\leq l_z+a,y=l_y+2a-d_4-h_{42},t\geq 0\end{cases} \tag{3.127b}$$

$$\frac{\partial p(x,y,z,t)}{\partial z}=\begin{cases}-\rho_0\frac{\partial^2 w_{52}(x,y,t)}{\partial t^2};a\leq x\leq l_x+a,a\leq y\leq l_y+a,z=d_5+h_{52},t\geq 0\\ \rho_0\frac{\partial^2 w_{62}(x,y,t)}{\partial t^2};a\leq x\leq l_x+a,a\leq y\leq l_y+a,z=l_z+2a-d_6-h_{62},t\geq 0\end{cases} \tag{3.127c}$$

The influence of radiating panels motion on the acoustic field in the fluid surrounding the active casing from the exterior is incorporated into the model by the nonhomogeneous Neumann boundary conditions for the homogeneous acoustic wave equation (3.121):

$$\frac{\partial p_s(x,y,z,t)}{\partial x}=\begin{cases}\rho_s\frac{\partial^2 w_{11}(y,z,t)}{\partial t^2};a\leq yl_{\leq}y+a,a\leq z\leq l_z+a,x=-h_{11},t\geq 0\\ -\rho_s\frac{\partial^2 w_{21}(y,z,t)}{\partial t^2};a\leq y\leq l_y+a,a\leq z\leq l_z+a,x=l_x+2a+h_{21},t\geq 0\end{cases} \tag{3.128a}$$

$$\frac{\partial p_s(x,y,z,t)}{\partial y}=\begin{cases}\rho_s\frac{\partial^2 w_{31}(x,z,t)}{\partial t^2};a\leq x\leq l_x+a,a\leq z\leq l_z+a,y=-h_{31},t\geq 0\\ -\rho_s\frac{\partial^2 w_{41}(x,z,t)}{\partial t^2};a\leq x\leq l_x+a,a\leq z\leq l_z+a,y=l_y+2a+h_{41},t\geq 0\end{cases} \tag{3.128b}$$

$$\frac{\partial p_s(x,y,z,t)}{\partial z}=\begin{cases}\rho_s\frac{\partial^2 w_{51}(x,y,t)}{\partial t^2};a\leq x\leq l_x+a,a\leq y\leq l_y+a,z=-h_{51},t\geq 0\\ -\rho_s\frac{\partial^2 w_{61}(x,y,t)}{\partial t^2};a\leq x\leq l_x+a,a\leq y\leq l_y+a,z=l_z+2a+h_{61},t\geq 0\end{cases} \tag{3.128c}$$

Remark 3.7 (Interactions in the model)
Model derived, on the one hand reflects the influence of casing structure vibration on the acoustic fields in fluids being in a direct contact with the casing structure. There are fluid-structure interactions of vibroacoustic character that are modelled by boundary conditions (3.124), (3.125), (3.127) *and* (3.128) *for acoustic wave equations* (3.96), (3.118) *and* (3.121), *respectively. On the other hand, it reflects the influence of considered fluids fields on the structural vibration of the casing structure. There are fluid-structure interactions of acoustic-vibration character that are modelled by functions* $f_{1i}(\eta_{1i},\eta_{2i},t)$ and $f_{2i}(\eta_{1i},\eta_{2i},t)$ *given by* (3.100) *and* (3.101), *respectively.*

Remark 3.8 (Models complexity)
It should be emphasized that although the models of both types of casings are described by the partial differential equations of the same type the double-panel casing model turns out to be much more complicated compared to the single-panel casing

model. The single-panel casing model is described by the system of 7 coupled partial differential equations of higher orders while the double-panel casing model is described by the system of 20 coupled partial differential equations of higher orders. This is the direct consequence of the fact that the number of subsystems that has been singled out in case of double-panel casing for the purpose of model synthesis is much greater compared to single-panel casing.

3.3.3 LIGHTWEIGHT CASING

In this section so called *lightweight* casing is considered (Fig 2.6). As in Section 3.3.1, casing is assumed to be of a rectangular parallelepiped shape consisting of single-panel thin flexible walls. However, in contrast to single-panel rigid casing it is assumed that the flexible walls are interconnected directly to each other without rigid frame forming self-supporting construction (Fig 2.6).

The walls are assumed to be able to radiate sound when made to vibrate. Inside the casing there is an acoustic field due to enclosed certain source of noise (e.g. machine, device, another appliance, etc.).

The notation used within this section (system of coordinates, walls notation, assumptions, etc.) is generally the same as in Section 3.3.1 (Fig. 3.4).

To derive a model of the *lightweight* casing the vibroacoustic system under discussion is split into simpler subsystems. Then, mathematical models for all subsystems are formulated. Finally, by incorporating interactions between the subsystems, their models are interconnected to obtain a complete model of the overall active casing. In the adopted approach the following most important physical phenomena are taken into account:

- acoustic field in the fluid enclosed within the casing due to primary noise source operation and actuator-forced vibration of casing walls.
- dynamics of casing single-panel walls excited both acoustically by the enclosed and surrounding fluids, and mechanically by the actuators.
- fluid-structure interactions including:
  - interactions between casing walls and the acoustic field in the fluid enclosed in the interior of the casing.
  - interactions between casing walls and the acoustic field in fluid surrounding the casing from the outside.
- structure-structure interactions including structural interactions between directly adjacent walls of the casing.
- influence of the fluid loading effect on walls' vibration.
- influence of point operating actuators on walls' vibration.

Model synthesis begins with the analysis of single-panel walls dynamics.

Dynamics of walls excited by the acoustic fields and actuators

In general, the active casing consists of six *light-weight thin flexible single-panel* walls that are connected directly to each other forming self-supporting construction.

Therefore, in the system under discussion structural interactions between directly adjacent walls cannot be neglected. In the case the adjacent walls significantly influence one another. Point-operating actuators (exciters) are bonded to all walls, except the wall at the bottom of the casing (numbered as the 5th wall), i.e. $K_5 = 0$. Masses of the actuators and sensors as well as their size are neglected, what is justified as they can be much smaller compared to the masses and size of corresponding walls. If heavier actuators are used, e.g. electrodynamic ones, it can be necessary to include their mass, what will complicate the model further.

To describe the behaviour of each wall basically it is necessary to model its motion. Moreover, the model has to incorporate the structural interactions between the adjacent walls as well as their interactions with acoustic field enclosed within the interior of the casing and surrounding the casing from the exterior. As it was aforementioned, in contrast to rigid casings considered in Sections 3.3.1 and 3.3.2 in case of lightweight casing adjacent walls significantly influence one another. To incorporate this phenomenon into the model, the motions of each casing wall in three mutually orthogonal directions, i.e., transverse and in-plane longitudinal and shear motion have to be taken into account. Therefore, the dynamics of each wall has to be described by the system of three coupled second order with respect to time partial differential equations.

To make the exposition as concise as possible the description of two selected adjacent walls is presented. Consider two adjacent walls denoted by 2 and 5 (Fig. 3.9). The dynamics of wall 5 is described by the following system of coupled partial differential equations:

$$D_5\left(\frac{\partial^4}{\partial x^4}+2\frac{\partial^4}{\partial x^2\partial y^2}+\frac{\partial^4}{\partial y^4}\right)w_5(x,y,t)+R_5\left(\frac{\partial^4}{\partial x^4}+2\frac{\partial^4}{\partial x^2\partial y^2}+\frac{\partial^4}{\partial y^4}\right)\frac{\partial w_5(x,y,t)}{\partial t}$$
$$+\rho_5 h_5\frac{\partial^2 w_5(x,y,t)}{\partial t^2}=-p(x,y,h_5,t)+p_s(x,y,0,t)$$
$$(x,y)\in\Omega_3,\ t>0, \tag{3.129}$$

$$\left(\frac{\partial^2}{\partial x^2}+\frac{1-\nu_5}{2}\frac{\partial^2}{\partial y^2}\right)w_{5x}(x,y,t)+R_5\left(\frac{\partial^2}{\partial x^2}+\frac{1-\nu_5}{2}\frac{\partial^2}{\partial y^2}\right)\frac{\partial w_{5x}(x,y,t)}{\partial t}$$
$$-\frac{(1-\nu_5^2)\rho_2}{E_5}\frac{\partial^2 w_{5x}(x,y,t)}{\partial t^2}=-\frac{1+\nu_5}{2}\frac{\partial^2}{\partial x\partial y}w_{5y}(x,y,t),$$
$$(x,y)\in\Omega_3,\ t>0, \tag{3.130}$$

$$\left(\frac{\partial^2}{\partial x^2}+\frac{1-\nu_5}{2}\frac{\partial^2}{\partial y^2}\right)w_{5y}(x,y,t)+R_5\left(\frac{\partial^2}{\partial x^2}+\frac{1-\nu_5}{2}\frac{\partial^2}{\partial y^2}\right)\frac{\partial w_{5y}(x,y,t)}{\partial t}$$
$$-\frac{(1-\nu_5^2)\rho_2}{E_5}\frac{\partial^2 w_{5y}(x,y,t)}{\partial t^2}=-\frac{1+\nu_5}{2}\frac{\partial^2}{\partial x\partial y}w_{5x}(x,y,t),$$
$$(x,y)\in\Omega_3,\ t>0. \tag{3.131}$$

In turn, the dynamics of wall 2 is described by the following system of coupled partial differential equations:

$$D_2\left(\frac{\partial^4}{\partial y^4}+2\frac{\partial^4}{\partial z^2\partial y^2}+\frac{\partial^4}{\partial z^4}\right)w_2(y,z,t)+R_2\left(\frac{\partial^4}{\partial y^4}+2\frac{\partial^4}{\partial z^2\partial y^2}+\frac{\partial^4}{\partial z^4}\right)\frac{\partial w_2(y,z,t)}{\partial t}$$
$$+\rho_2 h_2\frac{\partial^2 w_2(y,z,t)}{\partial t^2}=p(l_x-h_2,y,z,t)-p_s(l_x,y,z,t)+u_2(y,z,t)$$
$$(y,z)\in\Omega_1,\ t>0, \tag{3.132}$$

$$\left(\frac{\partial^2}{\partial z^2}+\frac{1-\nu_2}{2}\frac{\partial^2}{\partial y^2}\right)w_{2z}(y,z,t)+R_2\left(\frac{\partial^2}{\partial z^2}+\frac{1-\nu_2}{2}\frac{\partial^2}{\partial y^2}\right)\frac{\partial w_{2z}(y,z,t)}{\partial t}$$
$$-\frac{(1-\nu_2^2)\rho_2}{E_2}\frac{\partial^2 w_{2z}(y,z,t)}{\partial t^2}=-\frac{1+\nu_2}{2}\frac{\partial^2}{\partial z\partial y}w_{2y}(y,z,t),$$
$$(y,z)\in\Omega_1,\ t>0, \tag{3.133}$$

$$\left(\frac{\partial^2}{\partial z^2}+\frac{1-\nu_2}{2}\frac{\partial^2}{\partial y^2}\right)w_{2y}(y,z,t)+R_2\left(\frac{\partial^2}{\partial z^2}+\frac{1-\nu_2}{2}\frac{\partial^2}{\partial y^2}\right)\frac{\partial w_{2y}(y,z,t)}{\partial t}$$
$$-\frac{(1-\nu_2^2)\rho_2}{E_2}\frac{\partial^2 w_{2y}(y,z,t)}{\partial t^2}=-\frac{1+\nu_2}{2}\frac{\partial^2}{\partial z\partial y}w_{2z}(y,z,t)$$
$$(y,z)\in\Omega_1,\ t>0. \tag{3.134}$$

where:

$$D_i=\frac{E_i h_i^3}{12(1-\nu_i^2)},\ i=1,2,\ldots,6 \tag{3.135}$$

$$(\eta_{1i},\eta_{2i})=\begin{cases}(y,z)\in\Omega_1 & \text{for } i=1,2,\\(x,z)\in\Omega_2 & \text{for } i=3,4,\\(x,y)\in\Omega_3 & \text{for } i=5,6,\end{cases}\quad \eta_{3i}=\begin{cases}x,\ i=1,2,\\y,i=3,4,\\z,i=5,6,\end{cases}$$

$$u_i(\eta_{1i},\eta_{2i},t)=\sum_{m=1}^{K_i}B_{\mathrm{im}}F_{\mathrm{im}}(t)\delta(\eta_{1i}-\eta_{1im})\delta(\eta_{2i}-\eta_{2im}),\ i=1,2,\ldots,6, \tag{3.136}$$

$$\delta(\eta_{1i}-\eta_{1im})\delta(\eta_{2i}-\eta_{2im})=\begin{cases}\delta(y-y_{\mathrm{im}})\delta(z-z_{\mathrm{im}}),\ i=1,2,\\\delta(x-x_{\mathrm{im}})\delta(z-z_{\mathrm{im}}),i=3,4,\\\delta(x-x_{\mathrm{im}})\delta(y-y_{\mathrm{im}}),i=5,6,\end{cases} \tag{3.137}$$

$$\eta_{1im}=\begin{cases}y_{\mathrm{im}},\ i=1,2,\\x_{\mathrm{im}},i=3,4,5,6,\end{cases},\eta_{2im}=\begin{cases}z_{\mathrm{im}},\ i=1,2,3,4,\\y_{\mathrm{im}},i=5,6,\end{cases} \tag{3.138}$$

Remark 3.9 (Dependence of in-plane motions)
It should be pointed out the that in-plane motions of wall strongly dependent one another. It means that longitudinal motion depends on shear motion and vice versa. Actually the solution of Eq. (3.131), $w_{5y}(x,y,t)$, *say longitudinal displacement appears in the right-hand side of Eq.* (3.130), *so it can be treated as a stimulus for equation describing shear motion. On the other hand, the solution of Eq.* (3.130), $w_{5x}(x,y,t)$, *say shear displacement appears in the right-hand side of Eq.* (3.131), *so it can be treated as a stimulus for equation describing longitudinal motion.*

Equations (3.129) and (3.132) describe flexural (transverse) vibrations while Eqs. (3.130), (3.131) and (3.133), (3.134) (longitudinal and shear) in-plane vibrations of abutting walls 5 and 2, respectively. Functions $w_i(\eta_{1i},\eta_{2i})$ describe transverse displacements while $w_{in}(\eta_{1i},\eta_{2i})$ $(i=1,2\ldots,6,n=x,y,z)$ describe in-plane (longitudinal and shear) displacements in n-direction of the ith wall from the reference and stress-free state.

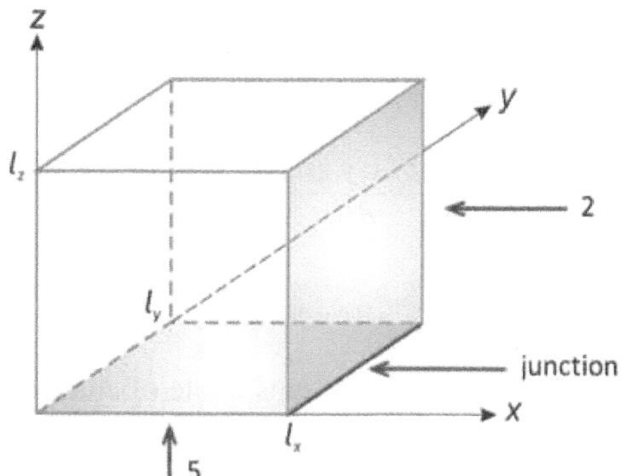

Figure 3.9 A lightweight casing with the system of coordinates and walls 2 and 5 with junction.

Since, Eqs. (3.129) to (3.134) are second order with respect to time, two initial conditions are formulated for each of Eq.:

$$w_2(y,z,0)=w_2^0(y,z),\quad w_5(x,y,0)=w_5^0(x,y), \tag{3.139a}$$

$$w_{2z}(y,z,0)=w_{2z}^0(y,z),\quad w_{5x}(x,y,0)=w_{5x}^0(x,y), \tag{3.139b}$$

$$w_{2y}(y,z,0)=w_{2y}^0(y,z),\quad w_{5y}(x,y,0)=w_{5y}^0(x,y), \tag{3.139c}$$

$$\left.\frac{w_2(y,z,t)}{\partial t}\right|_{t=0}=w_2^1(y,z),\quad \left.\frac{w_5(x,y,t)}{\partial t}\right|_{t=0}=w_5^1(x,y), \tag{3.139d}$$

$$\left.\frac{w_{2z}(y,z,t)}{\partial t}\right|_{t=0}=w_{2z}^1(y,z),\quad \left.\frac{w_{5x}(x,y,t)}{\partial t}\right|_{t=0}=w_{5x}^1(x,y), \tag{3.139e}$$

$$\left.\frac{w_{2y}(y,z,t)}{\partial t}\right|_{t=0} = w_{2y}^{1}(y,z)\,, \quad \left.\frac{w_{5y}(x,y,t)}{\partial t}\right|_{t=0} = w_{5y}^{1}(x,y)\,, \tag{3.139f}$$

where functions $w_i^k(\eta_{1i},\eta_{2i})$, $w_{in}^k(\eta_{1i},\eta_{2i})$ $(i=1,2\ldots,6,,k=0,1,n=x,y,z)$ are assumed to be given. Functions $w_i^0(\eta_{1i},\eta_{2i})$ and $w_{in}^0(\eta_{1i},\eta_{2i})$ describe the initial transverse and in-plane displacements of the ith wall from the reference and stress-free state, whereas $w_i^1(\eta_{1i},\eta_{2i})$ and $w_{\text{in}}^1(\eta_{1i},\eta_{2i})$ represents the initial distribution of transverse and in-plane velocities on the surface of the ith wall.

Function $u_i(\eta_{1i},\eta_{2i},t)$ $(i=1,2,\ldots,6,j=1,2)$ given by (3.136) describes control corresponding to point-operating actuators bonded to the surface of the ith wall.

Acoustic field in the fluid enclosed in the casing

The behaviour of acoustic field inside the casing through a homogeneous, viscous, isotropic and compressible fluid is described by the nonhomogeneous acoustic wave equation [142, 48]:

$$\begin{aligned}&\left(\frac{\partial^2}{\partial x^2}+\frac{\partial^2}{\partial y^2}+\frac{\partial^2}{\partial z^2}\right)p(x,y,z,t)\\&+\mu_0\left(\frac{\partial^2}{\partial x^2}+\frac{\partial^2}{\partial y^2}+\frac{\partial^2}{\partial z^2}\right)\frac{\partial p(x,y,z,t)}{\partial t}-\frac{1}{c_0^2}\frac{\partial^2 p(x,y,z,t)}{\partial t^2}=-\rho_0\sum_{k=1}^{K_s}\frac{\partial q_k}{\partial t}\,,\\&(x,y,z)\in\Omega_0,\ t>0,\ c_0^2=\frac{\gamma P_0}{\rho_0}.\end{aligned} \tag{3.140}$$

The right-hand side of Eq. (3.140) represents a rate of change of mass flux per unit area. It is assumed that the compressible fluid contained within the active casing is excited by acoustic waves generated by K_s sources of primary noise of strengths q_k $(k=1,2,\ldots,K_s)$.

Since Eq. (3.140) is second order with respect to time, two initial conditions are formulated:

$$p(x,y,z,0)=p^0(x,y,z)\,,(x,y,z)\in\Omega_0, \tag{3.141a}$$

$$\left.\frac{\partial p(x,y,z,t)}{\partial t}\right|_{t=0}=p^1(x,y,z)\,,(x,y,z)\in\Omega_0, \tag{3.141b}$$

where functions $p_0(x,y,z)$ and $p_1(x,y,z)$ are assumed to be given. Function $p_0(x,y,z)$ describes the initial distribution of acoustic pressure and $p_1(x,y,z)$ represents the initial distribution of the rate of change of acoustic pressure in the fluid enclosed within the casing.

Acoustic field in the fluid surrounding the casing

The behaviour of acoustic field outside the casing through a homogeneous, viscous, isotropic and compressible fluid is described by the homogeneous acoustic wave equation [142, 48]:

$$\left(\frac{\partial^2}{\partial x^2}+\frac{\partial^2}{\partial y^2}+\frac{\partial^2}{\partial z^2}\right)p_s(x,y,z,t)$$
$$+\mu_s\left(\frac{\partial^2}{\partial x^2}+\frac{\partial^2}{\partial y^2}+\frac{\partial^2}{\partial z^2}\right)\frac{\partial p_s(x,y,z,t)}{\partial t}-\frac{1}{c_s^2}\frac{\partial^2 p_s(x,y,z,t)}{\partial t^2}=0,$$
$$(x,y,z)\in\Omega_0^{'},\ t>0, c_s^2=\frac{\gamma_s P_s}{\rho_s}, \tag{3.142}$$

where c_s is the frequency-independent speed of sound, P_s is the mean fluid pressure, ρ_s is the mean fluid density, γ_s is the adiabatic bulk modulus of the fluid, $p_s(x,y,z,t)$ is the sound pressure outside the casing at point $(x,y,z)\in\Omega_0^{'}$, and time $t>0$.

Since Eq. (3.140) is second order with respect to time, two initial conditions are formulated:

$$p_s(x,y,z,0)=p_s^0(x,y,z),(x,y,z)\in\Omega_0^{'}, \tag{3.143a}$$

$$\left.\frac{\partial p_s(x,y,z,t)}{\partial t}\right|_{t=0}=p_s^1(x,y,z),(x,y,z)\in\Omega_0^{'}, \tag{3.143b}$$

where functions $p_s^0(x,y,z)$ and $p_s^1(x,y,z)$ are assumed to be given. Function $p_s^0(x,y,z)$ describes the initial distribution of acoustic pressure and $p_s^0(x,y,z)$ represents the initial distribution of the rate of change of acoustic pressure in the fluid enclosed within the casing.

Interactions in the system

Taking into account mechanical construction of the lightweight casing two basic type of interactions can be distinguished in the system:

- fluid-structure interactions that describes how the fluids being in a direct contact with mechanical structure of the casing interact each other. They can be divided into:
    - interactions of vibroacoustic character and,
    - interactions of acoustic-vibration character.
- structure-structure interactions that describe how abutting walls interact one another at the wall/wall junction.

Fluid-structure interactions

The fluids pressures on the surface of casing walls are the agents by which the fluids influence the structural motion of the casing walls. This type of fluid-structure interaction is referred to as an interaction of acoustic-vibration character. This phenomenon was incorporated into the model by adding term depending on $p(^\circ,^\circ,^\circ,t)$, and $p_s(^\circ,^\circ,^\circ,t)$, in the right-hand side of Eqs. (3.129) and (3.132) describing transverse motion of casing walls. They are the stimuli (forcing term) for Eqs. (3.129) and (3.132). The terms $p_s(^\circ,^\circ,^\circ,t)$ and $p(^\circ,^\circ,^\circ,t)$ represent a distributed transverse force per unit area acting on the corresponding walls surface and stimulating walls vibrations due to the acoustic fields in the fluids enclosed within the interior of the active casing as well as in the fluid in the casing exterior, respectively. The sign of particular component in the corresponding term depends on direction of the field force that a fluid exerts on the surface of walls being the result of positive increase in pressure of appropriate fluid.

On the other hand, casing walls that can be made to vibrate by actuators become sources of sound, and the sound emitted by each wall contributes to the acoustic fields in all fluids being in direct contact with active casing, i.e.:

- acoustic field in the fluid enclosed in the interior of the active casing,
- acoustic field in the fluid surrounding the active casing from the exterior.

This type of fluid-structure interaction is referred to as an interaction of vibro-acoustic character. The normal surface accelerations of the casing walls are the agents by which the casing structure influence the acoustic fields in the fluids being in contact with the casing structure. The influence of walls motion on the acoustic fields in the fluids enclosed within the interior of the casing and surrounding the casing from the exterior is incorporated into the model by the boundary conditions for acoustic wave Eqs. (3.140) and (3.142). The influence of walls motion on the acoustic field in the fluid enclosed in the interior of the active casing is incorporated into the model by the nonhomogeneous Neumann boundary conditions for the nonhomogeneous acoustic wave equation (3.140):

$$\frac{\partial p(x,y,z,t)}{\partial x} = \begin{cases} -\rho_0 \frac{\partial^2 w_1(y,z,t)}{\partial t^2}; h_3 \leq y \leq l_y - h_4, h_5 \leq z \leq l_z - h_6, x = h_1, t \geq 0 \\ +\rho_0 \frac{\partial^2 w_2(y,z,t)}{\partial t^2}; h_3 \leq y \leq l_y - h_4, h_5 \leq z \leq l_z - h_6, x = l_x - h_2, t \geq 0 \end{cases} \tag{3.144a}$$

$$\frac{\partial p(x,y,z,t)}{\partial y} = \begin{cases} -\rho_0 \frac{\partial^2 w_3(x,z,t)}{\partial t^2}; h_1 \leq x \leq l_x - h_2, h_5 \leq z \leq l_z - h_6, y = h_3, t \geq 0 \\ +\rho_0 \frac{\partial^2 w_4(x,z,t)}{\partial t^2}; h_1 \leq x \leq l_x - h_2, h_5 \leq z \leq l_z - h_6, y = l_y - h_4, t \geq 0 \end{cases} \tag{3.144b}$$

$$\frac{\partial p(x,y,z,t)}{\partial z} = \begin{cases} -\rho_0 \frac{\partial^2 w_5(x,y,t)}{\partial t^2}; h_1 \leq x \leq l_x - h_2, h_3 \leq y \leq l_y - h_4, z = h_5, t \geq 0 \\ +\rho_0 \frac{\partial^2 w_6(x,y,t)}{\partial t^2}; h_1 \leq x \leq l_x - h_2, h_3 \leq y \leq l_y - h_4, z = l_z - h_6, t \geq 0 \end{cases} \tag{3.144c}$$

The influence of walls motion on the acoustic field in the fluid surrounding the active casing from the exterior is incorporated into the model by the nonhomogeneous Neumann boundary conditions for the homogeneous acoustic wave equation (3.142):

$$\frac{\partial p_s(x,y,z,t)}{\partial x} = \begin{cases} \rho_s \frac{\partial^2 w_1(y,z,t)}{\partial t^2}; 0 \leq y \leq l_y, 0 \leq z \leq l_z, x = 0, t \geq 0 \\ -\rho_s \frac{\partial^2 w_2(y,z,t)}{\partial t^2}; 0 \leq y \leq l_y, 0 \leq z \leq l_z, x = l_x, t \geq 0 \end{cases} \tag{3.145a}$$

$$\frac{\partial p_s(x,y,z,t)}{\partial y} = \begin{cases} \rho_s \frac{\partial^2 w_3(x,z,t)}{\partial t^2}; 0 \le x l_{\le} x, 0 \le z \le l_z, y = 0, t \ \ge 0 \\ -\rho_s \frac{\partial^2 w_4(x,z,t)}{\partial t^2}; 0 \le x \le l_x, 0 \le z \le l_z, y = l_y, t \ \ge 0 \end{cases} \tag{3.145b}$$

$$\frac{\partial p_s(x,y,z,t)}{\partial z} = \begin{cases} \rho_s \frac{\partial^2 w_5(x,y,t)}{\partial t^2}; 0 \le x \le l_x, 0 \le y \le l_y, z = 0, t \ \ge 0 \\ -\rho_s \frac{\partial^2 w_6(x,y,t)}{\partial t^2}; 0 \le x \le l_x, 0 \le y \le l_y, z = l_z, t \ \ge 0 \end{cases} \tag{3.145c}$$

Structure-structure interactions

In contrast to rigid casings, in case of lightweight casing that forms a self-supporting structure by direct connections between adjacent walls without rigid frame interactions between abutting walls are significant and cannot be neglected. Therefore, to construct a reliable model in case of this vibroacoustic system interactions between abutting walls have to be incorporated into the model. Wall-wall interactions are incorporated into the model by synthesizing appropriate boundary conditions for partial differential equations describing transverse and in-plane motions of abutting casing walls.

The boundary conditions for the system under consideration can be formulated by applying the following conditions:

- continuity of walls displacements at the wall/wall junction,
- continuity of walls slope (rotation) at the wall/wall junction,
- equilibrium of forces at the wall/wall junction,
- equilibrium of moments at the wall/wall junction.

In order to apply above-mentioned conditions to derive boundary conditions the following forces and moments acting in the walls 2 and 5 are identified and described as follows:

- the shear forces due to bending Q_{5xx}, Q_{5yy}, Q_{2zz}, Q_{2yy}:

$$Q_{5yy} = -D_5 \left(\frac{\partial^3 w_5(x,y,t)}{\partial x^3} + \frac{\partial^3 w_5(x,y,t)}{\partial x \partial y^2} \right), \tag{3.146a}$$

$$Q_{5xx} = -D_5 \left(\frac{\partial^3 w_5(x,y,t)}{\partial y^3} + \frac{\partial^3 w_5(x,y,t)}{\partial x^2 \partial y} \right), \tag{3.146b}$$

$$Q_{2yy} = -D_2 \left(\frac{\partial^3 w_2(y,z,t)}{\partial z^3} + \frac{\partial^3 w_2(y,z,t)}{\partial z \partial y^2} \right), \tag{3.146c}$$

$$Q_{2zz} = -D_2 \left(\frac{\partial^3 w_2(y,z,t)}{\partial y^3} + \frac{\partial^3 w_2(y,z,t)}{\partial y \partial z^2} \right), \tag{3.146d}$$

- the net vertical shear forces (transverse shearing forces) V_{5x}, V_{5y}, V_{2z}, V_{2y}:

$$V_{5y} = Q_{5yy} + \frac{\partial M_{5xy}}{\partial y} = -D_5 \left(\frac{\partial^3 w_5(x,y,t)}{\partial x^3} + (2 - \nu_5) \frac{\partial^3 w_5(x,y,t)}{\partial x \partial y^2} \right), \quad (3.147a)$$

$$V_{5x} = Q_{5xx} + \frac{\partial M_{5xy}}{\partial x} = -D_5 \left(\frac{\partial^3 w_5(x,y,t)}{\partial y^3} + (2 - \nu_5) \frac{\partial^3 w_5(x,y,t)}{\partial x^2 \partial y} \right), \quad (3.147b)$$

$$V_{2y} = Q_{2yy} + \frac{\partial M_{2zy}}{\partial y} = -D_2 \left(\frac{\partial^3 w_2(y,z,t)}{\partial z^3} + (2 - \nu_2) \frac{\partial^3 w_2(y,z,t)}{\partial z \partial y^2} \right), \quad (3.147c)$$

$$V_{2z} = Q_{2zz} + \frac{\partial M_{2zy}}{\partial z} = -D_2 \left(\frac{\partial^3 w_2(y,z,t)}{\partial y^3} + (2 - \nu_2) \frac{\partial^3 w_2(y,z,t)}{\partial z^2 \partial y} \right), \quad (3.147d)$$

- the in-plane longitudinal forces N_{5xx}, N_{5yy}, N_{2zz}, N_{2yy}:

$$N_{5yy} = \frac{E_5 h_5}{1 - \nu_5^2} \left(\frac{\partial w_{5x}(x,y,t)}{\partial x} + \nu_5 \frac{\partial w_{5y}(x,y,t)}{\partial y} \right), \quad (3.148a)$$

$$N_{5xx} = \frac{E_5 h_5}{1 - \nu_5^2} \left(\frac{\partial w_{5y}(x,y,t)}{\partial y} + \nu_5 \frac{\partial w_{5x}(x,y,t)}{\partial x} \right), \quad (3.148b)$$

$$N_{2yy} = \frac{E_2 h_2}{1 - \nu_2^2} \left(\frac{\partial w_{2z}(y,z,t)}{\partial z} + \nu_2 \frac{\partial w_{2y}(y,z,t)}{\partial y} \right), \quad (3.148c)$$

$$N_{2zz} = \frac{E_2 h_2}{1 - \nu_2^2} \left(\frac{\partial w_{2y}(y,z,t)}{\partial y} + \nu_2 \frac{\partial w_{2z}(y,z,t)}{\partial z} \right), \quad (3.148d)$$

- the in-plane shear force N_{5xy}, N_{2yz}:

$$N_{5xy} = \frac{E_5 h_5}{2(1 + \nu_5)} \left(\frac{\partial w_{5x}(x,y,t)}{\partial y} + \frac{\partial w_{5y}(x,y,t)}{\partial x} \right), \quad (3.149a)$$

$$N_{2yz} = \frac{E_2 h_2}{2(1 + \nu_2)} \left(\frac{\partial w_{2z}(y,z,t)}{\partial y} + \frac{\partial w_{2y}(y,z,t)}{\partial z} \right), \quad (3.149b)$$

- the twisting moments M_{5xy}, M_{2yz} :

$$M_{5xy} = -D_5 (1 - \nu_5) \frac{\partial^2 w_5(x,y,t)}{\partial x \partial y}, \quad (3.150a)$$

$$M_{2yz} = -D_2 (1 - \nu_2) \frac{\partial^2 w_2(y,z,t)}{\partial y \partial z}, \quad (3.150b)$$

- the bending moments M_{5xx}, M_{5yy}, M_{2xx}, M_{2yy} :

$$M_{5yy} = -D_5\left(\frac{\partial^2 w_5(x,y,t)}{\partial x^2} + \nu_5\frac{\partial^2 w_5(x,y,t)}{\partial y^2}\right), \tag{3.151a}$$

$$M_{5xx} = -D_5\left(\frac{\partial^2 w_5(x,y,t)}{\partial y^2} + \nu_5\frac{\partial^2 w_5(x,y,t)}{\partial x^2}\right), \tag{3.151b}$$

$$M_{2yy} = -D_2\left(\frac{\partial^2 w_2(y,z,t)}{\partial z^2} + \nu_2\frac{\partial^2 w_2(y,z,t)}{\partial y^2}\right), \tag{3.151c}$$

$$M_{2zz} = -D_2\left(\frac{\partial^2 w_2(y,z,t)}{\partial y^2} + \nu_2\frac{\partial^2 w_2(y,z,t)}{\partial z^2}\right). \tag{3.151d}$$

Fig. 3.10 shows all displacements, forces and moments that have to be taken into account in the formulation of boundary condition at the junction between walls 2 and 5. Taking into account spatial distribution of displacements shown in Fig. 3.10, it can be easily seen that the requirement of continuity of walls displacements at the wall/wall junction leads to the following boundary conditions:

$$w_5(x,y,t) = w_{2z}(y,z,t), \tag{3.152a}$$

$$w_{5x}(x,y,t) = w_2(y,z,t), \tag{3.152b}$$

$$w_{5y}(x,y,t) = w_{2y}(y,z,t), \tag{3.152c}$$

all for $x = l_x$, $z = 0$, $y \in (0, l_y)$, $t > 0$.

Taking into account the continuity of plates slope (rotation) at the wall/wall junction the next boundary condition can be constructed:

$$\left.\frac{\partial w_5(x,y,t)}{\partial x}\right|_{x=l_x} = \left.\frac{\partial w_2(y,z,t)}{\partial z}\right|_{z=0} \quad \text{for} \quad y \in (0,l_y),\ t > 0. \tag{3.153}$$

Taking into account spatial distribution of forces shown in Fig. 3.10, on the basis of equilibrium of forces at the wall/wall junction the following boundary conditions are formulated:

$$V_{5y}(x,y,t) = N_{2yy}(y,z,t), \tag{3.154a}$$

$$N_{5yy}(x,y,t) = -V_{2y}(y,z,t), \tag{3.154b}$$

$$N_{5xy}(x,y,t) = N_{2yz}(y,z,t), \tag{3.154c}$$

all for $x = l_x$, $z = 0$, $y \in (0, l_y)$, $t > 0$.

Finally, taking into account spatial distribution of moments shown in Fig. 3.10, on the basis of equilibrium of moments at the wall/wall junction the following boundary condition is formulated:

$$M_{5yy}(x,y,t) = -M_{2yy}(y,z,t) \tag{3.155}$$

for $x = l_x$, $z = 0$, $y \in (0, l_y)$, $t > 0$.

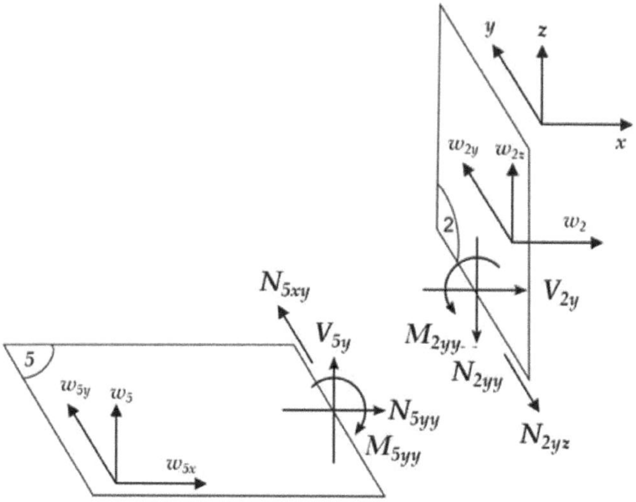

Figure 3.10 Displacements, forces and moments at the junction between abutting walls 2 and 5.

Taking into account Eqs. (3.147a), (3.147c), (3.148a), (3.148c), (3.149a), (3.149b), (3.151a), (3.151c) describing forces and moments appearing in Eqs. (3.152)–(3.155) boundary conditions can be rewritten in a straightforward form:

$$w_5(x,y,t) = w_{2z}(y,z,t), \tag{3.156a}$$

$$w_{5x}(x,y,t) = w_2(y,z,t), \tag{3.156b}$$

$$w_{5y}(x,y,t) = w_{2y}(y,z,t), \tag{3.156c}$$

$$\left.\frac{\partial w_5(x,y,t)}{\partial x}\right|_{x=l_x} = \left.\frac{\partial w_2(y,z,t)}{\partial z}\right|_{z=0}, \tag{3.156d}$$

$$D_5\left(\frac{\partial^3 w_5(x,y,t)}{\partial x^3} + (2-\nu_5)\frac{\partial^3 w_5(x,y,t)}{\partial x \partial y^2}\right)\Bigg|_{x=l_x}$$
$$= -\frac{E_2 h_2}{1-\nu_2^2}\left(\frac{\partial w_{2z}(y,z,t)}{\partial z} + \nu_2 \frac{\partial w_{2y}(y,z,t)}{\partial y}\right)\Bigg|_{z=0}, \tag{3.156e}$$

$$\frac{E_5 h_5}{1-\nu_5^2}\left(\frac{\partial w_{5x}(x,y,t)}{\partial x} + \nu_5 \frac{\partial w_{5y}(x,y,t)}{\partial y}\right)\Bigg|_{x=l_x}$$
$$= D_2\left(\frac{\partial^3 w_2(y,z,t)}{\partial z^3} + (2-\nu_2)\frac{\partial^3 w_2(y,z,t)}{\partial z \partial y^2}\right)\Bigg|_{z=0}, \tag{3.156f}$$

$$\frac{E_5h_5}{2(1+\nu_5)}\left(\frac{\partial w_{5x}(x,y,t)}{\partial y}+\frac{\partial w_{5y}(x,y,t)}{\partial x}\right)\Bigg|_{x=l_x}$$
$$=\frac{E_2h_2}{2(1+\nu_2)}\left(\frac{\partial w_{2z}(y,z,t)}{\partial y}+\frac{\partial w_{2y}(y,z,t)}{\partial z}\right)\Bigg|_{z=0}, \quad (3.156g)$$

$$D_2\left(\frac{\partial^2 w_2(y,z,t)}{\partial z^2}+\nu_2\frac{\partial^2 w_2(y,z,t)}{\partial y^2}\right)\Bigg|_{z=0}$$
$$=-D_5\left(\frac{\partial^2 w_5(x,y,t)}{\partial x^2}+\nu_5\frac{\partial^2 w_5(x,y,t)}{\partial y^2}\right)\Bigg|_{x=l_x}, \quad (3.156h)$$

all for $x = l_x$, $z = 0$, $y \in (0, l_y)$, $t > 0$.

Eqs. (3.156) determines boundary conditions that relate to one pair of casing walls 2–5.

Remark 3.10 (Model of the whole casing)
Considerations presented take into account only one of twelve pairs of casing walls (Fig. 3.9). To construct the complete model for the whole lightweight casing the considerations presented should be repeated with respect to all other pairs of casing walls (wall/wall junctions).

Remark 3.11 (Model complexity)
The complexity of the model of lightweight single-panel casing is comparable with the model of double-panel rigid casing. In both cases the dynamics of each flexible wall of the casing is described by the system of three coupled partial differential equations. Both models are described by the system of 20 coupled partial differential equations. In case of rigid double-panel casing PDEs describing each wall are second order with respect to time and two of them are fourth order with respect to spatial variables while one is second order. In case of lightweight single-panel casing all PDEs describing each wall are once again second order with respect to time and two of them are second order with respect to spatial variables while one fourth order. So the model of lightweight casing seems to be slightly simpler as far as the complexity of PDEs describing the model is concerned. However, boundary conditions for the model of lightweight casing are more complicated compared to those formulated in the model of double-panel rigid casing.

3.4 NUMERICAL MODELLING

Simulation that can be treated as application of the model to the problem follows modelling. In this context, the act of simulating something first requires that a model to be developed what is reported in Sections 3.2 and 3.3. Models developed in this sections represent the key characteristics of vibroacoustic systems being analysed. Simulation, that is the imitation of the operation of a real-world system over time, is also strongly related to answering the question of deriving of an effective strategy to solve the equations describing the model. In many cases, general purpose software packages (e.g. Matlab, Mathematica, etc.) may be used to implement this strategy in practice. Beyond this, it may be necessary to write dedicated software in cases where the mathematical model under consideration involves nonstandard equations.

In case under consideration both strategies are taken into account. The general purpose software packages are used to solve the task of simulating model of the whole casing and composite structures. Optimization problems aimed at shaping of structural and acoustic responses of vibroacoustic systems under discussion (Section 4.2) as well as optimization of actuators and sensors arrangement and optimization of error microphones (Section 6.2) are solved by original dedicated software implemented by authors for these purposes.

Results reported in Section 4.4.1 related to functionally-graded materials are obtained by simulations performed in ANSYS software package. Details regarding employed numerical modelling of vibrating plate in ANSYS are given in Section 3.4.1.

For producing trustworthy results showing the operation of the whole active casing over time a numerical model in COMSOL Multiphysics is built. Section 3.4.2 is devoted to this.

3.4.1 NUMERICAL SIMULATIONS OF A SINGLE PLATE

Vibrating Panel Modelling

This section uses first order deformation theory and finite element methods to model a structural casing for vibroacoustic computation. One side of the casing is bounded by a fully clamped vibrating panel while the other sides are bounded by the casing walls as illustrated in Fig. 3.11.

Fig. 3.12 represents a vibrating structural panel with thickness h and at a mid-surface distance z, placed in an acoustic enclosure medium. By applying the first-order shear deformation theory and taking the vibrating structural panel as a shell (i.e., the thickness of the vibrating panel is far smaller than its length and width), the displacement field can be established by the following equations

$$\overline{U}(x,y,z,t) = \overline{u}(x,y,t) + z\beta_x(x,y,t) \tag{3.157}$$

$$\overline{V}(x,y,z,t) = \overline{v}(x,y,t) + z\beta_y(x,y,t) \tag{3.158}$$

$$\overline{W}(x,y,z,t) = \overline{w}(x,y,t) \tag{3.159}$$

Eqs. (3.157)–(3.159) show the displacement functions $\overline{u}$, $\overline{v}$ and $\overline{w}$ at the mid-surface along x, y and z directions, respectively. Also, the symbols $\beta_x(x,y,t) = -\frac{\partial \overline{w}}{\partial x}$ and $\beta_y(x,y,t) = -\frac{\partial \overline{w}}{\partial y}$ denote rotations of the normal in yz and xz planes, respectively with time (t). By applying the linear small strain-displacement relations along the middle plane surface, the deformations are defined and expressed as

$$\varepsilon_x = \frac{\partial \overline{u}}{\partial x} - z\frac{\partial^2 \overline{w}}{\partial x^2} \quad ; \quad \varepsilon_y = \frac{\partial \overline{v}}{\partial y} - z\frac{\partial^2 \overline{w}}{\partial y^2} \tag{3.160}$$

$$\gamma_{xy} = \frac{\partial \overline{u}}{\partial y} + \frac{\partial \overline{v}}{\partial x} - 2z\frac{\partial^2 \overline{w}}{\partial x^2} \tag{3.161}$$

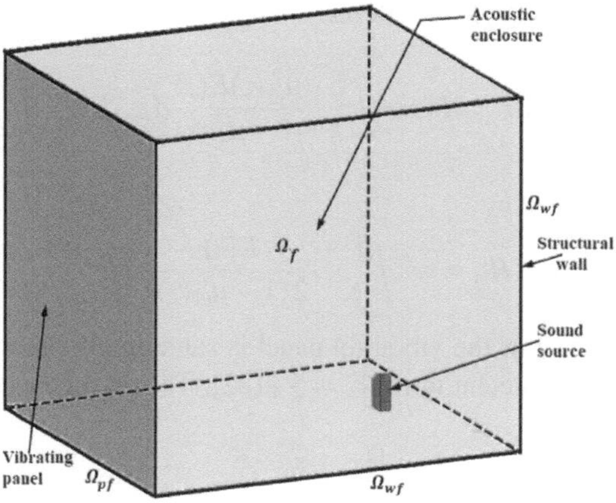

Figure 3.11 Diagrammatic illustration of a typical structural casing for vibroacoustic modelling.

$$\gamma_{xz} = \frac{\partial \overline{w}}{\partial x} \quad ; \quad \gamma_{yz} = \frac{\partial \overline{w}}{\partial y} \tag{3.162}$$

where the strains ε_x, ε_y and γ_{xy} are defined as the in-plane meridional, circumferential and shearing strain components, respectively. The transverse shear terms along xz and yz plane are γ_{xz} and γ_{yz}, respectively. Owing to the thin-walled structural characteristics, it is assumed that the curvature of the shell structure is negligible such that $\chi_x = \chi_y = \chi_{xy} = 0$. Let $\overline{N}$ be the in-plane force resultant, $\overline{M}$ the bending and twisting moment force resultant and $\overline{R}$ be the transverse shear. Describing the stress $\sigma(x,y)$ in relationship to its Poisson ratio μ_{ef}, in components form, it is given by

$$\begin{Bmatrix} \overline{N}_x \\ \overline{N}_y \\ \overline{N}_{xy} \end{Bmatrix} = A^c \begin{bmatrix} 1 & \mu_{ef} & 0 \\ \mu_{ef} & 0 & 0 \\ 0 & 0 & \frac{1-\mu_{ef}}{2} \end{bmatrix} \begin{Bmatrix} \frac{\partial \overline{u}}{\partial x} \\ \frac{\partial \overline{v}}{\partial y} \\ \frac{\partial \overline{u}}{\partial y} + \frac{\partial \overline{v}}{\partial x} \end{Bmatrix} + B^c \begin{bmatrix} 1 & \mu_{ef} & 0 \\ \mu_{ef} & 0 & 0 \\ 0 & 0 & \frac{1-\mu_{ef}}{2} \end{bmatrix} \begin{Bmatrix} -\frac{\partial^2 \overline{w}}{\partial x^2} \\ -\frac{\partial^2 \overline{w}}{\partial y^2} \\ -2\frac{\partial^2 \overline{w}}{\partial x \partial y} \end{Bmatrix} \tag{3.163}$$

$$\begin{Bmatrix} \overline{M}_x \\ \overline{M}_y \\ \overline{M}_{xy} \end{Bmatrix} = B^c \begin{bmatrix} 1 & \mu_{ef} & 0 \\ \mu_{ef} & 0 & 0 \\ 0 & 0 & \frac{1-\mu_{ef}}{2} \end{bmatrix} \begin{Bmatrix} \frac{\partial \overline{u}}{\partial x} \\ \frac{\partial \overline{v}}{\partial y} \\ \frac{\partial \overline{u}}{\partial y} + \frac{\partial \overline{v}}{\partial x} \end{Bmatrix} + F^c \begin{bmatrix} 1 & \mu_{ef} & 0 \\ \mu_{ef} & 0 & 0 \\ 0 & 0 & \frac{1-\mu_{ef}}{2} \end{bmatrix} \begin{Bmatrix} -\frac{\partial^2 \overline{w}}{\partial x^2} \\ -\frac{\partial^2 \overline{w}}{\partial y^2} \\ -2\frac{\partial^2 \overline{w}}{\partial x \partial y} \end{Bmatrix} \tag{3.164}$$

$$\begin{Bmatrix} \overline{R}_x \\ \overline{R}_y \end{Bmatrix} = H^c \begin{bmatrix} 1 & 0 \\ 0 & 1 \end{bmatrix} \begin{Bmatrix} \frac{\partial \overline{w}}{\partial x} \\ \frac{\partial \overline{w}}{\partial y} \end{Bmatrix} \tag{3.165}$$

Where the terms A^c, B^c, F^c and H^c correspond to the stiffness coefficients and can be realized using the following relations

$$A_{ij}^c = \int_{-\frac{h}{2}}^{\frac{h}{2}} \frac{E(z)}{1-\mu_{\text{ef}}^2(z)} dz \ ; \ B_{ij}^c = \int_{-\frac{h}{2}}^{\frac{h}{2}} \frac{\mu_{\text{ef}}^2(z)E(z)}{1-\mu_{\text{ef}}^2(z)} dz; \ F_{ij}^c = \int_{-\frac{h}{2}}^{\frac{h}{2}} \frac{\mu_{\text{ef}}^2(z)E(z)}{1-\mu_{\text{ef}}^2(z)} dz \tag{3.166}$$

$$H_{ij}^c = \varkappa^c \int_{-\frac{h}{2}}^{\frac{h}{2}} \frac{E(z)}{2(1-\mu_{\text{ef}}(z))} dz \tag{3.167}$$

Assuming the property of the vibrating panel is same in all directions, a shear correction factor with coefficient term $\varkappa^c = \frac{5}{6}$ [182]. The total energy functional (Π_T) is

$$\Pi_T = \Xi_\varepsilon - \mathbb{K} \tag{3.168}$$

where the strain energy Ξ_ε over the domain bounded by Ω_{pf} is defined as

$$\Xi_\varepsilon = \frac{1}{2}\int_{\Omega_{\text{pf}}} \left[\int_{-\frac{h}{2}}^{\frac{h}{2}} (\overline{N}_x \varepsilon_x + \overline{N}_y \varepsilon_y + \overline{N}_{\text{xy}} \gamma_{\text{xy}} + \overline{R}_x \gamma_{\text{xz}} + \overline{R}_y \gamma_{\text{yz}}) dz \right] dydx \tag{3.169}$$

and the kinetic energy symbol $\mathbb{K}$ over the domain also bounded by Ω_{pf} is defined as

$$\mathbb{K} = \frac{1}{2}\int_{\Omega_{\text{pf}}} \left[\overline{I}_0 \left(\dot{\overline{u}}^2 + \dot{\overline{v}}^2 + \dot{\overline{w}}^2 \right) + \overline{I}_2 \left(\dot{\beta}_x^{\,2} + \dot{\beta}_y^{\,2} \right) \right] d\Omega_{\text{pf}} \tag{3.170}$$

where $\dot{\overline{u}}$, $\dot{\overline{v}}$ and $\dot{\overline{w}}$ are velocity functions at the mid-plane with $\overline{I}_0$ and $\overline{I}_2$ as the moment of inertial which are defined as

$$\overline{I}_0 = \int_{-\frac{h}{2}}^{\frac{h}{2}} \overline{\rho}(z) dz \tag{3.171}$$

$$\overline{I}_1 = \int_{-\frac{h}{2}}^{\frac{h}{2}} z\,\overline{\rho}(z) dz \tag{3.172}$$

$$\overline{I}_2 = \int_{-\frac{h}{2}}^{\frac{h}{2}} z^2 \overline{\rho}(z) dz \tag{3.173}$$

By adopting the principle of virtual displacement, one can conveniently write the equation of motion of the vibrating panel as

$$-\omega^2 [\mathbf{M_p}] \{\overline{\mathbf{u}}\} + [\mathbf{K_p}] \{\overline{\mathbf{u}}\} = 0 \tag{3.174}$$

where the matrix symbol $[\mathbf{M_p}]$ denotes the mass matrix and $[\mathbf{K_p}]$ denotes the stiffness matrix. The vector symbol $\{\overline{\mathbf{u}}\}$ represents the generalized displacement with ω as the natural frequency. Eq. (3.174) can be modified by substituting the time function and expressed as

$$\left([\mathbf{K_p}] - \omega^2 [\mathbf{M_p}] \right) \{\overline{\mathbf{u}}\} = 0 \tag{3.175}$$

At the midplane of the square thin-walled panel, a set of nodes $x_r\{r = 1, \ldots, \xi\}$ can be used to obtain the finite element parts. The expression of the displacement takes the form

$$\bar{\mathbf{u}} = \begin{bmatrix} \bar{u} \\ \bar{v} \\ \bar{w} \\ \beta_x \\ \beta_y \end{bmatrix} = \sum_{r=1}^{\xi} \mathbb{N}_r \begin{bmatrix} \bar{u} \\ \bar{v} \\ \bar{w} \\ \beta_x \\ \beta_y \end{bmatrix} e^{\mathrm{i}\omega \mathrm{t}} = \sum_{k=1}^{\xi} \mathbb{N}_r(x)\bar{\mathbf{u}}^{\mathbf{T}} e^{\mathrm{i}\omega \mathrm{t}} \tag{3.176}$$

Where the notation $\mathbb{N}_r$ denotes the standard displacement shape functions. The bending component displacement ($\overline{w}$) with relation to the shape function can be expressed as

$$\overline{w}(x,y) = \mathbb{N}_r^T \overline{w} \tag{3.177}$$

for which $\Delta\overline{w}$ represent the bending degree of freedom vector. The stiffness matrix $\mathbf{K_p}$ is defined by two components $\mathbf{K_p^e}$ and $\mathbf{K_p^b}$ namely—stiffnesses resulting from extension and that from bending, respectively. It is expressed as

$$\mathbf{K_p} = \mathbf{K_p^e} + \mathbf{K_p^b} \tag{3.178}$$

Over the domain, these stiffness matrices are respectively discretized as

$$\mathbf{K}_{\mathrm{rt}}^e = \int_{\Omega_{\mathrm{pf}}} \mathbf{B}_r^{e^T} \mathbf{D} \mathbf{B}_t^e d\Omega_{\mathrm{pf}} \tag{3.179}$$

$$\mathbf{K}_{\mathrm{rt}}^b = \int_{\Omega_{\mathrm{pf}}} \mathbf{B}_r^{b^T} \mathbf{A} \mathbf{B}_t^b d\Omega + \int_{\Omega_{\mathrm{pf}}} \mathbf{B}_r^{b^T} \mathbf{B} \mathbf{B}_t^e d\Omega + \int_{\Omega_{\mathrm{pf}}} \mathbf{B}_r^{e^T} \mathbf{B} \mathbf{B}_t^b \mathrm{d}\Omega \tag{3.180}$$

where the symbol $\mathbf{B}^e$ denotes the extensional strain matrix and the symbol $\mathbf{B}^b$ denotes the strain matrix resulting from bending. These terms are defined as

$$\mathbf{B_r^e} = \begin{bmatrix} 0 & 0 & 0 & \frac{\partial \mathbb{N}_r}{\partial x} & 0 \\ 0 & 0 & 0 & 0 & \frac{\partial \mathbb{N}_r}{\partial y} \\ 0 & 0 & 0 & \frac{\partial \mathbb{N}_r}{\partial y} & \frac{\partial \mathbb{N}_r}{\partial x} \end{bmatrix} \tag{3.181}$$

$$\mathbf{B_t^b} = \begin{bmatrix} \frac{\partial \mathbb{N}_r}{\partial x} & 0 & 0 & 0 & 0 \\ 0 & \frac{\partial \mathbb{N}_r}{\partial y} & 0 & 0 & 0 \\ \frac{\partial \mathbb{N}_r}{\partial y} & \frac{\partial \mathbb{N}_r}{\partial x} & 0 & 0 & 0 \end{bmatrix} \tag{3.182}$$

Moreover, the expression of the mass matrix over the domain (Ω_{pf}) of the structural panel is

$$\mathbf{M}_{\mathrm{rt}} = \int_{\Omega_{\mathrm{pf}}} \mathbf{G}_r^T \overline{\mathbf{m}}\, \mathrm{G_t} d\Omega_{\mathrm{pf}} \tag{3.183}$$

where

$$\mathbf{G}_r = \begin{bmatrix} \mathbb{N}_r & 0 & 0 & 0 & 0 \\ 0 & \mathbb{N}_r & 0 & 0 & 0 \\ 0 & 0 & \mathbb{N}_r & 0 & 0 \\ 0 & 0 & 0 & \mathbb{N}_r & 0 \\ 0 & 0 & 0 & 0 & \mathbb{N}_r \end{bmatrix} \tag{3.184}$$

and

$$\overline{\mathbf{m}} = \begin{bmatrix} \bar{I}_0 & 0 & 0 & \bar{I}_1 & 0 \\ 0 & \bar{I}_0 & 0 & 0 & \bar{I}_1 \\ 0 & 0 & \bar{I}_0 & 0 & 0 \\ \bar{I}_1 & 0 & 0 & \bar{I}_2 & 0 \\ 0 & \bar{I}_1 & 0 & 0 & \bar{I}_2 \end{bmatrix} \tag{3.185}$$

The $\mathbf{K}_\mathbf{p}^e$, $\mathbf{K}_\mathbf{p}^b$ and $\mathbf{M}_{\mathrm{rt}}$ matrices can be solved numerically using Gauss integration procedure while the solution for the matrices **A, B, D** and $\overline{\mathbf{m}}$ can be obtained with less computation.

Acoustic Enclosure Modelling

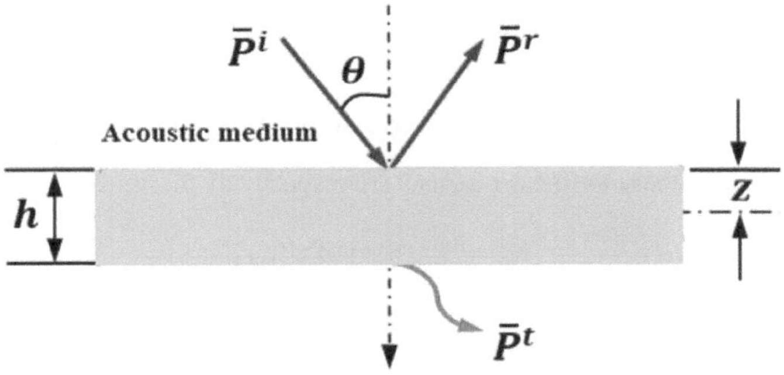

Figure 3.12 Schematic diagram of vibrating panel showing sound waves across its thickness.

To model the acoustic enclosure, all the acoustic pressures including the incident pressure wave $\overline{P}^i$, reflected pressure wave $\overline{P}^r$ and transmitted pressure wave $\overline{P}^t$ are represented in the acoustic enclosure domain. This acoustic enclosure region is modelled as a homogenous elastic flow property having density ρ_f with speed of sound c_s. Given that boundary domain of the acoustic field is Ω_f, the Helmholtz equation of time-harmonic excitation is written as

$$\nabla^2 \phi_i - \frac{1}{c_f^2}\frac{\partial^2 \phi_i}{\partial t^2} = 0 \quad , \quad \nabla = \begin{Bmatrix} \frac{\partial}{\partial x} \\ \frac{\partial}{\partial y} \\ \frac{\partial}{\partial z} \end{Bmatrix} \tag{3.186}$$

For which the symbol ϕ_i denotes the velocity potential and the symbol, ∇ denote the gradient operator. The dynamic pressure is expressed in the form

$$\overline{P} = -\rho_f \frac{\partial \phi_i}{\partial t} \tag{3.187}$$

The velocity of the acoustic enclosure fluid is $\mathbf{v} = \nabla\phi_i$ and the velocity Neumann boundary condition is $n_f^T \nabla \phi_i = \mathbf{v}_n$. Where n_f represents the outward unit normal vector of the acoustic enclosure domain and $\mathbf{v}_n$ is the outward normal velocity. Taking into consideration the time harmonic excitation of Eq. (3.186), with the time dependence ($e^{\mathrm{i}\omega t}$), Helmholtz equation could be adjusted to obtain the following equation

$$\nabla^2 \phi_i - \frac{\omega^2}{c_f^2}\phi_i = 0 \tag{3.188}$$

Also, considering Fig. 3.12, the acoustic pressure waves can be defined according to the forms outlined by [53] given as

$$\overline{P}^i(x,y,z) = \overline{Q}_0^i e^{j\omega t - \psi_{\mathrm{ix}} x - \psi_{\mathrm{iy}} y - \psi_{\mathrm{iz}} z} \tag{3.189}$$

$$\overline{P}^r(x,y,z) = \overline{Q}_0^r e^{j\omega t - \psi_{\mathrm{ix}} x - \psi_{\mathrm{iy}} y - \psi_{\mathrm{iz}} z} \tag{3.190}$$

$$\overline{P}^t(x,y,z) = \overline{Q}_0^t e^{j\omega t - \psi_{\mathrm{ix}} x - \psi_{\mathrm{iy}} y - \psi_{\mathrm{iz}} z} \tag{3.191}$$

The symbols $\overline{Q}_0^i$, $\overline{Q}_0^r$ and $\overline{Q}_0^t$ are the amplitude of the incident, reflected and transmitted waves, respectively. Also, the notations ψ_{ix}, ψ_{iy} and ψ_{iz} are the wavenumbers in the x, y and z directions which can be expressed with respect to their incident angle θ and their azimuth angle α [5]. Given that the amplitude which the incident wave makes is unity. Also, taking into account the quantity $\overline{Q}_0^r$ and $\overline{Q}_0^t$ as well as the displacement and rotational terms at the mid-plane, the total displacement field ($\overline{\mathbf{u}}_\mathbf{T}$) due to vibration and acoustic response can be calculated as

$$\overline{\mathbf{u}}_\mathbf{T} = [\overline{u}\ \ \overline{v}\ \overline{\mathrm{w}}\ \ \beta_x\ \ \beta_y\ \ \overline{Q}_0^r\ \ \overline{Q}_0^t]^T \tag{3.192}$$

Structural Wall – Acoustic Enclosure Modelling

The coupled region of the enclosed structural wall—acoustic enclosure domain is the interface between the acoustic enclosure region of the casing and the internal structural wall of the casing. Representing the domain of the element connecting the interface of these regions as, Ω_{wf}, and assuming this element is bounded by four nodes with two degrees of freedom for each of the nodes. The relationship between the cubic shape function and the discretized elemental displacement ($\mathbb{U}^e$) can be written as [64]

$$\mathbb{U}^e = \mathbb{N}_{\mathrm{uk}} \mathbf{u}_k^e, \qquad k = 1,\ 2,\ 3,\ 4 \tag{3.193}$$

Where the term $\mathbb{N}_{\mathrm{uk}}$ represents the cubic shape function of the elemental normal displacement. The acoustic pressure at the interface of the two regions acts on the four nodes of the element and is written as

$$p^e = \mathbb{N}_{\mathrm{fk}} p_k^{eT},\ \ k = 1,\ 2,\ 3,\ 4 \tag{3.194}$$

where the symbol $\mathbb{N}_{\mathrm{fk}}$ represents the linear shape functions evaluated at each elemental node. The elementary coupling matrix ($\mathscr{K}_{\mathrm{wf}}^e$), bounded by Ω_{wf}, is related to the cubic and linear shape functions according to the equation

$$\mathscr{K}_{\mathrm{wf}}^e = \int_{\Omega_{\mathrm{wf}}} \mathbb{N}_{\mathrm{uk}}{}^T \mathbb{N}_{\mathrm{fk}} J_f d\Omega_{\mathrm{wf}} \tag{3.195}$$

Vibrating Panel – Acoustic Enclosure Modelling

The internal surface of vibrating structural panel placed along the boundary of the structural casing (i.e., Fig. 3.11) interacts and acts as a velocity exciter for the acoustic enclosure. The matrix equation of Eq. (3.175) for the vibrating panel bounded by Ω_{pf} can be modified as

$$[m_p]\{\ddot{u}_f\}+[c_p]\{\dot{u}_f\}+[k_p]\{u_f\}=\{f_f\} \tag{3.196}$$

where m_p, c_p and k_p are the vibrating panel mass, damping and stiffness matrices, respectively. Also, the term u_f denotes nodal displacement with the corresponding nodal velocity ($\dot{u}_f$) and acceleration ($\ddot{u}_f$) of the vibrating panel at the fluid surface; and the term f_f denotes the nodal force due to the acoustic fluid in the enclosure region which can be expressed as

$$f_f=\rho_f\mathscr{K}_{\mathrm{pf}}\dot{\phi}_i \tag{3.197}$$

where ρ_f is the density of the acoustic enclosure fluid and $\dot{\phi}_i$ is the nodal velocity potential between the interface of the vibrating panel and the acoustic enclosure. The coupling matrix $\mathscr{K}_{\mathrm{pf}}$ bounded by Ω_{pf} can be expressed as

$$\mathscr{K}_{\mathrm{pf}}=\int_{\Omega_{\mathrm{pf}}}\mathbb{N}_p{}^T n_f\mathbb{N}_f J_p d\Omega_{\mathrm{pf}} \tag{3.198}$$

where $\mathbb{N}_p$ and $\mathbb{N}_f$ are respectively the nodal shape functions on the interface of the vibrating panel having Jacobian matrix J_p and the effect of the acoustic enclosure fluid having n_f as the outward unit normal vector along the acoustic enclosure domain.

3.4.2 NUMERICAL SIMULATIONS OF AN ACTIVE CASING

A numerical model built in COMSOL Multiphysics is presented in this section. It encompasses the following steps in the modelling workflow:

- defining geometry of the system,
- defining material properties,
- defining the physics that describe specific phenomena in the system,
- solving the numerical model.

With the model simulation implemented in the *COMSOL Multiphysics* software, the system of the coupled partial differential equations describing the model is solved by the finite element method.

Model set up in COMSOL Multiphysics

Taking into account the assumptions introduced and the model formulation described in Section 3.3.2, a numerical model is built in *COMSOL Multiphysics* and solved using finite element method. As is illustrated in Fig. 3.7, the whole system includes twelve flexible thin panels and eight acoustic fields. To build the numerical model for simulation, first the geometry and material for all these structural and acoustic fields are defined. Presented simulation assumes a symmetrical cubic casing, i.e. the twelve thin walls' panels are totally the same, including material, dimensions and connections with the rigid frame. The eight acoustic domains use identical material properties. All the geometry parameters are as follows:

$$l_x = l_y = l_z = 0.42\ m,\quad a = 0.09\ m,\quad h_{ij} = 0.001\ m,\quad d_i = 0.05\ m \\ (i = 1,2,\ldots,6,\ j = 1,2)\,. \tag{3.199}$$

The material properties of the panels (steel) are given by:

$$E_{ij} = 200\ \text{GPa},\quad \rho_{ij} = 7850\ \text{kg/m}^3,\quad \nu_{ij} = 0.3,\quad R_{ij} = 0.0001\ \text{N}\cdot\text{m}\cdot\text{s} \\ (i = 1,2,\ldots,6,\ j = 1,2)\,. \tag{3.200}$$

The acoustic fields (air, 20, 1 atm) satisfy:

$$c_i = c_0 = c_s = 343\ \text{m/s},\ \rho_i = \rho_0 = \rho_s = 1.2\ \text{kg/m}^3,\quad \mu_i = \mu_0 = \mu_s = 2.5\times 10^{-10}\ \text{s} \\ (i = 1,2,\ldots,6)\,. \tag{3.201}$$

Fig. 3.13 shows the geometry built in COMSOL *Multiphysics* to simulate the whole system including all the fields. Figs. 3.13–3.22 were generated by Ms. L. Liu, within joint research reported in [178]. As is shown, considering the symmetricity of the casing, the acoustic field in the exterior is modelled using a sphere, whose centre is located at the geometric centre of the casing.

Appropriate physics is then imposed to the corresponding geometry to match the model developed in Section 3.3.2. For the acoustic fields, viscous fluid model is used, and the viscosity parameters are set to match the absorption parameter in Eq. (3.201). On the exterior boundary of the spherical acoustic field, perfectly matched layers (PML) [9], which effectively absorbs the wave by imposing a complex-valued coordinate transformation, are used to simulate the non-reflecting boundary conditions. For the panels, shell model is used. Though COMSOL *Multiphysics* uses elements of the Mindlin-Reissner type instead of the implemented Kirchhoff-Love plate theory, the results have little difference since the panel is as thin as 1 mm and 1/420 of the length. The viscous damping coefficient indicated in Eq. (3.200) is imposed by setting up a frequency dependent structural loss factor. The boundary conditions introduced in Eq. (3.73) are also implemented. The boundary restraint coefficients are assumed to take the value $\xi_{ij} = 22.9\ (i = 1,2,\ldots,6, j = 1,2)$. The setting $\xi_{ij} = 22.9$ corresponds to prescribing the rotational stiffness along the panel edge l_{ki} as $D\frac{\xi_{ij}}{l_{\text{ki}}} = 10^3\ \text{N}{\cdot}\text{m/m}\ (i = 1,2,\ldots,6, j = 1,2, k = 1,2)$. The value of boundary restraint coefficient has been chosen to analyse the system behaviour in case of imperfect fastening of panels' edges to the rigid frame. For the fluid-structure interactions, they are assured by specifying the panels as the acoustic-structure boundaries.

Finally, the study and the excitation are determined, and the mesh is built up according to the frequency range of interest. As shown in Fig. 3.7, the simulation

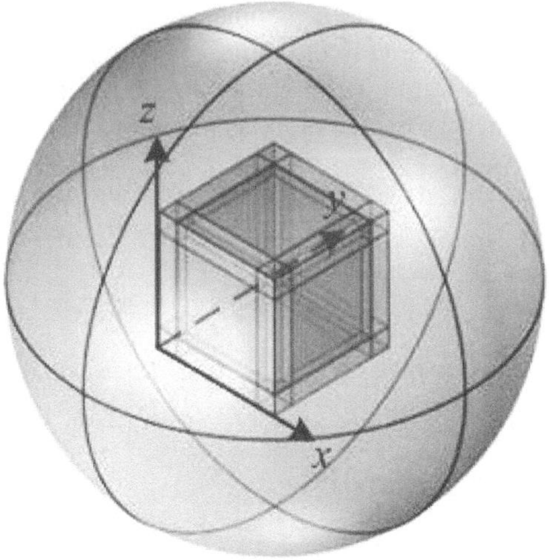

Figure 3.13 Model geometry of the vibroacoustic system [178]. Reprinted from Mechanical Systems and Signal Processing, 152:107371, J. Wyrwal, M. Pawelczyk, L. Liu, and Z. Rao, "Double-panel active noise reducing casing with noise source enclosed inside-modelling and simulation study", Copyright (2022), with permission from Elsevier.

studies the response of the system under the excitation of a primary source inside the casing. In practice, the sound source is modelled by a monopole of the strength $q = 10^{-3}$ m^3/s, exciting the system in the frequency domain in the range from 30 Hz to 350 Hz. Considering the mesh convergence in high frequencies, the maximum size of the (quadratic) shell element is set as 0.024 m, i.e. 18 elements per side of the panel, and the counterpart of the (quadratic) acoustic element is 0.196 m, i.e. 5 elements per wavelength of the acoustic wave at 350 Hz. The convergence at low frequencies is also checked because of the perfectly matched layers.

Simulation of selected models

In this section, the model of double-panel casing developed in Section 3.3.2 that is described by the system of the 20 coupled partial differential equations is simulated in COMSOL *Multiphysics*. The simulation is performed to prove the model correctness, i.e. convergence, appropriate spatial and temporal discretization, and correct implementation of all physical aspects describing phenomena that are to be explored. Finally, the results are interpreted in relation to the physical vibroacoustic system that is described by the developed model in the context of results reasonableness. On the one hand, results are interpreted in relation to the observed physical reality of the system under discussion and on the other hand they are used to analyse the system behaviour and consequently to understand the system behaviour better.

Two independent simulations are carried out depending on the location of the primary noise source being the source of undesirable noise that is to be attenuated. The first simulation is performed assuming the monopole noise source is placed at the casing centre (0.3 m, 0.3 m, 0.3 m) (noise source symmetrical with respect to casing walls). In case of second simulation the monopole noise source is moved to the point (0.4 m, 0.4 m, 0.4 m) (noise source asymmetrical with respect to casing walls). Transverse displacement is studied for all panels and sound pressure level is investigated for all acoustic fields in the system under discussion in case of both simulations.

Noise source at the casing centre

Assuming that the sound source is placed at the casing centre (at the point (0.3 m, 0.3 m, 0.3 m)) and that the casing is symmetrical cubic, the response of the system is also symmetric. It is symmetric about the typical nine symmetry planes of a cube. In the given coordinate system, the equations of the symmetry planes are given by:

$$x = 0.3 \text{ m},\ y = 0.3 \text{ m},\ z = 0.3 \text{ m} \tag{3.202a}$$

$$x - y = 0,\ y - z = 0,\ x - z = 0 \tag{3.202b}$$

$$x + y - 0.6 \text{ m} = 0,\ y + z - 0.6 \text{ m} = 0, x + z - 0.6 \text{ m} = 0 \tag{3.202c}$$

Taking into account the symmetry in the response of the casing in the case under discussion, the responses of the six double-panel walls are identical. Fig. 3.14 shows transverse displacement of the incident and radiating panels for different exciting frequencies in the range from 30 Hz to 350 Hz with the step of 1 Hz. Both the amplitude at the maximum displacement point and the amplitude at the central point of each panel are used to draw the graph. Different vibration strengths can be observed for the incident and radiating panels at most frequencies, however both of them exhibit resonant at 52 Hz, 102 Hz, 162 Hz, 178 Hz and 271 Hz. The vibration patterns of the panels at these resonant frequencies are shown in Fig. 3.15. To distinguish the difference between vibration patterns depicted in Figs. 3.15(a) and (b), and the corresponding difference between Figs. 3.15(c) and (d), the cross-section of the wall 5 (on the bottom) looking from x axis direction is additionally plotted for Figs. 3.15(a)–(d). In these plots the relative motion between the incident and radiating panels can be observed that plays a crucial role in the correct interpretation of the results.

Figs. 3.15(a) and (b) indicate two coupling modes that relate to the first *in vacuo* mode of a single panel, whose natural frequency is 43.5 Hz (Fig. 3.22). For Fig. 3.15(a), incident and radiating panels vibrate towards or against the casing centre together. The acoustic field in the interior of the casing stiffens the panels, and thus the resonance frequency 52 Hz is higher than the natural frequency of the panel. Fig. 3.15(b) is different from Fig. 3.15(a) concerning the relationship between relative motion of the incident and radiating panels. While in Fig. 3.15(a) the two panels move in the same direction, in Fig. 3.15(b) they do the opposite. At 102 Hz, since the two panels always vibrate in the opposite direction, the acoustic field in between (in the cavity) always resists their vibration due to the change of air volume. Accordingly, this mode has a much higher frequency. To see the difference with respect to the acoustic field, Fig. 3.16 shows the sound pressure level of all acoustic fields in the system. At 52 Hz, from the interior to the exterior of the casing, the sound pressure level is decreasing. The acoustic fields between the incident and radiating panels are about 5 dB smaller than the interior. In contrast, the sound pressure level in these narrow cavities is amplified at 102 Hz and even approximately 13 dB higher than the interior of the casing. At first glance it seems surprising but in the context of the analysis carried out it is obvious that the result of strong coupling effect that can be observed in the double-panel walls at 102 Hz. The relative motion between the two panels strengthens the acoustic field and the air volume inside stiffens the vibration of the panels.

The difference between Figs. 3.15(c) and (d), that relate to resonant at 162 Hz and 178 Hz, is similar. But this time, the coupling effect is lighter, which benefits from the mode shape of a single panel. When some part of the panel vibrates forward and some other part vibrates backward, the panel itself can compensate some air volume change. As a result, the third and fourth resonances are closer. It can also be expected that, for higher modes, this effect will be even smaller.

Corresponding to Fig. 3.14, Fig. 3.17 shows the sound pressure level measured at point (0.3 m, 0.8 m, 0.3 m) and point (0.3 m, 1.1 m, 0.3 m), respectively 0.3 m and 0.5 m in front of the wall 4. As is expected, the sound pressure level changes in agreement with the maximum displacement of radiating panel shown in Fig. 3.14. Apart from the similar curve shapes, the frequencies of the peaks are coincident. Fig. 3.17 also illustrates that the sound pressure level decreases as the distance from the casing wall increases.

To have a better insight about each acoustic field, the sound pressure level for different acoustic fields at 52 Hz (corresponding to Fig. 3.16(a)) is separately depicted in Fig. 3.18. The structure of the active casing encloses 7 cavities. Inside each cavity, the sound field varies a little, within 1 dB, because the frequency is much lower than the first non-zero eigenfrequency of the corresponding rigid-wall cavity, which is 343 Hz for the interior of casing and 409 Hz for the interiors of casing walls. In the interior of the active casing (Fig. 3.18(a)), the maximum sound pressure level is seen at the centre since the noise source is placed there. For the interiors of the casing walls (Fig. 3.18(b)), around the centre of the incident panels has the maximum value. However, the situation will change for different frequencies, depending on the fluid-structure interactions among the panels and cavities. Considering the exterior acoustic field ((Fig. 3.18(c)), as the distance from the casing increases, the sound pressure level decreases in general. The maximum value is shown around the casing walls. Each wall is like a sound source and their radiated sounds interfere with each other.

Noise source moved from the centre

When the noise source is moved to point (0.4 m, 0.4 m, 0.4 m), the level of the symmetricity is decreased comparing to case when it is placed at the casing centre,

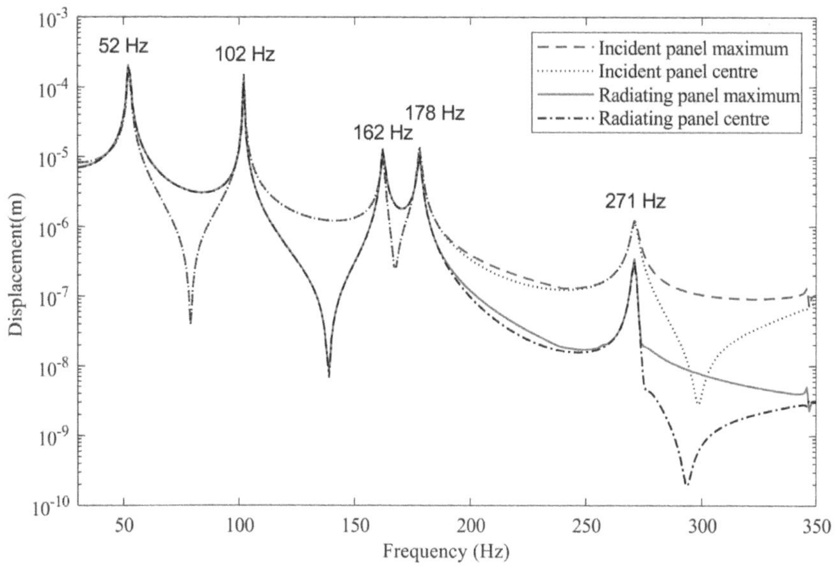

Figure 3.14 The transverse displacement of the incident and radiating panels [178]. Reprinted from Mechanical Systems and Signal Processing, 152:107371, J. Wyrwal, M. Pawelczyk, L. Liu, and Z. Rao, "Double-panel active noise reducing casing with noise source enclosed inside-modelling and simulation study", Copyright (2022), with permission from Elsevier.

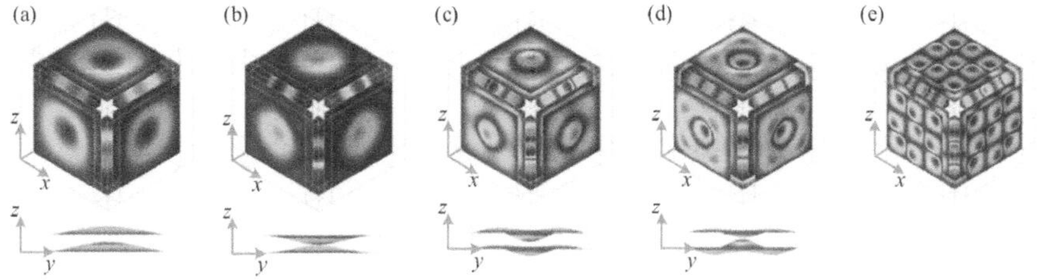

Figure 3.15 Field displacement for resonant frequencies: (a) 52 Hz, (b) 102 Hz, (c) 162 Hz, (d) 178 Hz, (e) 271 Hz [178]. Reprinted from Mechanical Systems and Signal Processing, 152:107371, J. Wyrwal, M. Pawelczyk, L. Liu, and Z. Rao, "Double-panel active noise reducing casing with noise source enclosed inside-modelling and simulation study", Copyright (2022), with permission from Elsevier.

although it is still symmetric with respect to the planes of Eq. (3.202b). As a result, some of the modes that are not symmetric with respect to the other planes can be excited. With a discussion in this case, the influence of the position of the primary noise source is present.

Fig. 3.19 shows the transverse displacement amplitudes of the maximum and central points of the two panels on wall 4 for different source frequencies, at the intervals

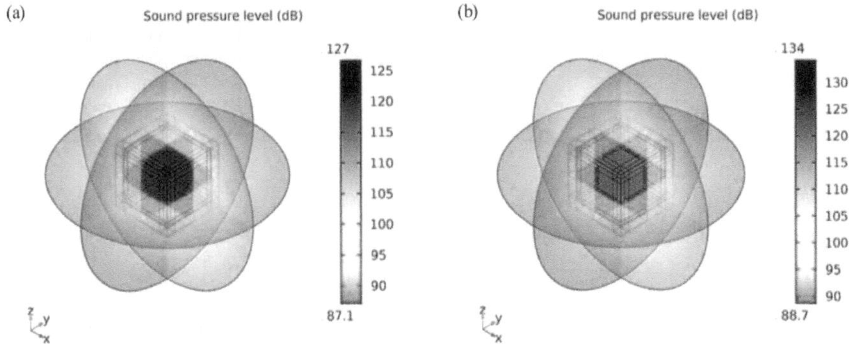

Figure 3.16 Sound pressure level for all acoustic fields under the resonant frequencies: (a) 52 Hz, (b) 102 Hz [178]. Reprinted from Mechanical Systems and Signal Processing, 152:107371, J. Wyrwal, M. Pawelczyk, L. Liu, and Z. Rao, "Double-panel active noise reducing casing with noise source enclosed inside-modelling and simulation study", Copyright (2022), with permission from Elsevier.

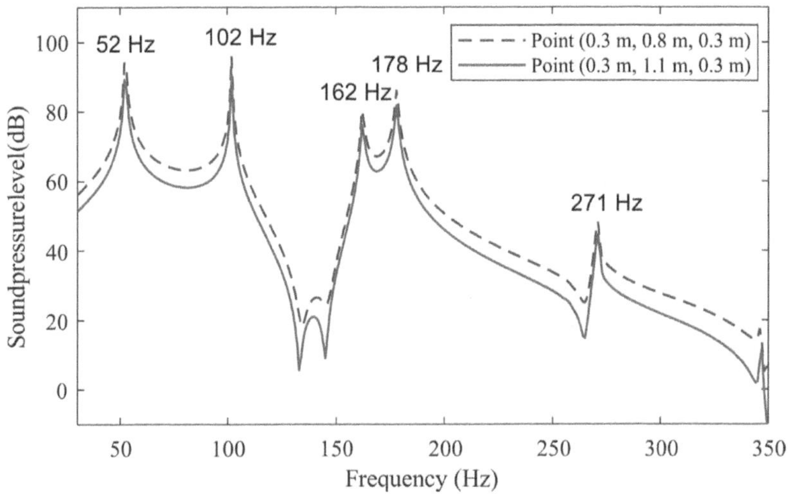

Figure 3.17 Sound pressure level at points (0.3 m, 0.8 m, 0.3 m) and (0.3 m, 1.1 m, 0.3 m), respectively 0.3 m and 0.5 m in front of the wall 4 [178]. Reprinted from Mechanical Systems and Signal Processing, 152:107371, J. Wyrwal, M. Pawelczyk, L. Liu, and Z. Rao, "Double-panel active noise reducing casing with noise source enclosed inside-modelling and simulation study", Copyright (2022), with permission from Elsevier.

of 1 Hz from 30 Hz to 350 Hz. Comparing Fig. 3.19 with Fig. 3.14, it can be observed that the panel centre serves as the maximum displacement point for fewer frequencies. Besides, with respect to the maximum displacement curves, additional peaks are shown at 43 Hz, 86 Hz, 130 Hz, 202 Hz, 258 Hz, 300 Hz and 347 Hz, and the

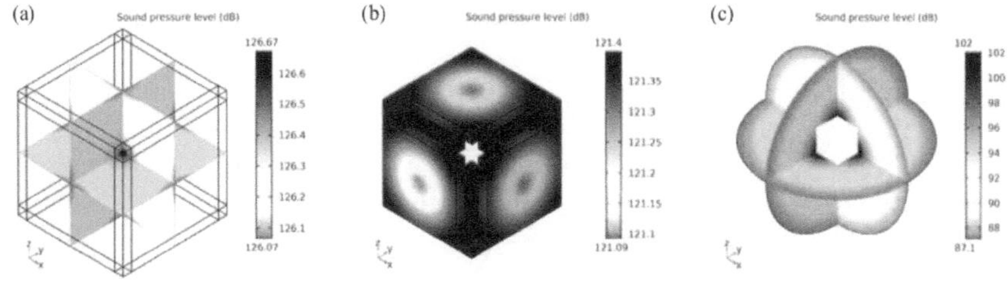

Figure 3.18 Sound pressure level for the acoustic fields at 52 Hz: (a) the interior of the active casing, (b) the interiors of the casing walls, (c) the exterior surrounding the active casing [178]. Reprinted from Mechanical Systems and Signal Processing, 152:107371, J. Wyrwal, M. Pawelczyk, L. Liu, and Z. Rao, "Double-panel active noise reducing casing with noise source enclosed inside-modelling and simulation study", Copyright (2022), with permission from Elsevier.

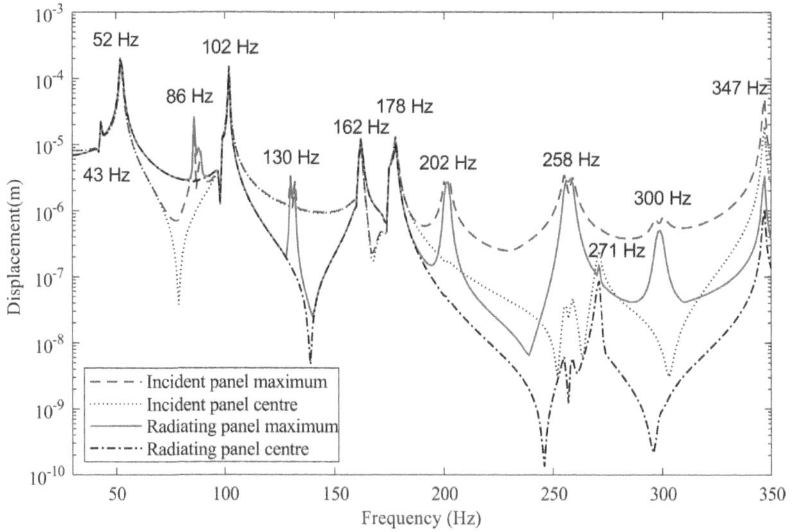

Figure 3.19 The transverse displacement of the incident and radiating panels on wall 4 [178]. Reprinted from Mechanical Systems and Signal Processing, 152:107371, J. Wyrwal, M. Pawelczyk, L. Liu, and Z. Rao, "Double-panel active noise reducing casing with noise source enclosed inside-modelling and simulation study", Copyright (2022), with permission from Elsevier.

resonance at 271 Hz is not so obvious. Under the symmetricity of this case, wall 2, wall 4 and wall 6 have identical response. They have the same smaller distance to the noise source and can be represented by wall 4. In opposite, wall 1, wall 3 and wall 5 can be represented by wall 3. Fig. 3.20 shows the transverse displacement of the two panels on wall 3, which is opposite to the wall 4. Comparison of Fig. 3.20

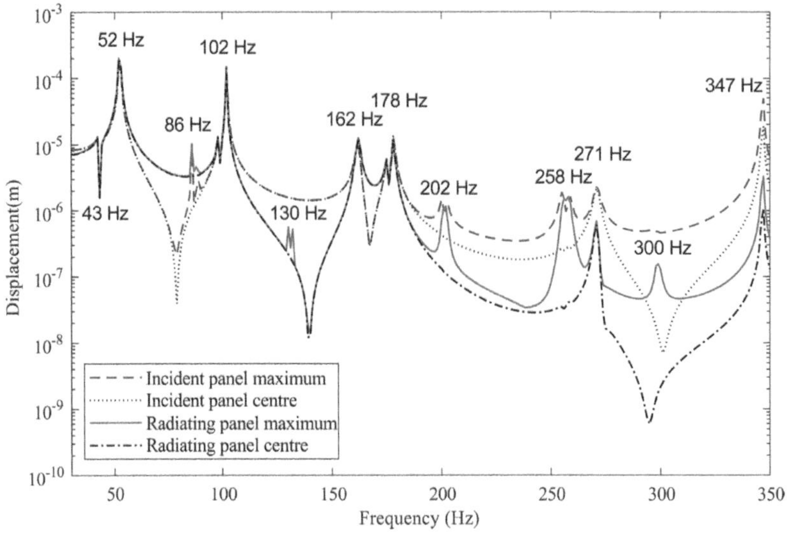

Figure 3.20 The transverse displacement of the incident and radiating panels on wall 3 [178]. Reprinted from Mechanical Systems and Signal Processing, 152:107371, J. Wyrwal, M. Pawelczyk, L. Liu, and Z. Rao, "Double-panel active noise reducing casing with noise source enclosed inside-modelling and simulation study", Copyright (2022), with permission from Elsevier.

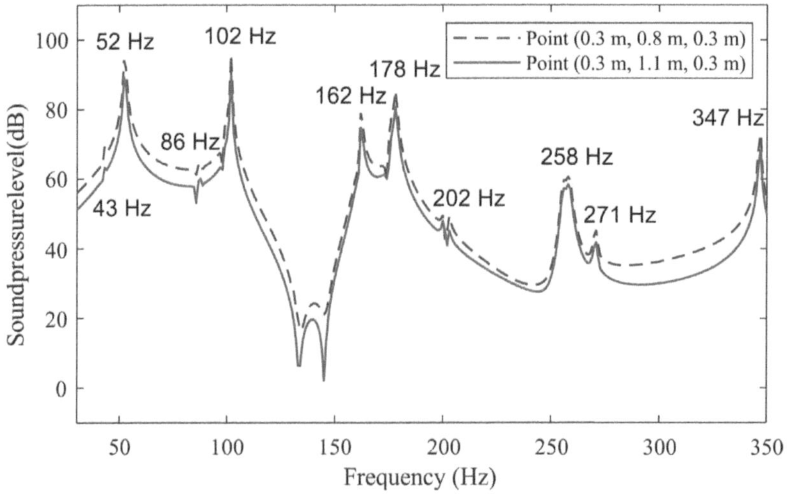

Figure 3.21 Sound pressure level at points (0.3 m, 0.8 m, 0.3 m) and (0.3 m, 1.1 m, 0.3 m), respectively 0.3 m and 0.5 m in front of the wall 4 [178]. Reprinted from Mechanical Systems and Signal Processing, 152:107371, J. Wyrwal, M. Pawelczyk, L. Liu, and Z. Rao, "Double-panel active noise reducing casing with noise source enclosed inside-modelling and simulation study", Copyright (2022), with permission from Elsevier.

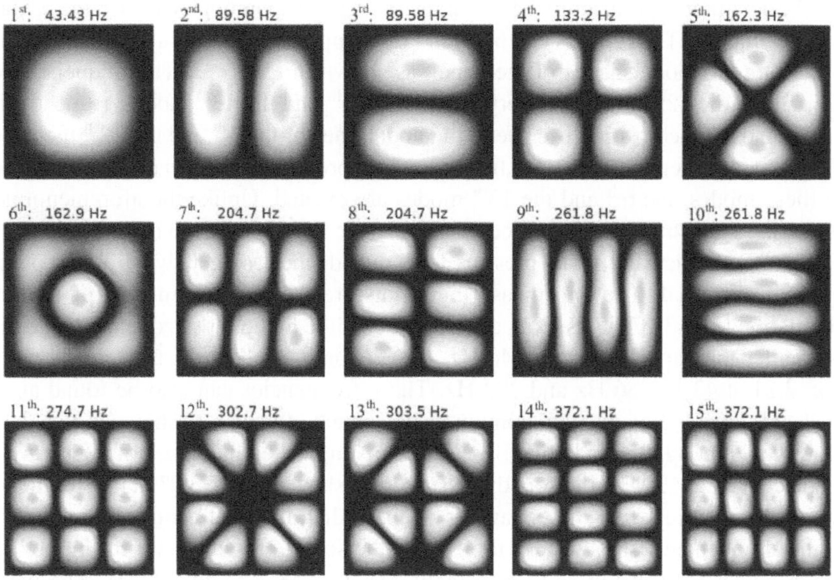

Figure 3.22 The first 15 *in vacuo* modes of a single panel [178]. Reprinted from Mechanical Systems and Signal Processing, 152:107371, J. Wyrwal, M. Pawelczyk, L. Liu, and Z. Rao, "Double-panel active noise reducing casing with noise source enclosed inside-modelling and simulation study", Copyright (2022), with permission from Elsevier.

and Fig. 3.19 reveals the different responses between these two opposite walls. For most resonances, the responses on the wall 4 are higher than the wall 3. Particularly, at 43 Hz, the vibration of wall 3 is suppressed while the vibration of wall 4 is amplified. One of the reasons is that the noise source is closer to wall 4. To understand the situation, knowing the *in vacuo* modes of the panel would be helpful. The first 15 natural frequencies and mode shapes of a single panel are listed and shown in Fig. 3.22. The frequency 43 Hz is close to the first natural frequency of the panel. When the sound field is at this frequency, it cannot provoke the resonances of all panels, because of the stiffening effect of the cavities (which has been discussed in detail in case of the first simulation, causing the resonances at 52 Hz and 102 Hz). However, when the noise source is moved from the casing centre to point (0.4 m, 0.4 m, 0.4 m), the six walls do not need to response alike, resonances of walls 2, 4 and 6 are provoked because they are closer to the sound source. Instead, the vibrations of other three walls are suppressed. For other frequencies, there are similar interactions among panels, cavities and the noise source. The peaks labelled in Fig. 3.19 (or Fig. 3.20) can all relate to the *in vacuo* modes of the panel (Fig. 3.22) except the one at 347 Hz: 43 Hz, 52 Hz, 102 Hz for the 1st mode; 86 Hz for the 2nd and 3rd modes; 130 Hz for the 4th mode; 162 Hz and 178 Hz for the 6th mode; 202 Hz for the 7th and 8th modes; 258 Hz for the 9th and 10th modes; 271 Hz for the 11th mode and 300 Hz for the 13th modes. It is noticed that there are no corresponding peaks for the 5th and the 12th modes. The reasons can be found by comparing the natural frequencies and observing the mode shapes of the panel, as shown in Fig. 3.22. The frequencies of the 5th and the 12th modes are very close to the 6th and the 13th modes, respectively, and sound source position is close to the nodal lines of the 5th and the 12th modes of the panel, so when the excitation frequency is approximate to the natural frequencies of these modes, the 6th and the 13th modes are excited. Unlike the aforementioned resonances are structure controlled, the resonance at 347 Hz is cavity controlled.

Fig. 3.21 depicts the sound pressure level outside the casing at two different points in front of wall 4. The two measuring points are the same as in case of the first simulation (Fig. 3.17). Figs. 3.21 and 3.17 look alike in the low frequency range that is smaller than approximately 240 Hz, except some small peaks that are shown in Fig. 3.21 at 43 Hz, 86 Hz and 202 Hz. These frequencies can also be found at the peaks of the radiating panel maximum displacement of wall 4, as shown in Fig. 3.19. However, although the radiating panel has a distinguished increase of the maximum displacement at 130 Hz, the sound pressure level shown in Fig. 3.21 at this frequency remains small, just like the behaviour of the radiating panel centre. Then, for the higher frequencies from 240 Hz to 350 Hz, the sound pressure level in Fig. 3.21 is much higher than in Fig. 3.17, especially around 258 Hz and 347 Hz. The peaks at these two frequencies of the sound pressure level are conform with the peaks of the maximum displacement on the radiating panel, but at 130 Hz and 300 Hz, the conformity is missing. To sum up, comparing the sound pressure level outside the casing (Fig. 3.21) and the displacement of the panels (Fig. 3.19), the peaks in Fig. 3.19 can be classified into three categories:

- The first category includes peaks at 43 Hz, 52 Hz, 102 Hz, 162 Hz, 178 Hz, 258 Hz and 271 Hz and 347 Hz. These peaks shown in Fig. 3.21 are controlled by the maximum displacement of the radiating panel, and the curve shapes around these peaks are similar to the curve of the radiating panel maximum in Fig. 3.19. Besides, Fig. 3.19 indicates that the displacement at the panel centre also reaches peak at these frequencies.
- The second category refers to the peaks at 86 Hz and 202 Hz. They can be observed in Fig. 3.21, but much smaller than the corresponding peaks on the radiating panel maximum displacement curve in Fig. 3.19. Meanwhile, the corresponding peaks are not shown in the radiating panel centre displacement.
- The third category is for the peaks at 130 Hz and 300 Hz. These peaks are not present in in Fig. 3.21. The discrepancy between the radiating panel maximum and centre displacements are also observed at these frequencies.

The agreement or disagreement between the panel maximum and the panel centre actually reflects the vibration pattern of the panel. Therefore, the sound pressure level outside the double-panel casing is closely related to the maximum displacement of the radiating panels as well as their vibrating patterns.

3.5 VALIDATION OF SELECTED MODELS

The previously developed model is validated in this section for selected cases. Results of simulations are compared with laboratory experiments. The model is validated by means of a comparison of natural frequencies, mode shapes and frequency response functions.

In experiments performed by the authors, measurements of natural frequencies and mode shapes are made with a laser vibrometer PDV-100 (measuring the normal plate velocity in a large number of points). A loudspeaker placed on the bottom of the casing is used as the noise source exciting the casing walls and making them vibrate. A random wideband signal is used as the excitation. Due to the close distance between the walls and the loudspeaker, not all of the eigenmodes in the considered frequency range are excited equally. Some of the less excited eigenmodes are even difficult to observe, because of vicinity of more excited ones.

3.5.1 RIGID CASING WALLS

In this subsection, the measurements and simulation results are presented, obtained for two configurations of rigid casing wall. In all performed experiments, single wall has been acoustically excited with a random wideband signal. The response of the plate has been measured in 400 uniformly distributed points (the distance between adjacent points was equal 20 mm). Remaining walls have been either removed or passively insulated to minimize the interference at this stage of the research.

3.5.1.1 Unloaded casing wall

First, an unloaded casing wall is evaluated. It is a 1 mm thick brushed aluminium plate. Due to the manufacturing process, the plate represents orthotropic properties—it follows from the analysis of initial 11 eigenmodes presented in Fig. 3.23 that despite square shape of the plate, e.g. 2nd and 3rd natural frequencies are not equal. Depending on the application, the wall could be satisfyingly approximated with an isotropic plate model. However, to obtain the highest modelling accuracy, the orthotropic Mindlin plate model is used. The parameters of the plate used for the purpose of modelling are as follows:

$$\begin{aligned} a &= 0.420\,\mathrm{m}, & b &= 0.420\,\mathrm{m}, & h &= 0.001\,\mathrm{m}, \\ E_x &= 91.2\,\mathrm{GPa}, & E_y &= 60.8\,\mathrm{GPa}, & G_{xy} &= 29.2\,\mathrm{GPa}, \\ \rho_p &= 2770\,{}^{\mathrm{kg}}/_{\mathrm{m}^3}, & \nu_x &= 0.3, & \kappa_x &= {}^5/_6. \end{aligned}$$

For the rigid casing walls, fully-clamped boundary conditions offer generally satisfactory results. However, in this case, boundary conditions elastically restrained against rotation are used. The stiffness coefficients of rotational springs have been assumed close to completely rigid (as it is for fully-clamped boundary conditions), but a minor softening of the springs offered better fitness to the real measurement data.

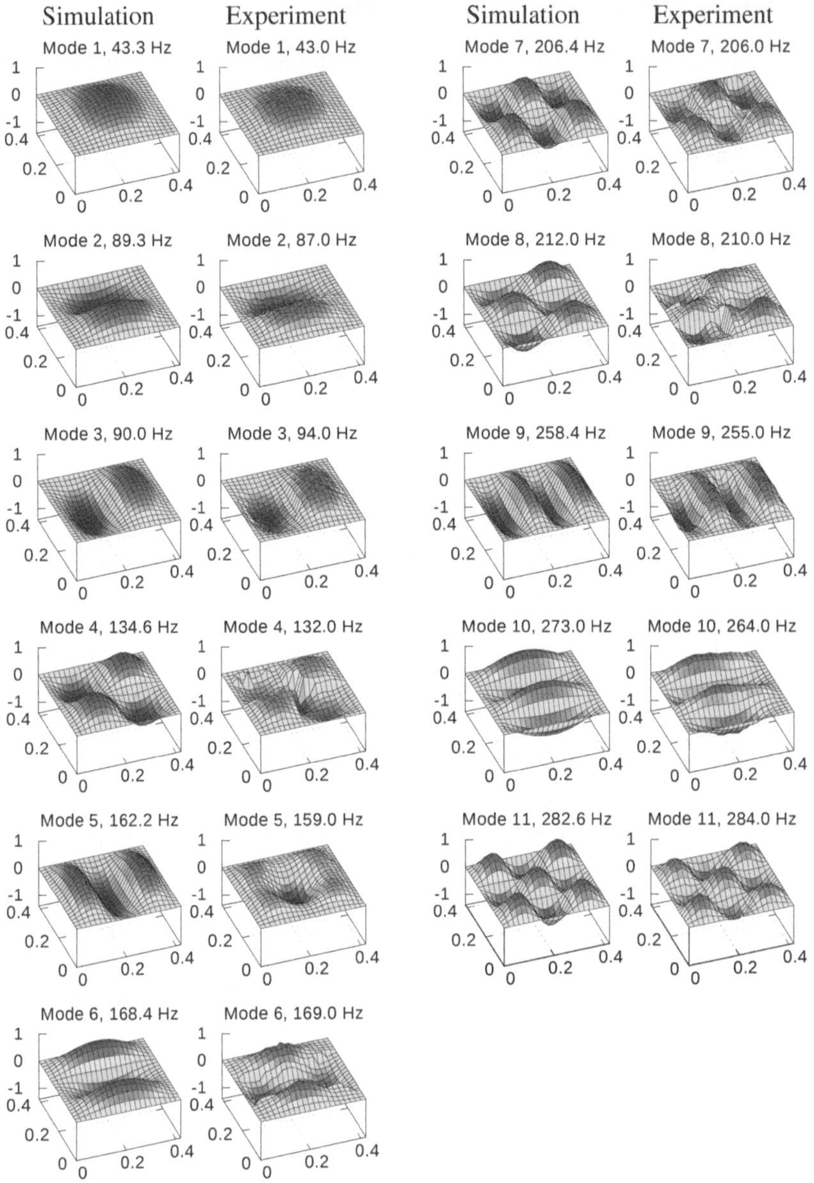

Figure 3.23 A comparison of initial 11 natural frequencies and mode shapes of rigid casing wall, calculated with the mathematical model and experimentally measured—1 mm thick aluminium unloaded plate [164]. Reprinted from "Modelling and control of device casing vibrations for active reduction of acoustic noise", PhD thesis, Silesian University of Technology, S. Wrona.

Assessing the fitness of theoretically calculated natural frequencies and mode shapes to the real measurements of rigid casing wall (see Fig. 3.23), it can be considered as entirely sufficient. Therefore, a model evaluated in such way for the unloaded plate can now be used to investigate behaviour of the plate when the additional elements are attached to its surface.

3.5.1.2 Casing wall with additional mass

The next case is an evaluation of the casing wall with an additional mass attached. In this investigation, a mass of $m_{m,1} = 0.080$ kg is used. It is rigidly bonded to the plate surface at location $x_{m,1} = 0.340$ m and $y_{m,1} = 0.340$ m. Measurements were performed in the same manner as for the unloaded plate. Rotary inertia of the attached mass is taken into account. Theoretically calculated and experimentally measured natural frequencies and mode shapes are compared in Fig. 3.24. A high consistency of both results is obtained.

3.5.2 LIGHTWEIGHT CASING WALLS

In this subsection, the light-weight casing is under consideration. A variant of dimensions 500 mm × 630 mm × 800 mm with 1.5 mm of plate thickness is used. The casing with accompanying laboratory setup is shown in Fig. 3.25. During the measurements of the natural frequencies and mode shapes, the whole casing was excited together with a random signal by a loudspeaker placed inside. The response of the casing was measured in uniformly distributed points with the distance between the adjacent points equal 20 mm (as in the case of the rigid casing walls). Such dense grid resulted in many measurement points: 1280 points for the top wall, 1000 points for the front wall and 800 points for the left wall. Hence, a high accuracy of the measured mode shapes is obtained. They can be presented for the whole casing together, as in an example for frequency of 155 Hz given in Fig. 3.26. However, as it was discussed in previous chapters, the light-weight casing constitutes a three-dimensional structure with strong couplings, but due to its specific dynamical behaviour, it can be studied with each of its walls separately. Therefore, due to methodology of the modelling, theoretical calculations and measurements are further compared for each wall separately (right and back walls are omitted as they are symmetrical to the left and front walls, respectively).

The mathematical modelling of the loading of the casing walls with additional elements is validated in previous subsection. Therefore, to justify the application of the developed model for the light-weight casing, only results for the unloaded casing are presented.

3.5.2.1 Unloaded casing walls

Mathematical model parameters used here can be divided into two groups. First group includes dimensions and material properties of the plates that are known, and

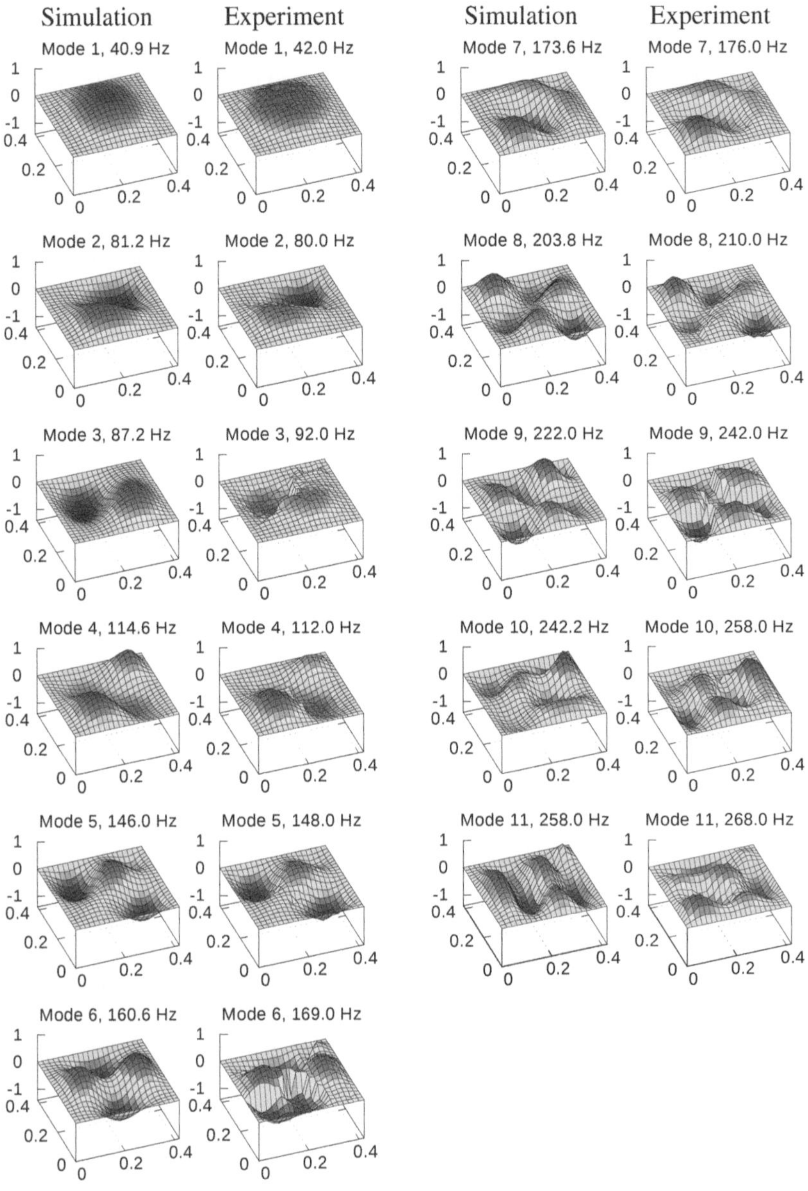

Figure 3.24 A comparison of initial 11 natural frequencies and mode shapes of rigid casing wall, calculated with the mathematical model and experimentally measured—1 mm thick aluminium plate with an additional mass of $m_{m,1} = 0.080$ kg mounted at $x_{m,1} = 0.340$ m and $y_{m,1} = 0.340$ m [164]. Reprinted from "Modelling and control of device casing vibrations for active reduction of acoustic noise", PhD thesis, Silesian University of Technology, S. Wrona.

in this research they are represented by following values:

$$h = 0.0015\ \mathrm{m},$$
$$E = 200\ \mathrm{GPa}, \qquad G = 76.9\ \mathrm{GPa},$$
$$\rho_p = 7850\,{}^{\mathrm{kg}}/_{\mathrm{m}^3}, \qquad \nu = 0.3, \qquad \kappa = {}^5/_6.$$

All of these values are common for all walls of the casing. The width and height of each plate (denoted by a and b, respectively) are defined by corresponding dimensions of the casing given earlier in the subsection.

The second group consists of parameters, which cannot be measured or calculated directly. Therefore, for the purpose of fitting the model to the behaviour of real whole vibrating structure, an optimization algorithm is used to identify them. In this research, such parameters are spring constants describing boundary conditions of each wall. Plate edges are assumed to be elastically restrained against both rotation and translation, hence there are four rotational spring constants k_{rx0}, k_{rx1}, k_{ry0}, k_{ry1}, and four translational spring constants k_{tx0}, k_{tx1}, k_{ty0}, k_{ty1}, defined as in Section 3.2.2,

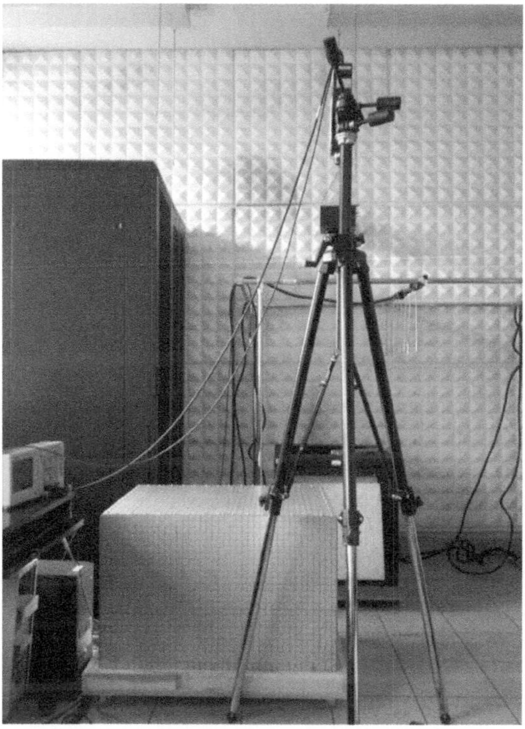

Figure 3.25 The laboratory setup with the laser vibrometer and the light-weight casing to measure mode shapes of the structure [164]. Reprinted from "Modelling and control of device casing vibrations for active reduction of acoustic noise", PhD thesis, Silesian University of Technology, S. Wrona.

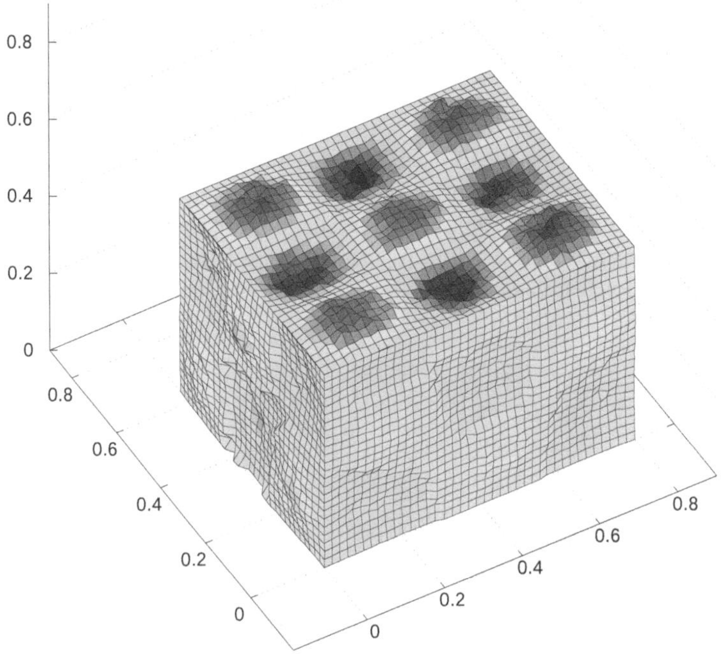

Figure 3.26 Three-dimensional visualization of experimentally measured mode shape of the whole light-weight casing, for an exemplary frequency of 155 Hz. All dimensions are given in [m] [164]. Reprinted from "Modelling and control of device casing vibrations for active reduction of acoustic noise", PhD thesis, Silesian University of Technology, S. Wrona.

to be identified for each casing wall (as the two pairs of walls are symmetrical—left and right, front and back—the same configuration is calculated for a given pair). The obtained results and process of identification procedure is described in more details in [169].

A comparison of initial 12 eigenmodes (simulated for optimal set of spring constants and measured experimentally) of the top, front and left wall are given in Figs. 3.27–3.29, respectively. All eigenmodes that were observed are consistent with the model, confirming the correctness of the model.

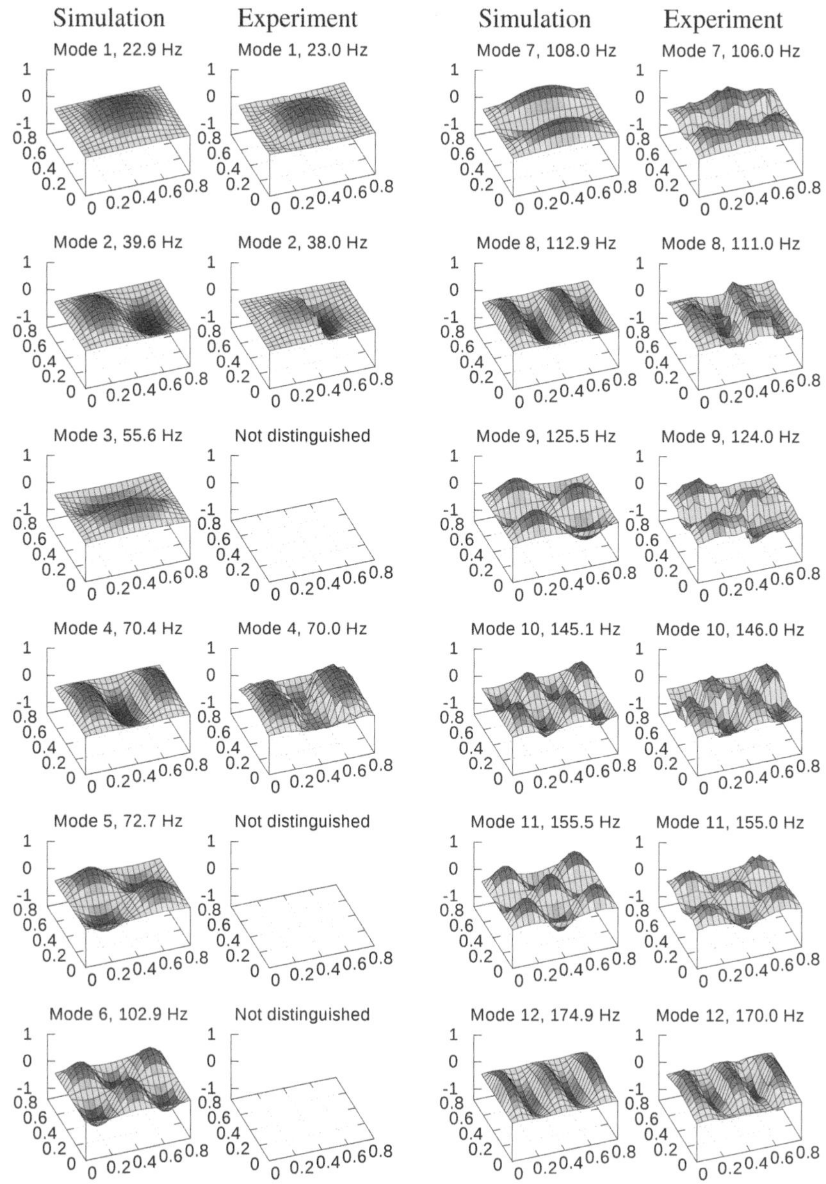

Figure 3.27 A comparison of initial 12 natural frequencies and modeshapes of the top wall of the light-weight casing, calculated with the mathematical model and experimentally measured [164]. Reprinted from “Modelling and control of device casing vibrations for active reduction of acoustic noise”, PhD thesis, Silesian University of Technology, S. Wrona.

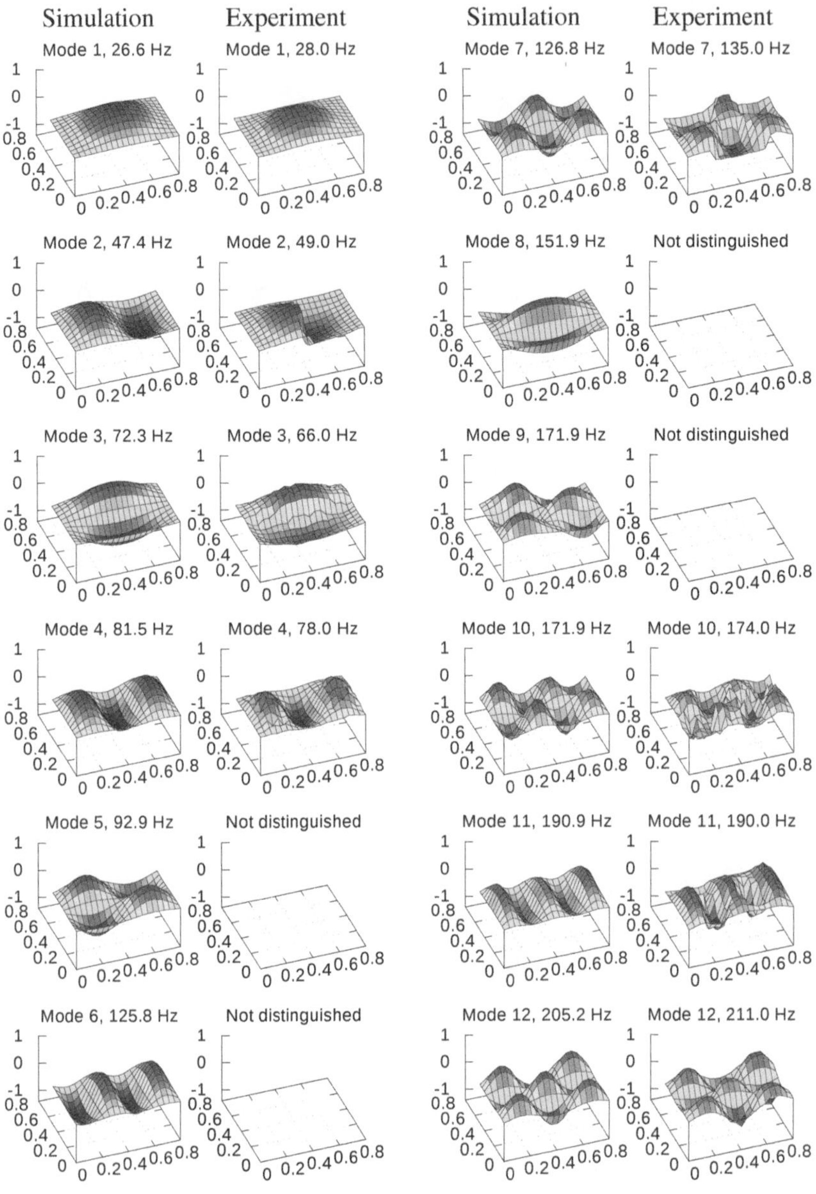

Figure 3.28 A comparison of initial 12 natural frequencies and mode shapes of the front wall of the light-weight casing, calculated with the mathematical model and experimentally measured [164]. Reprinted from "Modelling and control of device casing vibrations for active reduction of acoustic noise", PhD thesis, Silesian University of Technology, S. Wrona.

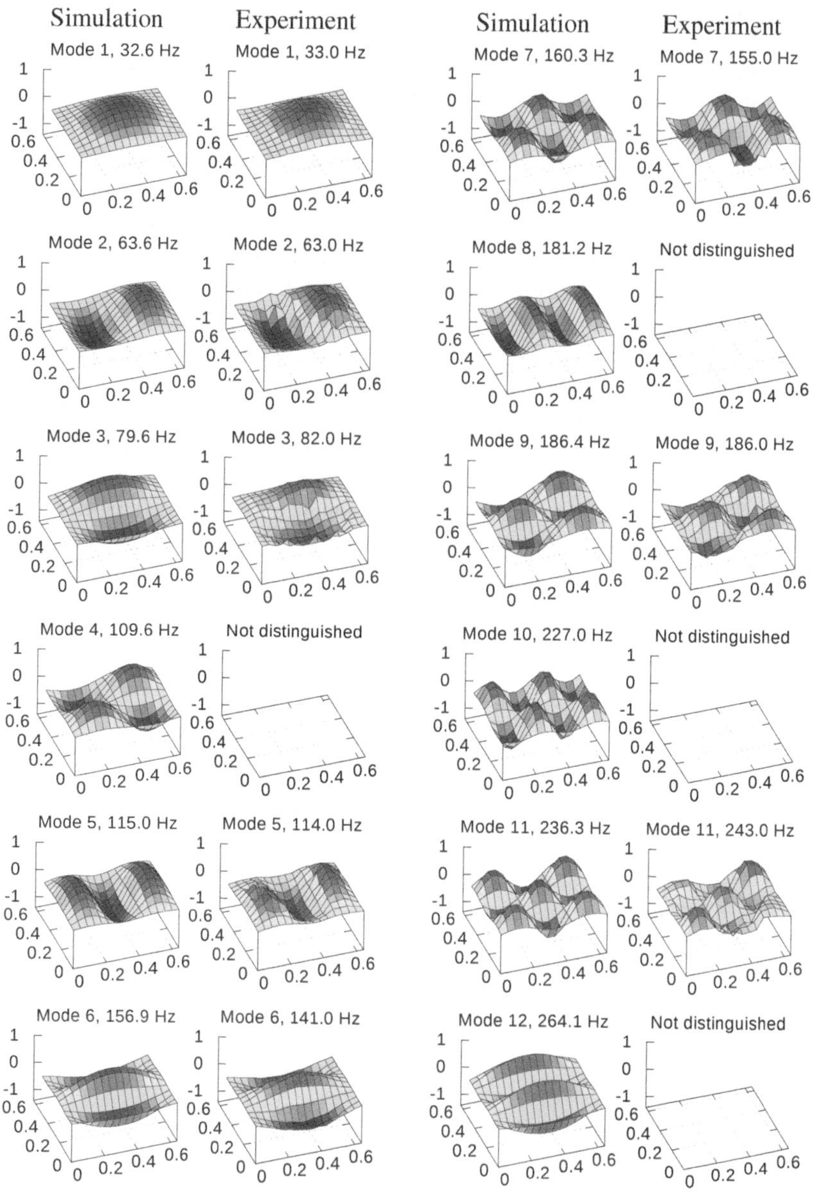

Figure 3.29 A comparison of initial 12 natural frequencies and mode shapes of the left wall of the light-weight casing, calculated with the mathematical model and experimentally measured [164]. Reprinted from "Modelling and control of device casing vibrations for active reduction of acoustic noise", PhD thesis, Silesian University of Technology, S. Wrona.

3.6 SUMMARY

The mathematical models of the device casing walls have been presented in this chapter, incorporating different approaches.

First, analytical model of individual casing walls has been developed. The model includes thin and thick plate theory, elastically restrained boundary conditions, structural thermoelastic damping model, and additional elements mounted to the casing surface—masses, ribs, actuators and sensors. These aspects are integrated in a state space model form, which facilitates further analysis and numerical simulation.

Second, analytical modelling of the whole casing has been presented. A number of interacting subsystems are specified, including acoustic medium inside cavities and surrounding the casing. For the light-weight casing, vibrations of a panel cause changes in the boundary conditions of the adjacent panels, what was also considered.

The developed theoretical models have been verified by simulation using multiphysics packages and experimentally in laboratory tests. Experimental measurements have also been used to fine tune the parameters of the mathematical models.

The model can be used to predict natural frequencies and mode shapes of a casing walls or other panels, and hence to optimize the structure parameters. If active control is concerned, it can be used also to calculate measures of controllability and observability of the system.

4 Passive control

4.1 INTRODUCTION

The first considered group of noise reduction methods are passive solutions, which are characterized by a complete lack of the need for external energy supply. In general, this group includes physical modifications to the noise source itself or barriers that separate it from the recipient. These modifications often lead to the optimal shaping of the vibroacoustic properties of the structure (e.g. shifting the natural frequencies or introducing additional damping) based on a previously prepared theoretical or numerical model [35, 171, 172]. These modifications can be made by adding metamaterials [32], mass, a damper, or increasing the stiffness of a selected structure component in an appropriate manner, provided, e.g. by an optimization algorithm. Modifications of this type may be introduced at the design stage and taken into account in production, or implemented individually for existing devices. Passive solutions can also include systems originally based on semi-active or even active solutions, but powered by energy harvesting within the system, which eliminates the need for external energy supply [105, 80, 37].

4.2 SHAPING OF STRUCTURAL AND ACOUSTIC RESPONSES

Plates exhibit a complicated frequency response, with multiple strong resonances and antiresonances. The ability to shape the frequency response of the plate as desired would therefore be highly beneficial in noise control applications, allowing to better fit to given noise spectrum. The passive transmission loss of the plate could be improved by appropriately shaping the frequency response, moving resonances in frequency domain away from dominant noise tones [176].

As described in [176], an arrangement of additional masses and ribs attached to the plate surface can be optimized in order to achieve a desired structural frequency response. The general rules for the addition of mass or stiffness are well known – additional masses reduce the natural frequencies of the plate, whereas stiffeners increase them. When these elements are used together at appropriate locations, they can be used to precisely alter the frequencies of multiple modes. Initially-developed methods have focused on natural frequencies of structural vibrations and have not taken into account their acoustic radiation efficiency [171, 172]. This should not be neglected, as even moderate alterations of the mode shapes can significantly affect their acoustic radiation power [176]. For example, for a noise barrier, instead of shifting structural resonances out of a frequency range of interest, their acoustic radiation can be reduced. This phenomenon is important for the applications under consideration, i.e. noise-reducing casings.

The limitations of the proposed method relate mainly to the maximum dimensions and masses of the added structures, but the values derived from the optimization

DOI: 10.1201/9781003273806-4

process are practical and can be employed for a wide class of applications. In active noise control applications structural sensors and actuators are modelled as additional masses. Their arrangement is optimized, together with that of passive masses and ribs, in order to maximize the controllability and observability of the system [179, 177], together with the frequency response shaping objectives. This requires the use of multi-objective cost functions. In this paper, accelerometers and inertial actuators are considered, however, the idea can also be applied to other kinds of transducers, e.g. piezoelectric patches [1, 49].

4.2.1 OPTIMIZATION ALGORITHM

This section describes the proposed method for optimizing the arrangement of additional elements mounted on a vibrating plate in order to shape its acoustic radiation, and to enhance controllability and observability measures, if necessary. An optimization problem with appropriate cost functions is defined, and an optimization algorithm is employed to find an optimal solution [176].

Optimization problem

The variables defined for the optimization problem correspond to the shape and placement of a predefined number of additional elements mounted on the surface of a vibrating plate. They are optimized in the sense of minimizing an arbitrarily-chosen cost function. Four kinds of elements are considered:

- **Additional masses**
  These are passive elements. A number of different items can be employed in this role, including custom-made weights and neodymium magnets. The total concentrated mass $m_{m,i}$ and position coordinates $(x_{m,i}, y_{m,i})$ are considered as optimization variables for the ith additional mass.
- **Ribs**
  These are passive elements. It is assumed for practical reasons that one of their dimensions is much greater than the others, and that their cross-sections are constant along this direction. Therefore, the length, location and orientation defined by coordinates $(x_{r0,i}, y_{r0,i})$ and $(x_{r1,i}, y_{r1,i})$ are considered as optimization variables for the ith rib. The rib material and cross-sectional dimensions are assumed here to be predefined, however, they could also be optimization variables if desired.
- **Actuators**
  In this paper inertial actuators are considered, although other types of actuators, such as PZT patches, PZT stacks or MFC patches, could also be employed and included in the model. The actuators are used to force vibration and generate sound with the vibrating plate, or to enhance its passive transmission loss. Their shape and mass are assumed to be predefined, hence their position coordinates $(x_{a,i}, y_{a,i})$ are the only optimization variables for the ith actuator.

- **Sensors**
  Accelerometers are used in this research as structural sensors to monitor vibrations or to provide signals for an active control algorithm. Other transducers, e.g. tensometers, MFC elements or PVDF films could also be considered and included in the model. Similarly, as in the case of actuators, only the position coordinates $(x_{s,i}, y_{s,i})$ are considered as optimization variables for the ith sensor.

Furthermore, the variables specified above are subject to various constraints. The number, weight and dimensions of the elements are limited due to plate shape and other practical constraints depending on the application. A further constraint is that the additional elements should not overlap. To summarize the formulated problem, the optimization algorithm is required to find a solution in a $(3N_m + 4N_r + 2[N_a + N_s])$-dimensional space.

Memetic algorithm

The multi-dimensional search space that follows from the described problem is very complicated, containing numerous local optima. Therefore, an efficient optimization algorithm should be employed in order to find a solution that satisfies the defined requirements. A Memetic Algorithm (MA) has been chosen for this task, which is a hybrid form of a population-based approach coupled with individual learning [130, 128]. MA combines advantages of global search and local refinement procedures, which enhance convergence to the local optima. The MA has been successfully used by the authors in previous applications, e.g. in [107, 106].

Fig. 4.1 shows a flowchart of the memetic algorithm used in this work. The algorithm starts with the *initialization* step, in which a given number of individuals (potential solutions) is randomly generated to create the initial population (a pool of possible solutions).b In the *evaluation* step, the fitness function is calculated for

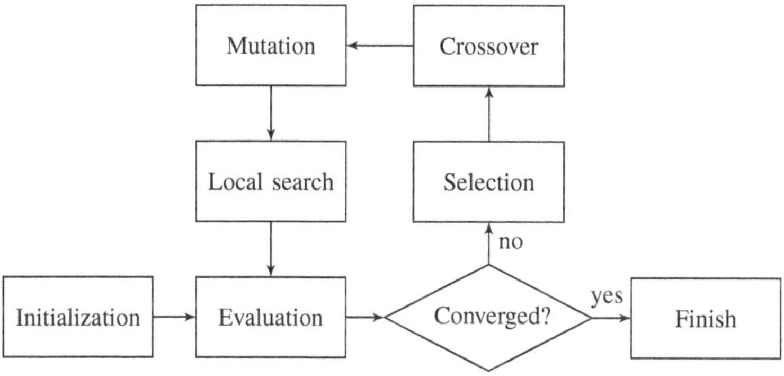

Figure 4.1 A memetic algorithm flowchart [176]. Reprinted from *Journal of Sound and Vibration*, 476:115285, S. Wrona, M. Pawelczyk, and X. Qiu, "Shaping the acoustic radiation of a vibrating plate", Copyright (2022), with permission from Elsevier.

each individual in the population. This part of the algorithm consumes most of the computation time, as it requires a complete model recalculation, including both vibration and acoustic radiation parts, for each fitness function evaluation. Hence, parallelized computation is advisable. Subsequently, the individuals are sorted based on their fitness values. The algorithm then checks whether the termination criteria have been met, i.e. whether a fixed maximum number of generations has been reached or the optimization has converged. Convergence is defined to have occurred when a predefined number of recent generations resulted in no improvement in the cost function. If any of the termination criteria are met, then the optimization process is finished. Otherwise, the process continues, as shown in Fig. 4.1, and the *selection* step is performed, where a subset of the existing population is chosen for reproduction depending on their associated fitness values. A fixed number of the fittest individuals are selected (survive) and the rest are rejected (die out). Then, in the *crossover* step, child solutions are generated by applying a crossover operator for a number of selected pairs of individuals (parents). Individuals with higher fitness values are more likely to be chosen as parents. Children join the population without replacing the parents at this stage, although this may happen at a subsequent stage depending on their fitness, as determined in the *evaluation* and *selection* steps. This process is equivalent to the elitist selection strategy, which allows a limited number of the best parents to carry over to the next generation, unaltered. To maintain population diversity, so as to counteract premature convergence and ensure a global search, the mutation operator is used in the *mutation* step. This operator is applied based on a predefined probability. In this implementation, the mutation results in the creation of an additional individual which joins the population, keeping the original individual unaltered. Then, a local search operator is employed to improve the individual fitness (the *local search* step). To maintain a balance between the degree of exploration and individual improvement, only a fraction of the population of individuals undergoes the learning process. The "hill climbing" technique is chosen as the individual learning strategy [130]. Afterwards, an *evaluation* is performed, and the process is repeated until one of the termination criteria is met.

4.2.2 OBJECTIVE FUNCTIONS

The cost function for the described problem should quantify the discrepancy between the desired and actual acoustic radiation of the plate. It can be expressed using the natural frequencies f_i and estimates of modal acoustic power P_i, where i stands for the mode number. The overall transmission in a specified frequency band can also be specified to be either enhanced or attenuated, depending on the role of the plate. Moreover, if an active control application is considered, measures of controllability and observability of the system, expressed in the form of diagonal elements of the controllability and observability Gramian matrices, $\lambda_{c,i}$ and $\lambda_{o,i}$, respectively, should also be introduced into the cost function, making it a multi-objective optimization problem. Therefore, in a general form, the cost function can be defined as [176]

$$J = f(f_i, P_i, \lambda_{c,i}, \lambda_{o,i}, \Theta), \tag{4.1}$$

where Θ is a vector representing the optimization variables of the additional elements, including their shapes and arrangement. Specific definitions of particular cost functions are presented in Section 4.2.3, where results for different scenarios are presented and discussed.

In the present research, the numbers of additional elements mounted to the plate are not a subject of optimization. They are predefined as initialization parameters. If the results obtained for particular numbers of elements are not satisfactory, the numbers should be increased. However, simpler configurations are usually more desirable. Therefore, if the goal is achieved, the possibility of reducing the number of elements should be explored—sometimes a satisfactory result can be obtained with fewer elements. This means that the final number of elements is a trade-off between the quality of the obtained result and the complexity of the solution. In the example cases reported in the following section, the choice of numbers of elements is made according to the guidelines described above. However, for the sake of brevity, this process is not discussed in detail for any particular scenario.

4.2.3 RESULTS

In the simulation studies, frequencies up to 400 Hz are considered, and first 12 eigenmodes of the plate are compared for each scenario. For the unloaded plate, frequency responses of the plate obtained from the model are shown in Fig. 4.2. The acoustic and structural vibration responses shown are (i) the mean sound pressure amplitude obtained by averaging over the measurement grid and (ii) the mean vibration amplitude obtained by averaging over the plate surface, respectively. Natural frequencies are given in Table 4.1. Additional masses are limited to a maximum value of 0.2 kg. The ribs are assumed to have a constant cross-section with a square shape, defined by dimensions:

$$e_{r,i} = h_{r,i} = 0.006 \text{ m}, \tag{4.2}$$

where $e_{r,i}$ is the width and $h_{r,i}$ is the height of the ith stiffener. Such a cross-section implies that the geometric properties of the stiffeners are given by:

$$A_{r,i} = e_{r,i} h_{r,i}, \quad I_{r,i} = \frac{1}{12} e_{r,i} \, h_{r,i}{}^{3} + e_{r,i} h_{r,i} \left(\frac{h + h_{r,i}}{2} \right)^{2}. \tag{4.3}$$

The material of the stiffeners is considered to be the same as that of the plate.

For each optimization, the population consisted of 300 individuals, the maximum number of generations was set to 15, and the probabilities of crossover, mutation and individual learning were 0.20, 0.30 and 0.06, respectively.

Minimization of the acoustic radiation in a wide frequency range

The first example considered is the minimization of the acoustic radiation in a wide frequency range. This objective is expressed using the following cost function

$$J_1 = \max_i P_i, \qquad \omega_i \leq \omega_{sp}, \tag{4.4}$$

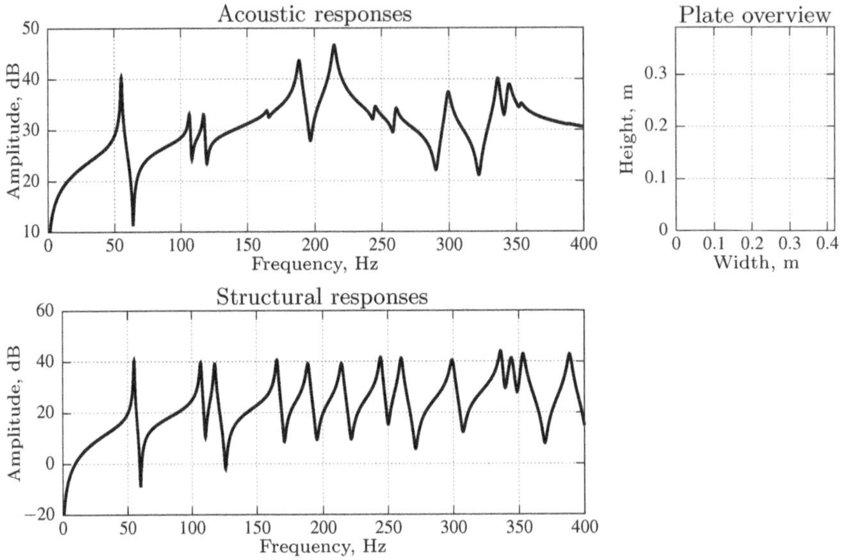

Figure 4.2 Acoustic and structural vibration responses of the unloaded plate [176]. Reprinted from *Journal of Sound and Vibration*, 476:115285, S. Wrona, M. Pawelczyk, and X. Qiu, "Shaping the acoustic radiation of a vibrating plate", Copyright (2022), with permission from Elsevier.

Table 4.1

Natural frequencies and estimates of modal acoustic power of the unloaded plate [176].

Mode	1	2	3	4	5	6	7	8	9	10	11	12
ω_i (Hz)	55.1	106.8	117.6	165.5	188.8	214.4	244.5	260.1	299.4	336.4	344.3	353.2
P_i (dB)	20.1	12.9	13.1	8.4	24.1	27.2	10.8	12.0	17.2	18.4	18.0	8.9

Reprinted from *Journal of Sound and Vibration*, 476:115285, S. Wrona, M. Pawelczyk, and X. Qiu, "Shaping the acoustic radiation of a vibrating plate", Copyright (2022), with permission from Elsevier.

where ω_{sp} is a set point frequency, limiting the frequency range of interest (only resonances with $\omega_i \leq \omega_{sp}$ are considered in the cost function). The cost function J_1 results in the minimization of the acoustic radiation of the most radiating mode. To evaluate the cost function, the set point frequency was set to $\omega_{sp} = 300$ Hz. Only passive elements were considered in this case. The numbers of additional masses and ribs were set to $N_m = 2$ and $N_r = 2$, respectively. Results of the optimization are given in Fig. 4.3 and Table 4.2, where circles and lines represent placements of additional masses and ribs, respectively.

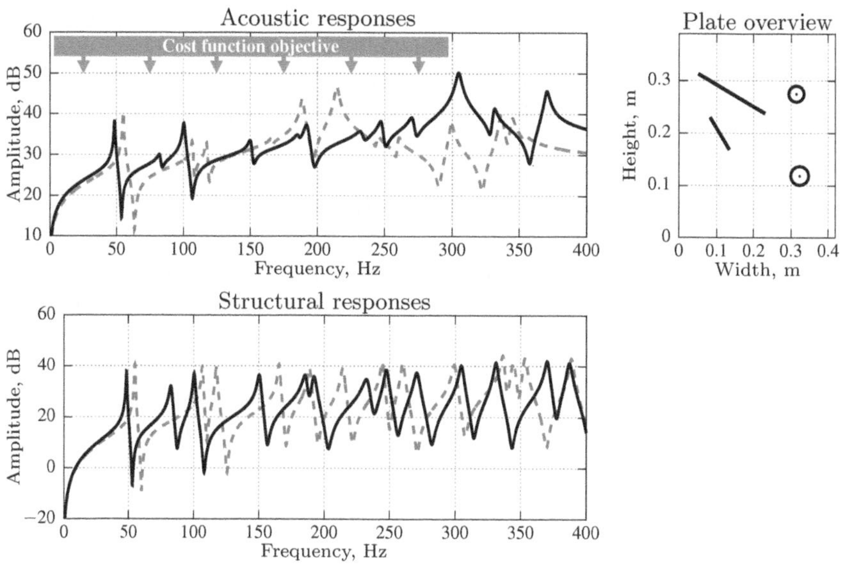

Figure 4.3 Acoustic and structural vibration responses of the plate, obtained for optimization index J_1 (solid line—plate with elements; dashed line—the unloaded plate). The plate overview with the additional elements is also shown (circles—additional masses; lines—ribs) [176]. Reprinted from *Journal of Sound and Vibration*, 476:115285, S. Wrona, M. Pawelczyk, and X. Qiu, "Shaping the acoustic radiation of a vibrating plate", Copyright (2022), with permission from Elsevier.

Table 4.2

Results of optimization of the cost function J_1 for $N_m = N_r = 2$, showing the natural frequencies and estimates of modal acoustic power of the loaded plate, together with the placement of additional elements [176].

Mode	1	2	3	4	5	6	7	8	9	10	11	12
ω_i (Hz)	48.5	82.7	100.5	150.5	185.8	192.3	233.0	247.7	270.7	304.7	331.0	370.6
P_i (dB)	19.2	11.7	19.2	14.2	12.8	19.4	15.8	18.1	19.0	30.2	19.9	24.7

Masses	$x_{m,i}$ (m)	$y_{m,i}$ (m)	$m_{m,i}$ (kg)	Ribs	$x_{r0,i}$ (m)	$y_{r0,i}$ (m)	$x_{r1,i}$ (m)	$y_{r1,i}$ (m)
1	0.323	0.119	0.200	1	0.231	0.238	0.050	0.313
2	0.314	0.275	0.146	2	0.134	0.169	0.082	0.230

Reprinted from *Journal of Sound and Vibration*, 476:115285, S. Wrona, M. Pawelczyk, and X. Qiu, "Shaping the acoustic radiation of a vibrating plate", Copyright (2022), with permission from Elsevier.

The algorithm achieved the solution with a cost function value of $J_{1,opt} = 19.4$ dB. As a result of the optimization, the amplitudes of the structural response were only slightly altered, whereas the acoustic response amplitudes were strongly reduced, reducing the amplitudes of the highest peaks in the considered frequency range by more than 7 dB.

It follows from the analysis of these results that the structural vibration resonances can be moved in the frequency domain with additional masses and ribs, but it is difficult to reduce their amplitudes. On the other hand, additional masses and ribs strongly affect the mode shapes, which play a large part in determining the acoustic radiation. Therefore the results confirm that the acoustic radiation capability of the plate can be significantly reduced with the use of additional elements, even over a wide frequency range.

It is noteworthy that, just above the considered frequency range, the tenth mode has a high estimate of the modal acoustic power. In such cases it may be desirable to increase the set point frequency ω_{sp} in order to avoid the proximity of such highly-radiating modes.

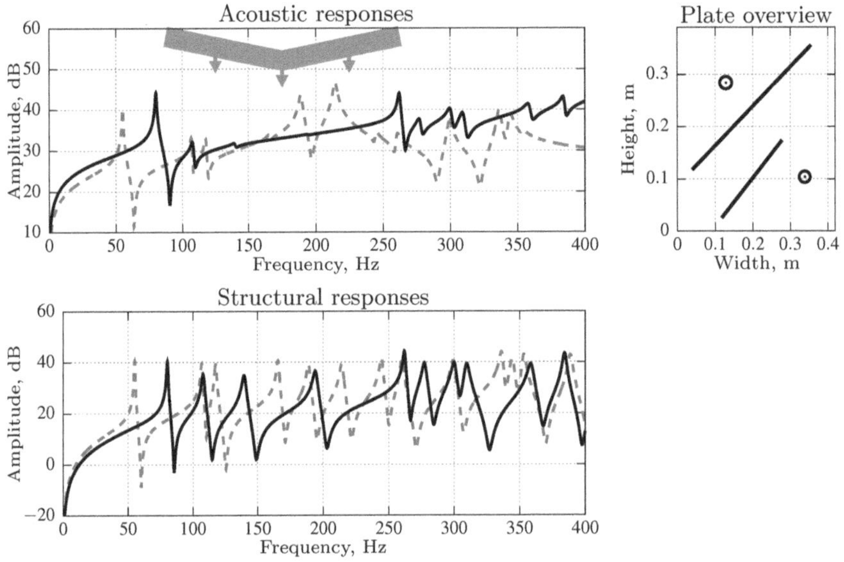

Figure 4.4 Acoustic and structural vibration responses of the plate, obtained for optimization index J_3 (solid line—plate with elements; dashed line—the unloaded plate). The plate overview with the additional elements is also shown (circles—additional masses; lines—ribs) [176]. Reprinted from *Journal of Sound and Vibration*, 476:115285, S. Wrona, M. Pawelczyk, and X. Qiu, "Shaping the acoustic radiation of a vibrating plate", Copyright (2022), with permission from Elsevier.

Minimization of the acoustic radiation for a narrower frequency bandwidth

The third scenario considers the minimization of the acoustic radiation for a narrower frequency band. This objective is expressed using the cost function

$$J_3 = \sum_i \left\{ \max \left[\left(1 - d_f \frac{|\omega_i - \omega_{sp}|}{\omega_{sp}} \right), 0 \right] \cdot P_i \right\}, \qquad i \in \{1,2,...,N\}, \tag{4.5}$$

where d_f determines the width of the chosen frequency band; and set point frequency ω_{sp} defines the centre of the band. The cost function J_3 results in minimization of the acoustic radiation of a set of modes, with the highest weighting given to the modes that are nearest to the set point frequency ω_{sp}. To evaluate the cost function, the set point frequency was set to ω_{sp} = 175 Hz and parameter d_f = 2. Only passive elements were considered. The numbers of additional masses and ribs were again set to $N_m = 2$ and $N_r = 2$, respectively. Results of the optimization are given in Fig. 4.4 and Table 4.3.

The algorithm achieved a solution with a cost function value of $J_{3,opt}$ = 10.2 dB. The solution found shows that the shaping method acted in two ways. First, the natural frequencies of the plate were moved as far away from the frequency band of interest as possible. Second, for those modes that could not be moved far enough, the estimates of modal acoustic power were greatly reduced. As a result, the acoustic response of the plate was flattened in the frequency band of interest, increasing its transmission loss significantly. The optimized solution therefore makes such a plate suitable for application as a noise barrier for the noise frequencies considered.

Table 4.3

Results of optimization of the cost function J_3 for $N_m = N_r = 2$, showing the natural frequencies and estimates of modal acoustic power of the loaded plate, together with the placement of additional elements [176].

Mode	1	2	3	4	5	6	7	8	9	10	11	12
ω_i (Hz)	80.2	107.9	139.9	194.1	262.0	277.4	300.2	309.6	358.7	384.3	425.0	446.0
P_i (dB)	24.0	13.6	8.0	3.0	21.9	16.2	19.4	18.9	20.0	19.9	23.8	35.4

Masses	$x_{m,i}$ (m)	$y_{m,i}$ (m)	$m_{m,i}$ (kg)	Ribs	$x_{r0,i}$ (m)	$y_{r0,i}$ (m)	$x_{r1,i}$ (m)	$y_{r1,i}$ (m)
1	0.338	0.104	0.075	1	0.117	0.026	0.278	0.174
2	0.128	0.284	0.092	2	0.039	0.118	0.355	0.356

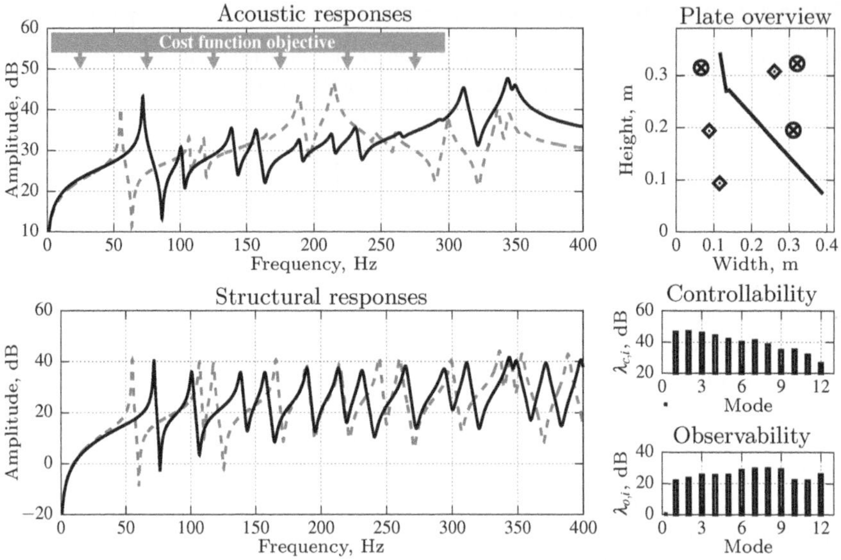

Figure 4.5 Acoustic and structural vibration responses of the plate, obtained for optimization index J_6 (solid line—plate with elements; dashed line—the unloaded plate). The plate overview with the additional elements is also shown (circles with "X" inside—actuators; diamonds—sensors; lines—ribs) [176]. Reprinted from *Journal of Sound and Vibration*, 476:115285, S. Wrona, M. Pawelczyk, and X. Qiu, "Shaping the acoustic radiation of a vibrating plate", Copyright (2022), with permission from Elsevier.

Hybrid passive-active acoustic barrier

A hybrid, passive-active acoustic barrier is optimized in this subsection. The objective in this scenario is expressed using the following cost function

$$J_6 = \prod_i \left\{ \frac{P_i}{\lambda_{c,i} \cdot \lambda_{o,i}} \right\}, \qquad \omega_i \leq \omega_{sp} \,. \tag{4.6}$$

The cost function J_6 leads to the simultaneous minimization of both acoustic radiation of modes within the considered frequency range, and the maximization of their controllability and observability. Structural sensors are introduced in this scenario, because the active barrier control system needs a feedback signal for its operation. To evaluate the proposed cost function, the set point frequency was set to $\omega_{sp} = 300$ Hz. Actuators, sensors and ribs were used as admissible additional elements, with $N_a = 3$, $N_s = 3$ and $N_r = 2$, respectively. The results of the optimization are shown in Fig. 4.5 and Table 4.4.

The algorithm achieves the solution with a cost function value of $J_{6,opt} = -54.0$ dB. The solution satisfies all of the stated objectives: (i) the estimates of modal acoustic power of resonances within the frequency range of interest are passively reduced (by more than 10 dB for certain frequency bands), while both (ii) controllability, and (iii) observability are ensured for all of the considered modes.

Table 4.4

Results of optimization of the cost function J_6 for $N_a = 3$, $N_s = 3$ and $N_r = 2$, showing the natural frequencies and estimates of modal acoustic power of the loaded plate, together with the placement of additional elements [176].

Mode	1	2	3	4	5	6	7	8	9	10	11	12
ω_i (Hz)	72.1	100.9	138.9	157.3	189.4	213.6	231.4	264.6	294.2	311.2	344.0	348.6
P_i (dB)	23.1	12.6	17.7	17.2	12.7	14.0	16.9	9.9	11.7	25.7	27.5	24.8
$\lambda_{c,i}$ (dB)	46.3	46.7	45.7	44.2	41.9	40.0	41.0	38.4	34.8	35.1	31.8	26.6
$\lambda_{o,i}$ (dB)	21.9	23.4	25.4	25.3	25.4	28.5	29.4	29.6	29.0	22.0	21.8	26.0

Actuators	$x_{a,i}$ (m)	$y_{a,i}$ (m)	$m_{a,i}$ (kg)
1	0.321	0.323	0.115
2	0.066	0.315	0.115
3	0.311	0.195	0.115

Ribs	$x_{r0,i}$ (m)	$y_{r0,i}$ (m)	$x_{r1,i}$ (m)	$y_{r1,i}$ (m)
1	0.133	0.267	0.116	0.344
2	0.138	0.273	0.389	0.074

Sensors	$x_{s,i}$ (m)	$y_{s,i}$ (m)	$m_{s,i}$ (kg)
1	0.115	0.094	0.010
2	0.087	0.194	0.010
3	0.261	0.308	0.010

The development of such hybrid solutions is a novel approach which requires more investigation. However, given its potential, it is worthy of further research.

It is noteworthy that it is only for the first mode that the acoustic power was not passively reduced by the optimized solution (see Fig. 4.5). This is due to the fact that the fundamental mode, by its nature, does not have any nodal lines, which makes it practically impossible to reduce its acoustic radiation by influencing the mode shape (virtually no alteration is possible for such a shape). Hence, the acoustic power of the fundamental mode can only be modified slightly by altering its frequency.

To summarize, the analysis of the results of the optimization algorithm shows that the structural vibration resonances can be adjusted in the frequency domain with additional masses and ribs, but that their magnitudes are difficult to influence. However, the addition of masses and ribs strongly affects the mode shapes, which play a significant part in determining the acoustic radiation. Therefore, the incorporation of the acoustic radiation model into the optimization process greatly extends the capabilities of the shaping approach. The transmission loss of a passive noise barrier can be enhanced by more than 7 dB in a chosen frequency band and, in the opposite case, the passive sound transmission of a plate can be enhanced by more than 10 dB.

Furthermore, the results of this passive optimization approach can be combined with optimization of the placement of actuators and sensors , resulting in hybrid passive-active control systems. The optimization process was successfully performed for both a driven sound source and a hybrid passive-active noise barrier. The passive transmission loss for the barrier was enhanced by more than 10 dB for certain frequency bands, while both controllability and observability were ensured for all of the modes considered.

There are few limitations of the method other than those related to the physical constraints imposed on the structure—limited dimensions, overall weight and selected type of elements. However, as shown by the examples discussed, the structural parameters of the elements considered are appropriate for the majority of applications. The proposed method could also be applied for more complex structures, such as panels of different shapes, and even whole casings of devices, if a model of the vibroacoustic system is available. The method allows for enhanced sound radiation or transmission loss properties of a structure, without affecting its outward appearance.

4.3 PASSIVE SHUNT SYSTEMS

The passive shunt circuits assume no need of external energy to work. It is based on the piezoelectric elements such as Lead Zirconate Titanate (PZT) or Macro Fiber Composite (MFC) structures. It is capable to transform the mechanical energy of the vibrating structure into the electrical charge. This energy is stored in internal capacitance of piezoelectric element and should be reused or dissipated. The easiest method is transform energy into the heat, by including resistor in circuit [55]. With internal capacitance of piezoelectric element and additional inductor a simple RLC circuit can be build. Such system is more efficient, however, it should be tuned to the specific frequency of vibration, i.e. the resistance and inductance values are optimized for the single mode. It is also sensitive for changes of environmental conditions, such as temperature, and requires inductance values higher than inductors existing in reality. Hence, a synthetic inductor, based on the operational amplifiers (opamps) is used [160]. However, it requires an external source of energy to supply opamps, which place this solution in group of active systems. An alternative approach is switching circuits, based on the Synchronized Switched Damping (SSD) idea. It assumes switching between open and short circuits, four times per each vibration cycle, when the displacement of vibrating structure occurs extremum [69]. A proper time is determined with the use of switching law. One of the most popular methods from SSD group is SSD on inductor (SSDI), which is RLC shunt circuit with switching [149]. This method can be implemented as passive or semi-active system, dependently on source of energy used to control the switches.

4.3.1 HARDWARE PLATFORM

The presented research is based on rigid frame casing described previously. Two different panels, i.e. 0.5 mm thick steel (Fig. 4.6(a)) with five MFCs and 1.0 mm thick aluminium (Fig. 4.6(b)) with nine MFCs were investigated as a front wall. To assure boundary conditions similar to fully clamped an additional clamping frame was used,

(a) Steel plate with 5 MFCs. (b) Aluminium plate with 9 MFCs.

Figure 4.6 Front wall panels with MFC elements attached.

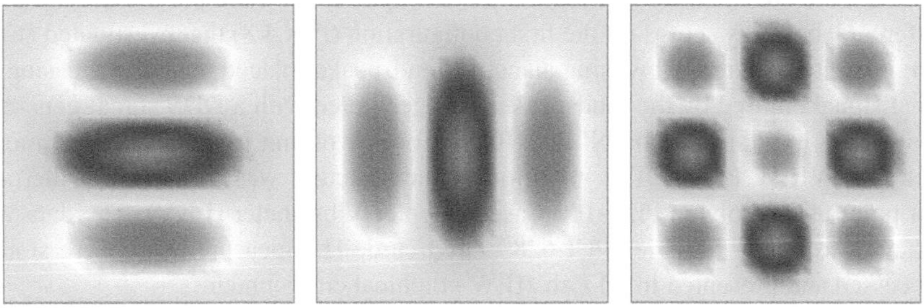

(a) (1,3) mode. (b) (3,1) mode. (c) (3,3) mode.

Figure 4.7 Investigated mode shapes of vibrating structure.

and the panels were mounted using twenty screws. The other—top and side—walls are made from 3 mm thick aluminium panels to enhance acoustic insulation of the casing. To assure mechanical coupling between the panel and piezoelectric elements as much as it is possible, the MFCs were attached to plate's surface using epoxy glue. To increase the efficiency of SSDI circuit, the MFC elements are connected in series, and their placement was determined by mode shapes of the vibrating structure. For the first experiment a (3,3) mode was investigated (Fig. 4.7(c)), and MFC elements were attached in places of anti-nodes compatible in phase. For the second experiment a (1,3) and a (3,1) modes were investigated (Figs. 4.7(a), and (b)). The MFCs were attached in places of every anti-node. To increase the voltage obtained from piezoelectric elements, in the places of anti-nodes with opposite phases a polarization of MFCs was inverted.

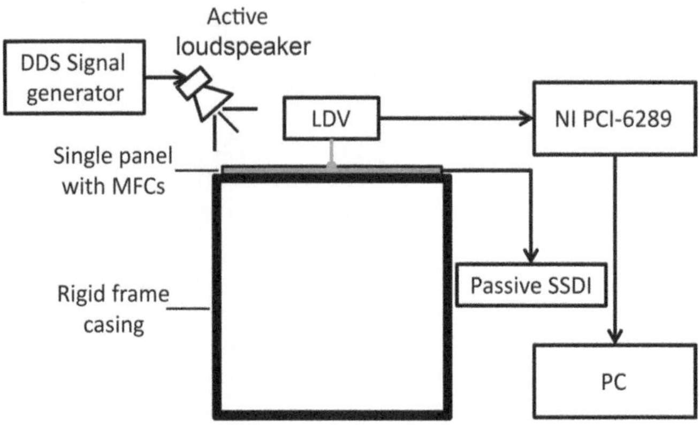

Figure 4.8 Measurement and control system for the first experiment.

The experiments are divided into two stages: with the front wall made of steel [150], and made of aluminium [104]. On the panel's surface five and nine MFC elements were attached, respectively. Both the experiments were performed using the different laboratory setups. In the first configuration (Fig. 4.8) the investigated structure is excited to vibrate by using an active loudspeaker placed outside the casing. A tonal signal emitted by the loudspeaker was generated with a DDS signal generator application implemented on NI myRIO platform, working in the standalone mode, using LabVIEW FPGA. The vibration of the structure was measured at the central point using Polytec PDV-100 Laser Doppler Vibrometer (LDV). The data was acquired using NI M-series PCI-6289 DAQ card. The measurements were stored, processed and presented in NI LabVIEW graphical environment.
The laboratory setup for the second experiment was similar (Fig. 4.9), however, with some modifications.

An active loudspeaker was placed inside the casing. NI myRIO was replaced by an external DDS signal generator and measurement system was based on the dSpace platform. Besides the vibrations measurement, the noise emitted outside the casing was measured using six microphones.
In Table 4.5 (x,y,z) coordinates of LDV and randomly distributed microphones are presented. The origin of the coordinate system is in the middle of front down edge of the investigated casing, which was also presented on the illustrative scheme (Fig. 4.10).

4.3.2 CONTROL STRUCTURE

In the presented research it was assumed that there are no mechanical interactions between the front and other walls. Modelling of the system including all of the couplings is much more complicated. The model of the front wall is based on the Kirchhoff-Love theory for thin plates [108]. Neglecting the damping, the transverse displacement η can be calculated using following equation:

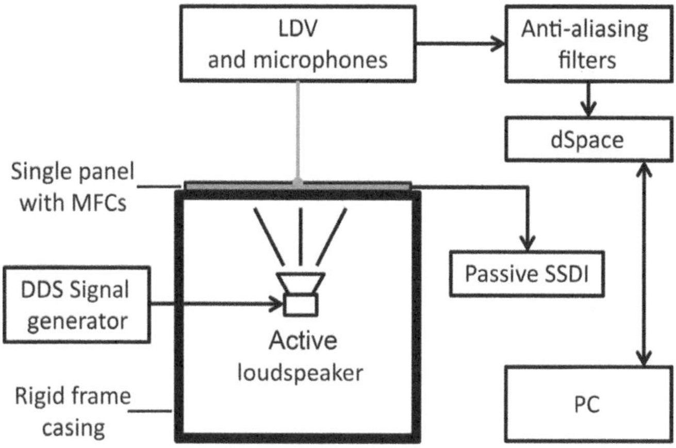

Figure 4.9 Measurement and control system for the second experiment.

Table 4.5
Sensors coordinates–all values are given in metres.

Axis/Sensor	M1	M2	M3	M4	M5	M6	V
x	0.37	0.02	0	−0.27	1.01	0.13	0
y	0.21	0.43	0.37	0.21	2.56	1.66	2.26
z	0.35	0.32	0.73	0.31	1.58	1.45	0.55

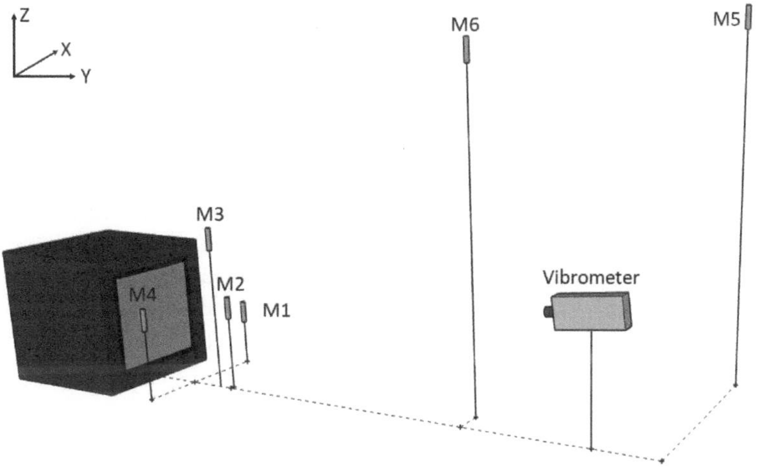

Figure 4.10 Scheme of placement of the casing, vibrometer and microphones.

Table 4.6
Main parameters of the investigated panels and MFCs.

Property	Steel panel	Aluminium panel	MFC
Dimensions (mm)	420× 420× 0.5	420× 420× 0.98	85× 14× 0.3
Mass (g)	692	463	2
Density (kg/m^3)	7850	2680	5440
Young's modulus (GPa)	200	70.5	30.3/15.8
Poisson's ratio	0.3	0.33	0.31

$$D\left(\frac{\partial^4\eta(x,y,t)}{\partial x^4}+2\frac{\partial^4\eta(x,y,t)}{\partial x^2\partial y^2}+\frac{\partial^4\eta(x,y,t)}{\partial y^4}\right)+\rho h\frac{\partial^2\eta(x,y,t)}{\partial t^2} = f_{ext}(x,y,t)+f_{mfc}(x,y,t), \quad (4.7)$$

where D is the bending stiffness per unit, ρ and h are density of plate, and plate thickness, respectively, and f_{ext} and f_{mfc} are distributed external forces coming from the primary excitation, and the distributed force generated by MFCs, respectively. The bending stiffness D can be expressed as:

$$D=\frac{Eh^3}{12(1-\nu^2)}, \quad (4.8)$$

where E and ν are Young's modulus and Poisson's ratio, respectively. The system has infinite number of dimensions which makes it difficult to control, however, in presented research the authors are focused only on the single vibrational mode. Hence, the complexity of model equation can be reduced and written as:

$$m\frac{\mathrm{d}^2x(t)}{\mathrm{d}t^2}+d\frac{\mathrm{d}x(t)}{\mathrm{d}t}+kx(t)=f_{exc}(t)+f_{MFC}(t), \quad (4.9)$$

where m, d, k and x are the effective mass, damping, stiffness, and displacement, respectively. f_{exc} and $f_{MFC}=\sum_{i=0}^{4} f_{MFC,i}(t)$ are effective external force, and force generated by all MFCs, respectively. The main parameters of the investigated steel and aluminium panels, and MFC elements are presented in Table 4.6.

In SSDI the piezoelectric element should generate force with the sign opposite to the velocity. The force generated by a piezoelectric element is equal to:

$$f_{MFC,i}(t)=-\alpha_i u_{MFC,i}(t), \quad (4.10)$$

where $u_{MFC,i}(t)$ is the voltage on ith MFC element, and α_i is a coefficient related to its efficiency.

With zero initial conditions, the force generated by the piezoelectric element is opposite to the velocity (and displacement). Each MFC generates current equal to:

$$i_i(t) = \alpha_i \frac{\mathrm{d}x(t)}{\mathrm{d}t}. \tag{4.11}$$

This current charges the MFC. When the terminals of the MFC are open and the internal MFC parallel resistance is discarded, and only internal MFC capacitance c_{MFC} is taken into account, the voltage on the MFC is equal to:

$$u_{MFC,i}(t) = \frac{1}{c_{MFC}} \int_{\tau=0}^{t} i_i(\tau)\mathrm{d}\tau = \frac{\alpha_i}{c_{MFC}} \int_{\tau=0}^{t} \frac{\mathrm{d}x(\tau)}{\mathrm{d}\tau}, \tag{4.12}$$

$$u_{MFC,i}(t) = \frac{\alpha_i}{c_{MFC}} (x(t) - x(0)). \tag{4.13}$$

For positive initial velocity the displacement is positive and the generated force is negative. After the first one-fourth cycle the sign of the velocity changes to negative and the piezoelectric voltage is in its maximum. The displacement is still positive, and the force is negative. The sign is the same as the sign of the velocity, which is not desired. So the sign of the force must be changed. Thus, the polarity of the voltage on the MFC element must be changed. In SSD systems the piezoelectric is discharged by shorting its terminals, the voltage drops to zero, and then the velocity recharges the piezoelectric with appropriate polarity. So, in SSD circuits the force generated by the piezoelectric is always opposite to the velocity and it is proportional to the displacement measured between the current position and the position in which the MFC is shorted. With all simplifying assumptions the MFC in SSD circuit acts as a spring, which is repositioned after each half-cycle so it always generates force that is opposite to the velocity.

SSDI circuits, instead of zeroing voltage on MFC element, reverse the voltage polarity using a LC resonant circuit (Fig. 4.11). The MFC element is used as the capacitor, and an inductor is added to the circuit. By shorting charged capacitor and inductor, the current starts flowing. When the capacitor is fully discharged all recovered energy is stored in the inductor. Then, the inductor acts as a current source and recharges the capacitor with a reversed polarity, and then unless this process is stopped, the voltage on capacitor induces the undesired current flow in the opposite direction. Stopping, when the capacitor is fully charged in opposite polarity, can be performed by opening the circuit or by allowing the current to flow in one direction only, which can be implemented by using a diode (Fig. 4.12). Unfortunately, some energy is lost on the diode. Additionally, some energy is lost on resistive elements [108].

The SSDI switching law can be written as:

If $f_{MFC}(t)\frac{\mathrm{d}x(t)}{\mathrm{d}t} > 0$ Then
    Switch Polarity;

The piezoelectric force and velocity should have opposite polarity, otherwise the

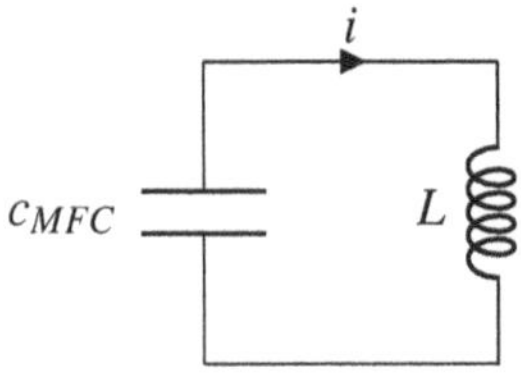

Figure 4.11 LC circuit used to change polarity of voltage on MFC.

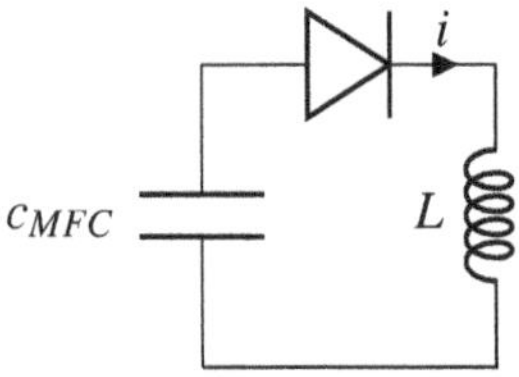

Figure 4.12 LC circuit with a diagonal diode used to stop oscillations after half-cycle.

piezoelectric force polarity must be changed. The piezoelectric force is not measured, but can be estimated using voltage, from Eq. (4.10). The switching law can be rewritten as:

```
If u_MFC(t) dx(t)/dt < 0 Then
        Switch Polarity;
```

This switching law requires measurement of both MFC voltage and velocity. It can be also rewritten as:

```
If dx(t)/dt > 0 Then
        Set u_MFC(t) to Positive;
Else
        Set u_MFC(t) to Negative;
```

The "Set $u_{MFC}(t)$ to Negative/Positive" can be implemented in hardware, by using two diodes and two switches (Fig. 4.13).

If SW1 is short and SW2 is open, then it is guaranteed that the voltage on MFC is non-positive (assuming ideal diode). If it is negative then the diode D2 is reverse biased. If the voltage is positive then diode D1 is forward biased, and the polarity switch occurs. If SW1 is open and SW2 is short, then it is guaranteed that the voltage on MFC is non-negative. To the switching law it can be rewritten as:

```
If dx(t)/dt > 0 Then
        SW1 ← Open, SW2 ← Short;
Else
        SW1 ← Short, SW2 ← Open;
```

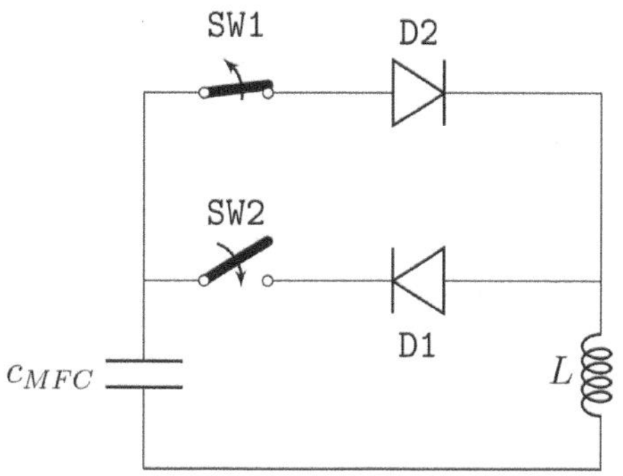

Figure 4.13 SSDI circuit with a pair of switches.

The velocity can be measured or estimated. Differentiation of Eq. 4.13 gives:

$$\frac{\mathrm{d}u_{MFC,i}(t)}{\mathrm{d}t} = \frac{\alpha_i}{c_{MFC}}\frac{\mathrm{d}x(t)}{\mathrm{d}t}. \tag{4.14}$$

Thus, the switching law can be rewritten as:

```
If du_MFC,i(t)/dt > 0 Then
        SW1 ← Open, SW2 ← Short;
Else
        SW1 ← Short, SW2 ← Open;
```

The above switching law effectively detects extreme values of $u_{MFC,i}(t)$. Alternatively, extreme values can be detected using a comparison between current value and maximum value in the near past. Such switching law can be implemented using a pair of transistors used as comparators and a pair of capacitors with diodes to detect maximum/minimum voltage (Fig. 4.14). It is a modified version of the solution proposed by Wei et al. [162]. This circuit implements the following pseudocode:

```
If u_MFC,i(t) > U_MAX(t) − U_BE Then
        SW1 ← Open;
Else
        SW1 ← Short;
If u_MFC,i(t) < U_MIN(t) + U_BE Then
        SW2 ← Open;
Else
        SW2 ← Short;
```

The U_{BE} is the threshold base-emitter voltage.

The circuit measures voltage on the piezoelectric element and based on the current value it detects the change of the velocity of structural vibrations. The transistors Q1, Q2 compare currently measured voltage values with minimal and maximal voltage

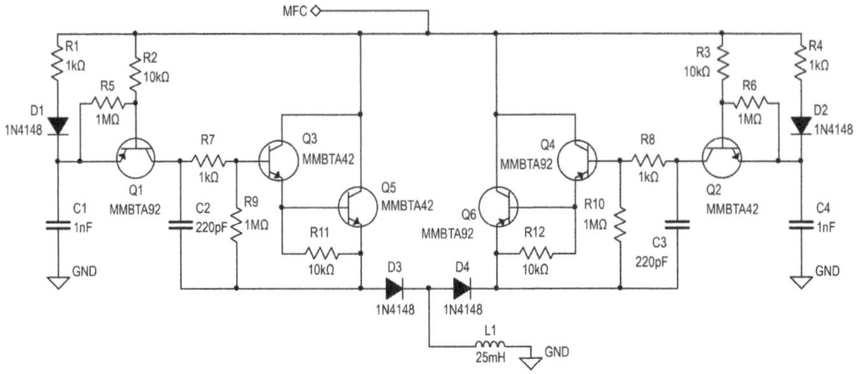

Figure 4.14 A scheme of passive SSDI electrical circuit [108]. Reprinted from "Synchronized switch damping on inductor for noise-reducing casing", K. Mazur, et al., 26th International Congress on Sound and Vibration, Montreal, Canada, 2019.

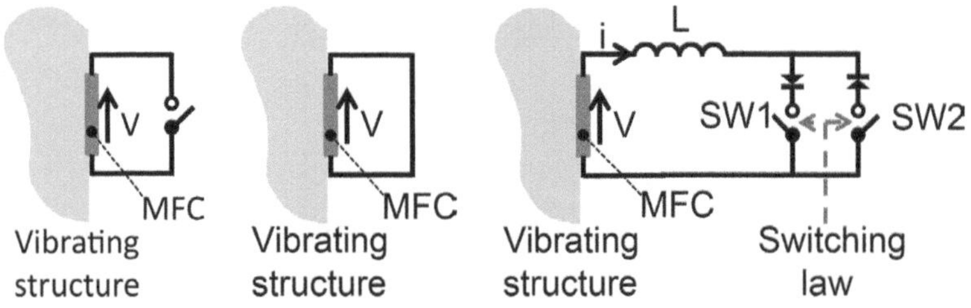

(a) Open. (b) Short. (c) SSDI.

Figure 4.15 Schematic representation of the investigated circuits.

values obtained using diodes D1, D2. A decrease of the voltage value is related to a change of the sign of the velocity and the circuit should be switched to discharge of the piezoelectric capacitance.

4.3.3 CONTROL RESULTS

In the first experiment three types of circuits were compared: open (Fig. 4.15(a)), short (Fig. 4.15(b)), and SSDI (Fig. 4.15(c)). In the second experiment the open case was not available and was omitted.

In the open circuit the MFC remains stiff and absorbs mechanical energy from the vibrating structure. In the short circuit the energy is dissipated from internal capacitance of the piezoelectric element similarly to discharging of the capacitor, and the MFC becomes flexible. The SSDI circuit switches between those two states [33]. Thus, with properly selected time of switching, the improvement of the vibration damping is expected.

Before the experiments the mode shapes and corresponding resonant frequencies of the steel and aluminium panels were identified. The first of the main experiments was

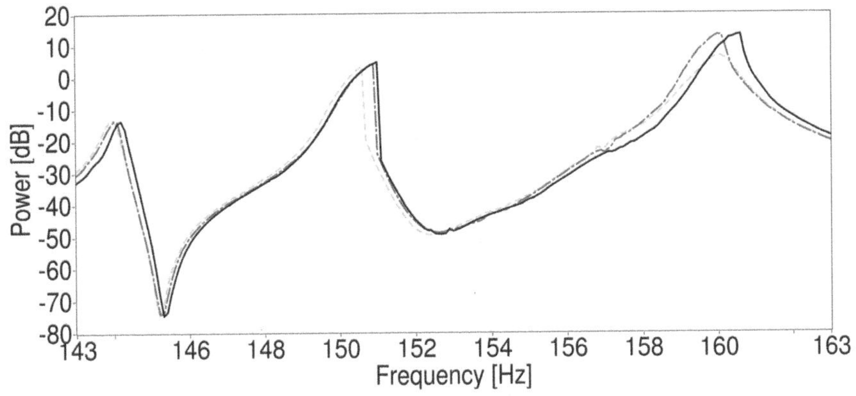

Figure 4.16 PSD of vibration signal measured using LDV at the central point of the vibrating structure for open (solid black), short (dot-dashed grey), and SSDI circuits (dashed grey) [150]. Adapted from "Semi-active reduction of device casing vibration using a set of piezoelectric elements", J. Rzepecki, et al., 20th International Carpathian Control Conference, Wieliczka, Poland, 2019.

focused on the (3,3) mode corresponding to frequency of 155 Hz. However, as the temperature in laboratory changes, the resonant frequencies also change. Hence, the excitation signal was emitted as a sequence of single tonal components from 143 to 163 Hz with 0.1 Hz step. In this experiment a (3,3) mode was observed at 160.1 Hz. The Power Spectral Density (PSD) is presented in Fig. 4.16.

The amplitudes for open and short circuits are similar, however, a frequency difference is observed. It may result from MFC properties in both states. In the open state MFC remains stiff, and the resonant frequency increases. In the short state the MFC is flexible, however, it is still present as an additional mass on the vibrating panel. For the SSDI a 3.5 dB vibration reduction is observed. In Fig. 4.17 the displacement of the vibrating structure for investigated frequency is presented. It is observed that the front panel's displacement for the short circuit is higher than for the open circuit. However, for the SSDI the displacement is around 18 % lower than for the open circuit.

Fig. 4.18 shows the measured signal powers from the second experiment. The shunt system provides up to 4.8 dB (at 162.2 Hz) velocity reduction. Above that frequency the noise reduction sharply drops to ca. 3.8 dB. For high frequencies the SSDI system increases vibrations. The SSDI system provides also noise reduction. Some noise reduction is observed on all microphones close to the casing: up to 2.8 dB at M1, up to 8.0 dB at M2, up to 3.4 dB at M3, up to 2.8 dB at M4. At distant microphones some coincidental noise reduction is observed for higher frequencies, but in the resonant band the peak noise reduction is equal to 0.2 dB at M5 and ab. 1.8 dB at M6.

Fig. 4.19 shows the measured velocity attenuation level for a different excitation levels at open circuit resonance. For low excitation levels the voltage on MFCs is

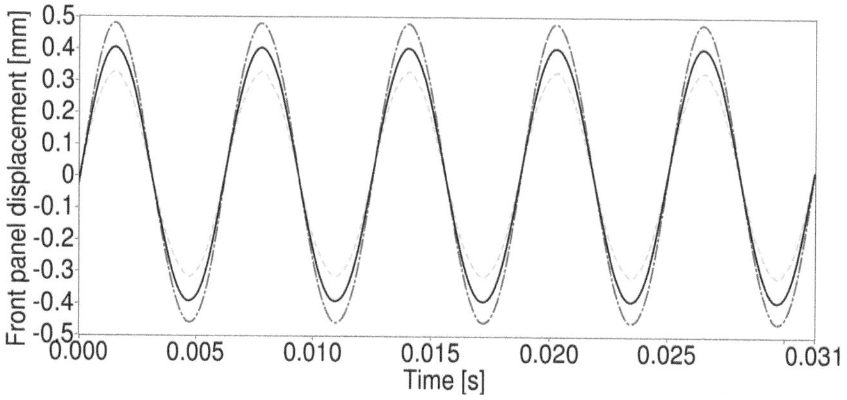

Figure 4.17 Displacement of the vibrating structure, measured at the central point, for open (black), short (dot-dashed grey), and SSDI circuits (dashed grey) [150]. Adapted from "Semi-active reduction of device casing vibration using a set of piezoelectric elements", J. Rzepecki, et al., 20th International Carpathian Control Conference, Wieliczka, Poland, 2019.

too small for the passive SSDI circuit. The reverse-polarity protection diodes, maxima detector and bipolar transistors used as the switch require high-enough voltage to operate correctly. The voltage drop on those elements causes additional losses and less energy after switching. At high excitation levels the voltage is high and a near-constant voltage drop on those element has less effect, and both the switching efficiency and performance increases. However, for excitation levels above -1 dB the performance drops. This performance drop is related to spurious switching. C2 and C3 capacitors try to address that problem. The values of C2 and C3 capacitors were chosen based on SPICE simulations [108]. Higher values could reduce that problem, but they also increase a capacitive load on the MFC elements and effectively reduce the voltage. Thus, the C2 and C3 are the compromise. Without them, the performance drops earlier, at levels above -7 dB.

4.4 COMPOSITE STRUCTURES

Apart from the conventional materials such as steel and aluminium typically utilized for constructing casings as well as their vibrating panels, composite structures have been found to be very useful materials for noise reduction especially in the high frequency regions (i.e. above 1000 Hz). For this reason, they are very suitable for vibro-acoustic problems involving passive structural control. Moreover, they have high internal damping over their conventional metallic counterpart which makes them attractive in industrial application involving the reduction of vibration due to resonance. Typical amongst these composites are the laminate composites and fibrous composites. However, investigation showed that they are less effective in transmitting sound losses at low frequencies than their conventional metallic counterpart. The probable reasons for this limitation could be attributed to the effect of mass law and

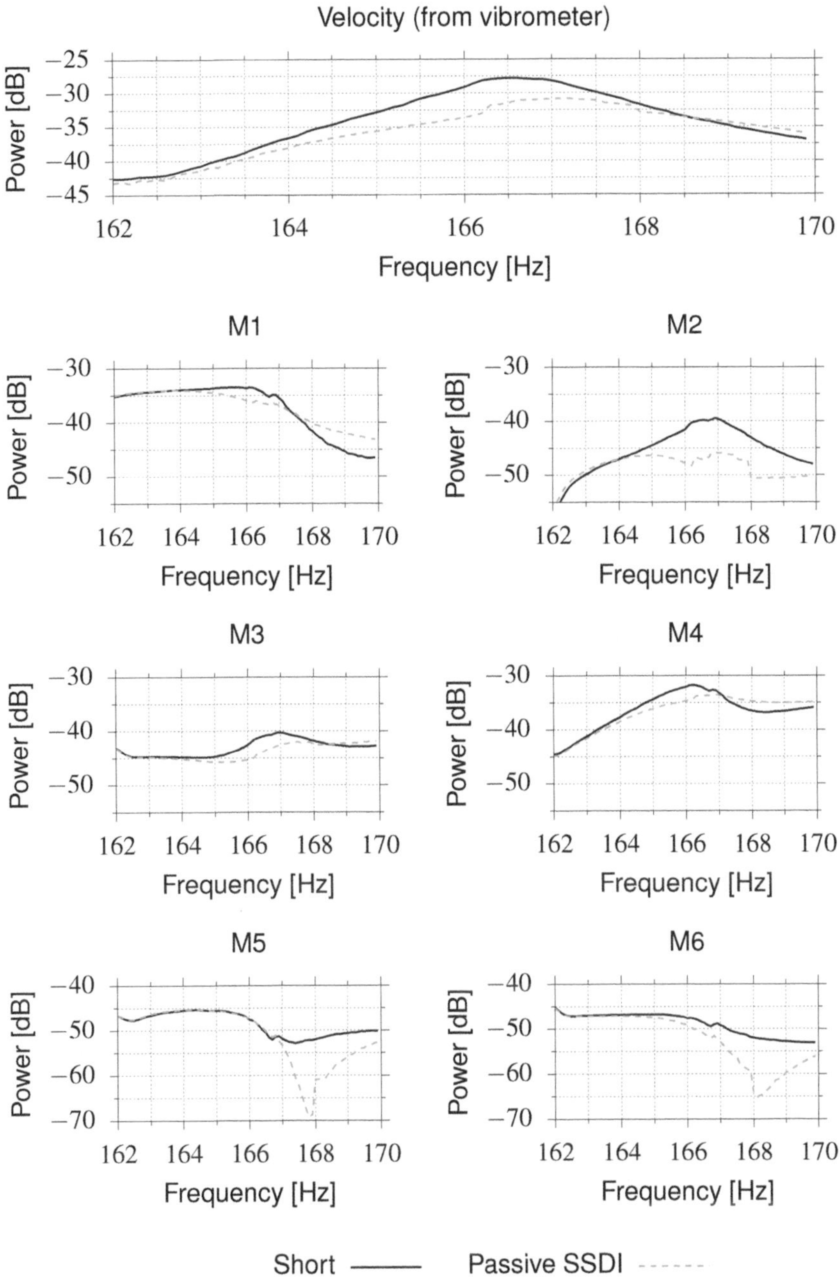

Figure 4.18 Measured signal powers for different excitation frequencies.

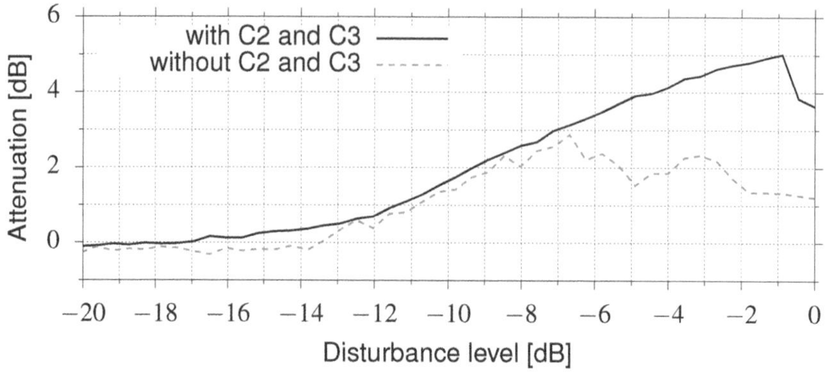

Figure 4.19 Vibration velocity attenuation level for different excitation levels at the open circuit resonance.

also, during vibroacoustic process, the coincident region of composite panel has the proclivity to excessively extend over higher frequencies.

Composite materials are lightweight and can be used for single-panel, double-panels, and multi-panels of vibrating structures. Single panels face-sheets are constructed by laminates and the face-sheets can be doubled or tripled, where the cores in between the face-sheets may be sandwiched or architecturally designed. The core may take the form of air (i.e. equivalent to spring connection), foam (or poroelastic material) or lattice/truss structures as shown in Fig. 4.20. The lattice structures which may be architecturally complex can also be made of composites and a very good way to fabricate this is through additive manufacturing or 3D printing. In Fig. 4.20, the incident and radiating face-sheets panels are made from composites with different cavity or core structures. During vibroacoustic analysis, three different sound transmission routes are taken into consideration. The first is the incident region which gives the incident and reflected sound waves. The second is the core or cavity region, in which there exist the positive and negative going sound waves being transmitted to the panels. Finally, in the radiating region, only one transmitted sound wave exists.

Composite double panel walls could have either orthotropic or anisotropic properties. The governing equation of an orthotropic sandwich panel with isotropic core have been formulated by Renji and Nair [148] as

$$D_{11}\frac{\partial^4 r}{\partial x^4}+2\left(D_{12}+2D_{66}\right)\frac{\partial^4 r}{\partial x^2 \partial y^2}+D_{22}\frac{\partial^4 r}{\partial y^4}=-\frac{1}{G}\left(D_{11}\frac{\partial^2 F}{\partial x^2}+D_{12}\frac{\partial^2 F}{\partial y^2}\right)+F \tag{4.15}$$

Where the terms D_{11}, D_{12}, D_{22} and D_{66} are the flexural rigidities of the composite laminates, G is the shear rigidity of the panel and F is the external force per unit area. However, due to the vibroacoustic excitation of the double panel composite structure, the governing equation can be obtained using the Hamilton's principle as applied by Li et al. [89]. The Hamilton's principle is written as

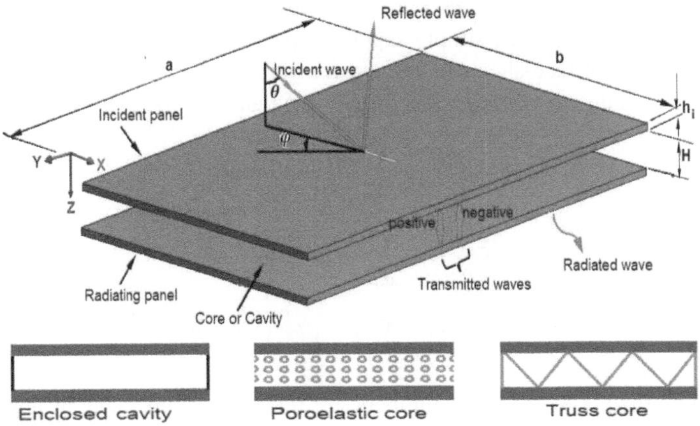

Figure 4.20 Illustration of vibroacoustic double-panel composite structure sandwiched with different cavity or core materials [63]. Reprinted from *Applied Sciences*, 10(4):1543, C. W. Isaac, M. Pawelczyk, and S. Wrona, "Comparative study of sound transmission losses of sandwich composite double panel walls", Copyright (2022), with permission from MDPI (STM).

$$\delta \int_{t_1}^{t_2} (K_e + W_e - U)\, dt = 0 \tag{4.16}$$

Where K_e is the kinetic energy relating to the displacements in the direction of x, y and mid-plane, given as

$$K_e = \frac{1}{2} \int_V \rho \left(\dot{u}^2 + \dot{v}^2 + \dot{r}^2\right) dV = 0 \tag{4.17}$$

The work of external load W_e is

$$W_e = \int_0^a \int_0^b \left(F + N_x \frac{\partial^2 r}{\partial x^2} + N_y \frac{\partial^2 r}{\partial y^2}\right) r dx dy \tag{4.18}$$

Where N_x and N_y are the thermal loads per unit width. Finally, the total strain energy U on the face-sheets and core can be written in compact form as

$$U = \frac{1}{2} \int_{V^{\text{ol}}} \{\sigma\}^T \{\varepsilon\}^T dV^{\text{ol}} = 0 \tag{4.19}$$

With components of the stress $\{\sigma\}^T$ and strain $\{\varepsilon\}^T$ respectively written as

$$\{\sigma\}^T = \left\{\sigma_{\text{fx}}\ \sigma_{\text{fy}}\ \sigma_{\text{cx}}\ \sigma_{\text{cy}}\ \tau_{\text{fxy}}\ \tau_{\text{cxy}}\ \tau_{\text{cxz}}\ \tau_{\text{cyz}}\right\} \tag{4.20}$$

$$\{\varepsilon\}^T = \left\{\varepsilon_{\text{fx}}\ \varepsilon_{\text{fy}}\ \varepsilon_{\text{cx}}\ \varepsilon_{\text{cy}}\ \gamma_{\text{fxy}}\ \gamma_{\text{cxy}}\ \gamma_{\text{cxz}}\ \gamma_{\text{cyz}}\right\} \tag{4.21}$$

Note, the subscripts f and c indicate face-sheet and core in their different planes, respectively. By substituting Eqs. (4.17)–(4.21) into Eq. (4.16), the governing equation can be obtained. The respective amplitudes can also be calculated using the free vibration analysis of the simply supported panels. Given the rotational angles of the upper and lower face-sheets in xoz and yoz planes as $\varnothing_1$and $\varnothing_2$, respectively; the modal functions due to vibration response of the panel can be expressed as

$$r(x,y,\ t) = \sum_{o,p} \xi_{\mathrm{op}}(x,y) A_{\mathrm{op}} e^{\mathrm{j}\omega \mathrm{t}} \tag{4.22}$$

$$s_1(x,y,\ t) = \sum_{o,p} \varnothing_{1,op}(x,y) A_{\mathrm{op}} e^{\mathrm{j}\omega \mathrm{t}} \tag{4.23}$$

$$s_2(x,y,\ t) = \sum_{o,p} \varnothing_{2,op}(x,y) A_{\mathrm{op}} e^{\mathrm{j}\omega \mathrm{t}} \tag{4.24}$$

The components of the wavenumbers are related to both incident angle θ and azimuthal angle φ. By applying the orthogonality condition and careful arrangement of the above equations, the amplitudes of the acoustic field can be evaluated with respect to their incident sound pressure (p_i) such that the transmission coefficient can be written as

$$\tau(\theta,\varphi) = \frac{Z_3}{Z_1} \qquad \text{where} \qquad Z_i = \frac{1}{2}\mathrm{Re}\iint_A p_i.v_i^{*} dA \qquad (i = 1,2,3) \tag{4.25}$$

Where v_i is the acoustic velocity of the incident sound wave, Re and the superscript $(*)$ represent the real part and the complex conjugate, respectively. The diffused transmission loss and the sound transmission loss can be, respectively written as

$$\tau_{\mathrm{df}} = \frac{\int_0^{2\pi}\int_0^{\theta_{\mathrm{lim}}} \tau(\theta,\varphi)\sin\Theta\cos\Theta d\Theta d\phi}{\int_0^{2\pi}\int_0^{\varphi_{\mathrm{lim}}} \sin\Theta\cos\Theta d\Theta d\phi} \tag{4.26}$$

$$STL = 10\log_{10}\frac{1}{\tau} \tag{4.27}$$

4.4.1 FUNCTIONALLY GRADED MATERIALS

Another special form of composite material is the Functionally Graded Material (FGM) which has shown great potential for constructing structural panels used for noise reduction and control.

The FGM are engineering materials which are manufactured in such a way that their microstructures are patterned or arranged in a graded form. Gradation of the microstructures can be laterally, axially or longitudinally patterned. The unique advantage of FGM over other types of composite materials are that they have very strong bonding coupled, as well as low stress concentration of the material due to the smooth variation at their interface. Fig. 4.21 compares a typical FGM and their composite laminate counterpart. The microstructures of the FGM are graded from a higher concentration to a lesser concentration to give it an improved mechanical property. The composite laminates on the other hand, have lower mechanical property which is caused by the interactions of the laminates on each interface.

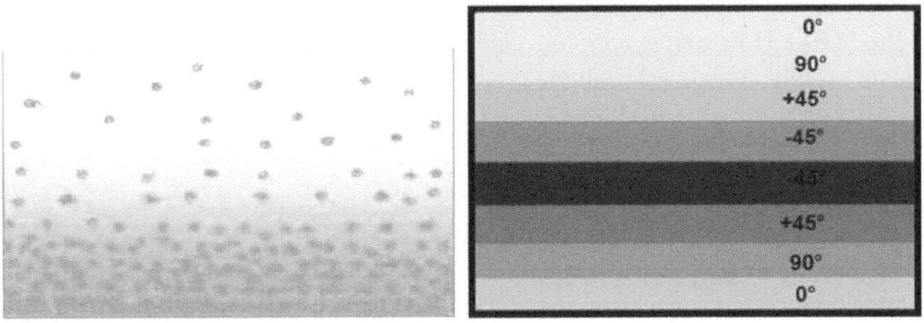

(a) (b)

Figure 4.21 Comparison between two kinds of composite structures used for vibroacoustic problems (a) FGM showing the microstructure with smooth graded variation and (b) composite laminates showing variations with sharp interfaces [63]. Reprinted from *Applied Sciences*, 10(4):1543, C. W. Isaac, M. Pawelczyk, and S. Wrona, "Comparative study of sound transmission losses of sandwich composite double panel walls", Copyright (2022), with permission from MDPI (STM).

Fig. 4.22 shows a typical vibroacoustic problem of a FGM panel. Just like any other noise vibrating panel, the FGM panel is placed inside a casing with a loudspeaker producing the sound wave. The incident sound wave passes through the graded thickness z and radiate sound wave at the radiating surface. In Fig. 4.22, the constituents of FGM are aluminium (Al) and ceramic alumina (Al_2O_3). In the construction of FGM panel, it is assumed that each graded part is governed by similar material properties such as the Poisson ratio, Young's modulus and density. For this reason, effective material properties have to be calculated. Eqs. (4.28), (4.29) and (4.30) gives the effective Young's modulus, density and Poisson ratio, respectively.

$$E_{\mathrm{ef}}(z) = (E_{\mathrm{cr}} - E_{\mathrm{Al}}) V_{\mathrm{cr}} + E_{\mathrm{Al}} \tag{4.28}$$

$$\rho_{\mathrm{ef}}(z) = (\rho_{\mathrm{cr}} - \rho_{\mathrm{Al}}) V_{\mathrm{cr}} + \rho_{\mathrm{Al}} \tag{4.29}$$

$$\mu_{\mathrm{ef}}(z) = (\mu_{\mathrm{cr}} - \mu_{\mathrm{Al}}) V_{\mathrm{cr}} + \mu_{\mathrm{Al}} \tag{4.30}$$

The effective material properties are linear expressions derived using the Voigt's rule of mixture. Where the terms E_{cr}, E_{Al} and V_{cr} are the Young's moduli of ceramic, aluminium and volume fraction of ceramics, respectively. Also, ρ_{cr} and ρ_{Al} are densities of the ceramic and aluminium, respectively; while μ_{cr} and μ_{Al} are the Poisson ratio of ceramic and aluminium, respectively. The subscript symbols, cr and Al, represent ceramic and aluminium, respectively. The volume fractions of ceramic and aluminium sum up to unity by applying the rule of mixture given by [64]

$$V_{cr} + V_{Al} = 1 \tag{4.31}$$

Various ways have been established to obtain the volume fraction of ceramics V_{cr} of Eq. (4.31). However, a very convenient way to obtain V_{cr} along the FGM panel's thickness is by applying the simple power law distribution calculated as [157]

$$V_{cr} = \left(\frac{1}{2} - \frac{z}{h}\right)^{i} \tag{4.32}$$

Where the term $z = \frac{h}{2}$ can be gotten by evaluating the distance from the mid-surface of the FGM panel as shown in Fig. 4.22. Moreover, the term i, typifies the volume fraction index which essentially governs the gradation profile along the thickness of the graded structural panel as illustrated in Fig. 4.23. The value of the fraction index can vary from zero (0) to infinity (∞). In Fig. 4.23, i varies from 0.2 to 20. The isotropic and homogeneous characteristics of the FGM is attained when the value of i is zero or infinity.

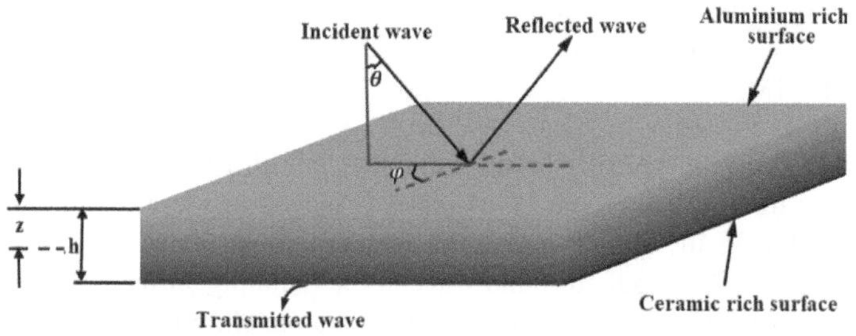

Figure 4.22 Illustration of vibroacoustic FGM panel with material property gradation along its thickness.

4.4.2 HYBRID COMPOSITES

Composite material typically consists of a reinforcement fibre and a matrix resin. The reinforcement fibre can be obtained naturally or through synthetic process. Hybrid composites, which include two or more composite materials architecturally joined together, can be used to form vibroacoustic panels for sound analyses and noise control. Consider a simple hybrid composites panel having the same matrix as illustrated in Fig. 4.24. The first composite material contains fibre a, while the second contains fibre b both having a hybrid property given as

$$Q_h = Q_a V_{ha} + Q_b V_{hb} \tag{4.33}$$

Where Q_a and Q_b are properties of fibre a and fibre b, respectively. As defined in the previous section, the volume fraction of fibre a, (V_{ha}) and fibre b (V_{hb}) also sum up to unity using the rule of mixture given as

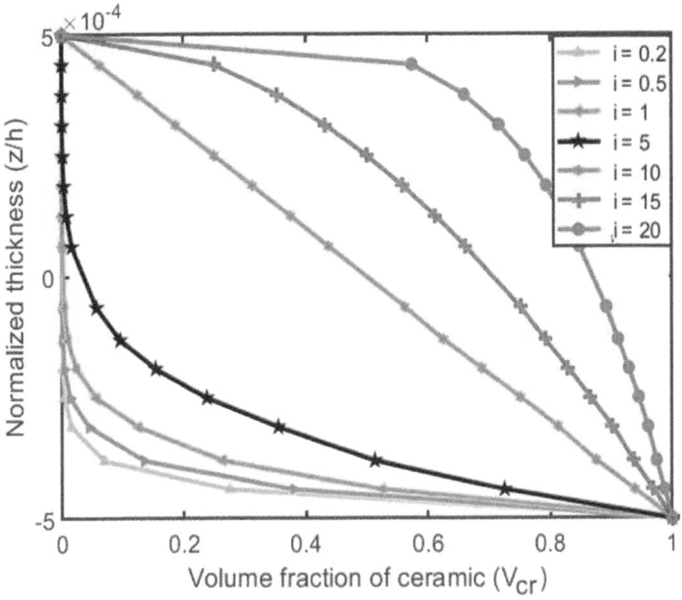

Figure 4.23 Plot of V_{cr} against $\frac{z}{h}$ for different i values of FGM panel according to Eq. (4.32).

$$V_{ha} + V_{hb} = 1 \tag{4.34}$$

As these volume fractions vary, the longitudinal modulus of the two fibres making up the hybrid composites also vary. Let E_a and E_b represent the respective longitudinal modulus of the two fibres. By the rule of hybrid mixture, the longitudinal modulus of the hybrid property can be written as

$$E_L = E_a V_{ha} + E_b V_{hb} + E_m V_m \tag{4.35}$$

Where the new term, E_m and V_m are the longitudinal modulus and volume fraction of the matrix, respectively.

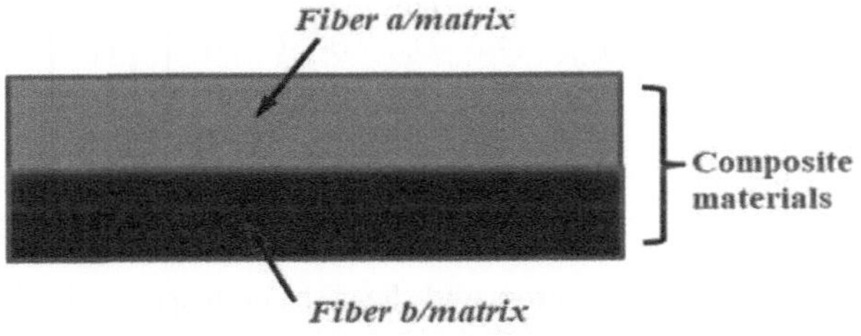

Figure 4.24 Schematic representation of a simple hybrid composites structure.

Using the Halpin-Tsai equation [3], the transverse modulus E_T for hybrid composites with respect to their arrangement coefficient κ can be written as

$$E_T = E_m\left(\frac{1+\kappa(\beta_a V_{\mathrm{ha}}+\beta_b V_{\mathrm{hb}})}{1-(\beta_a V_{\mathrm{ha}}+\beta_b V_{\mathrm{hb}})}\right) \tag{4.36}$$

Where the terms

$$\beta_a = \frac{\frac{E_a}{E_m}+1}{\frac{E_a}{E_m}+\kappa} \quad \text{and} \quad \beta_b = \frac{\frac{E_b}{E_m}+1}{\frac{E_b}{E_m}+\kappa} \tag{4.37}$$

In a similar way, for the hybrid composite, the effective Poisson ratio, density and Young's modulus according to the hybrid rule of mixture are respectively given as

$$\mu_{\mathrm{ef}} = \mu_a V_a + \mu_b V_b + \mu_m V_m \tag{4.38}$$

$$\rho_{\mathrm{ef}} = \rho_a V_a + \rho_b V_b + \rho_m V_m \tag{4.39}$$

$$E_{\mathrm{ef}} = E_L + E_T \tag{4.40}$$

Where μ_a, ρ_a and μ_b, ρ_b are the Poisson ratio, density of fibre a and Poisson ratio, density of fibre b, respectively. Also, μ_m and ρ_m are the Poisson ratio and density of the matrix resin, respectively. In a similar way of Eq. (4.36), the shear moduli, i.e. G_{12}, G_{13} and G_{23} can be derived using the Halpin-Tsai relation given by

$$G_T = G_m\left(\frac{1+\kappa(\beta_a V_{\mathrm{ha}}+\beta_b V_{\mathrm{hb}})}{1-(\beta_a V_{\mathrm{ha}}+\beta_b V_{\mathrm{hb}})}\right) \tag{4.41}$$

Where the terms

$$\beta_a = \frac{\frac{G_a}{G_m}+1}{\frac{G_a}{G_m}+\kappa} \quad \text{and} \quad \beta_b = \frac{\frac{G_b}{G_m}+1}{\frac{G_b}{G_m}+\kappa} \tag{4.42}$$

For orthotropic composite material whose fibres are along x, y and z axes, the stress (σ)-strain (ε) relationship is expressed as

$$\begin{Bmatrix}\sigma_x\\ \sigma_y\\ \sigma_z\\ \tau_{\mathrm{xy}}\\ \tau_{\mathrm{zx}}\\ \tau_{\mathrm{yz}}\end{Bmatrix} = \begin{bmatrix} D_{\mathrm{xx}} & D_{\mathrm{xy}} & D_{\mathrm{xz}} & 0 & 0 & 0\\ D_{\mathrm{yx}} & D_{\mathrm{yy}} & D_{\mathrm{yz}} & 0 & 0 & 0\\ D_{\mathrm{zx}} & D_{\mathrm{zy}} & D_{\mathrm{zz}} & 0 & 0 & 0\\ 0 & 0 & 0 & D_{\mathrm{ab}} & 0 & 0\\ 0 & 0 & 0 & 0 & D_{\mathrm{cd}} & 0\\ 0 & 0 & 0 & 0 & 0 & D_{\mathrm{ef}}\end{bmatrix} \begin{Bmatrix}\varepsilon_x\\ \varepsilon_y\\ \varepsilon_z\\ \gamma_{\mathrm{xy}}\\ \gamma_{\mathrm{zx}}\\ \gamma_{\mathrm{yz}}\end{Bmatrix} \tag{4.43}$$

Where τ and γ are the shear stress and shear strain, respectively. The strain components of Eq. (4.43) can be obtained using the linear small-strain elastic theory. Also, D is the constitutive material matrix. For orthotropic composite material under plane stress Eq. (4.43) can be reduced to

$$\begin{Bmatrix} \sigma_x \\ \sigma_y \\ \tau_{xy} \end{Bmatrix} = \frac{1}{1-\mu_{xy}\mu_{yx}} \begin{bmatrix} E_x & \mu_{yx}E_x & 0 \\ \mu_{xy}E_y & E_y & 0 \\ 0 & 0 & (1-\mu_{xy}\mu_{yx})G_{xy} \end{bmatrix} \begin{Bmatrix} \varepsilon_x \\ \varepsilon_y \\ \gamma_{xy} \end{Bmatrix} \tag{4.44}$$

Where the kinetic energy, work of external load and total strain energy of the hybrid composite can be obtained using Eqs. (4.17), (4.18) and (4.19), respectively, as discussed in the previous section.

4.5 SUMMARY

Solutions presented in this chapter are characterized by a common feature: they do not require any external source of energy. In many applications, it is not only attractive, but strictly required.

The first approach proposed shaping of structural and acoustic responses of noise barriers in order to enhance their transmission loss. If the spectrum of the noise is known, the responses of the barrier can be optimized accordingly. The alteration of responses is achieved by attaching additional passive components: masses or ribs.

The second approach employs shunt circuit better known for semi-active control systems, but in the proposed architecture the energy needed to switch MFCs properties is harvested from the vibrations of the barrier. Hence, no external power supply is needed.

The third solution is based on composite structures, which due to inhomogeneous structure offer valuable acoustic properties not available for regular materials. Considered materials include, inter alia, functionally graded materials and composite materials typically consisting of a reinforcement fibre and a matrix resin.

It is noteworthy that the discussed solutions can be employed together with other control systems, including both semi-active and active, creating a hybrid system.

5 Semi-active control

5.1 INTRODUCTION

People in modern society are often exposed to excessive acoustic noise and, as a result, a variety of noise reduction methods have been developed [68]. One such approach is to separate the recipients from the noise source using noise barriers. Commonly used passive barriers are generally ineffective for low-frequency noise (see Chapter 4). They also tend to be thick and heavy and introduce unwanted heat insulation. However, thanks to technological advancement, passive barriers can be complemented with or replaced by actively controlled barriers. They incorporate control sources that may be either acoustic, such as loudspeakers, or structural, such as vibration actuators. Such systems are most effective in the low-frequency range, where passive insulation fails. Actively controlled barriers have proven their effectiveness in a number of publications (see Chapter 6). However, when the availability of high-performance processors and energy sources is limited, it can be beneficial to adopt a semi-active solution instead. The semi-active barrier adjusts the characteristics of the structure itself with a small external energy supply [98]. It can offer considerable levels of noise reduction in a highly effective and economical manner.

5.2 SWITCHED MECHANICAL LINKS FOR DOUBLE PANELS

Double-panel structures can be used as a base to design semi-active noise barriers [175]. Double-wall structures have been widely used in aircraft fuselages, car doors and lightweight partition walls in buildings because they offer significantly higher passive transmission loss compared to equivalent single-wall structures. However, their acoustic performance deteriorates rapidly at low frequencies due to low order structural-acoustic resonances. So called mass-air-mass resonance is particularly responsible for the weak passive transmission loss [67, 183]. This is the frequency range where control systems can offer significant enhancement. This topic has gained a high interest in recent years. Langfeldt et al. considered a broadband low-frequency sound transmission loss improvement of double walls due to application of Helmholtz resonators [83]. De Melo Filho et al. studied dynamic mass based sound transmission loss prediction of vibroacoustic metamaterial double panels applied to the mass-air-mass resonance [39]. Mao investigated an improvement on sound transmission loss through a double-plate structure by using electromagnetic shunt damper [96]. Ma et al. analysed an active control of sound transmission through orthogonally rib stiffened double-panel structure [95]. In addition, double-panel structures enable application of a wide variety of control approaches, including (i) methods aimed at reduction of panel vibrations to reduce the acoustic radiation, and (ii) methods aimed at reduction of the acoustic response in the gap cavity to block the sound transmission. The first group consists of, e.g. tunable vibration

DOI: 10.1201/9781003273806-5

absorbers [158] and piezoelectric patches with shunt circuits. Gardonio et al. investigated a panel with self-tuning shunted piezoelectric patches for broadband flexural vibration control [51]. Billon et al. studied vibration isolation and damping using a piezoelectric flextensional suspension with a negative capacitance shunt [10]. Dal Bo et al. considered a smart panel with sweeping and switching piezoelectric patch vibration absorbers [36]. On the other hand, the second group includes, e.g. adaptive Helmholtz resonators [83, 97] and shunted loudspeakers located in the gap cavity between the panels [91].

The research presented in this section investigates a novel semi-active approach for double-panel noise barriers (as described in [175]). The proposed solution can significantly enhance the effective transmission loss, while being significantly lighter and requiring less space compared to vibration absorbers or Helmholtz resonators (which can be bulky when tuned for low frequencies). Meanwhile, it is less demanding in terms of system complexity as compared with a fully active approach. The solution is based on bistable links mounted between the incident and the radiating panels, which structurally couple (when turned on) or decouple (when turned off). Such semi-active links only require energy for switching between the states. They do not force vibration by themselves and hence the solution is semi-active. The structural couplings that have been introduced significantly alter the natural frequencies and mode shapes of the vibroacoustic system. This enables an adaptation of frequency-dependent transmission loss to the noise spectrum, which can be easily monitored, i.e. minimizing the radiation in the targeted frequency bands (especially in the vicinity of low-order resonances). Such an approach aims at an efficient reduction of transmission of non-stationary narrow-band noise, which is very common in real-life.

5.2.1 HARDWARE PLATFORM

This section presents an experimental setup of a double-panel barrier [175]. It evaluates both the designed semi-active actuators and the proposed approach to control. Photographs of the laboratory setup are shown in Fig. 5.1. The enclosure was built as a heavily reinforced-concrete box, of which the walls can be considered as acoustically rigid. A loudspeaker was placed inside the enclosure as a noise source. For the purpose of model verification, the loudspeaker was driven to generate band-limited white noise. Two rectangular steel plates were attached to the front of the enclosure at a distance of $L_{g,z} = 0.080$ m. The concrete walls of the box provided a high noise attenuation, hence most of the acoustic energy, which was transmitted outside the box, was transmitted through the panels. The dimensions of each panel area that was free to vibrate (i.e. the area inside the clamping frame) are 0.420 m $\times$ 0.390 m.

A scheme and a photograph of the setup of the semi-active link are shown in Fig. 5.2. In Fig. 5.2(a), each component described has the name of a panel to which it is attached denoted in parentheses, when the link is turned off (decoupled). The masses of ith semi-active link components are $m_{La,i} = 0.043$ kg and $m_{Lb,i} = 0.013$ kg. Both ends of the link are attached to the panels with neodymium magnets. To change the state of the link, the motor either rotates the brake counter-clockwise to hold the pin inside the main block (coupling panels) or rotates it clockwise to release the pin and allow it to move freely inside the main block. The motor only requires energy for a short time in order to switch between states, otherwise the system is self-locking. Pressure springs under the main block push the brake block down, ensuring that, when turned on, the brake holds the pin firmly. The pin is covered with a layer of PTFE to minimize friction when the link is turned off. Most of the components are

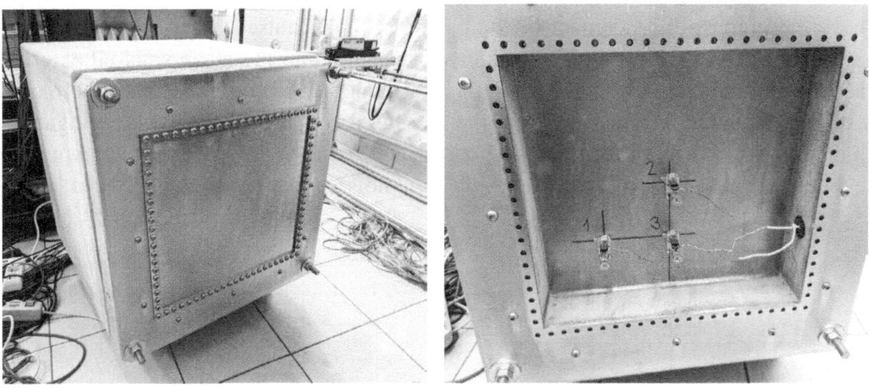

(a) (b)

Figure 5.1 (a) A photograph of the heavy concrete box and mounted panel *b*. (b) A photograph with dissembled panel *b* and three semi-active links mounted in the gap cavity [175]. Reprinted from *Mechanical Systems and Signal Processing*, 154:107542, S. Wrona, M. Pawelczyk, and L. Cheng, "Semi-active links in double-panel noise barriers", Copyright (2022), with permission from Elsevier.

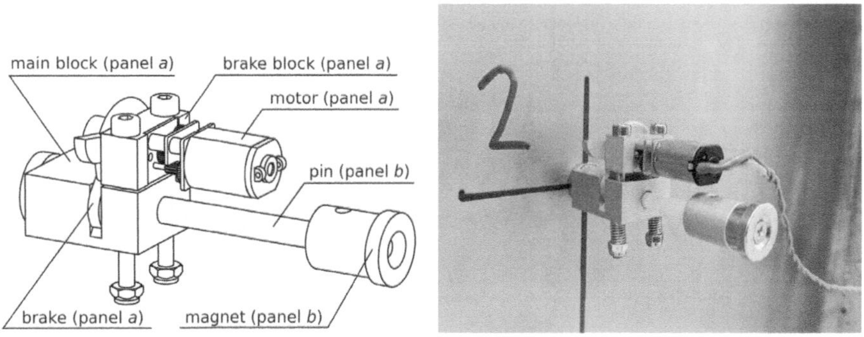

(a) (b)

Figure 5.2 (a) A scheme of the semi-active link. (b) A scheme and a photograph of the setup of the semi-active link [175]. Reprinted from *Mechanical Systems and Signal Processing*, 154:107542, S. Wrona, M. Pawelczyk, and L. Cheng, "Semi-active links in double-panel noise barriers", Copyright (2022), with permission from Elsevier.

made of stainless steel or aluminium, so the permanent magnets used for mounting the link to panels do not introduce additional forces acting on panels.

5.2.2 CONTROL ALGORITHM

The mass-air-mass resonance, when both panels vibrate out-of-phase with mode shapes (1,1), is particularly responsible for noise transmission through double-panel barriers [38, 39, 175]. Hence, the semi-active link was attached to the centres of the panels to have the highest impact on the mass-air-mass resonance (its mode shape have the highest displacement of the panels at their centres). The frequency characteristics of the mean vibration velocity of panel b and the mean squared external acoustic pressure are presented in Fig. 5.3. The adopted dB reference is equal to one, i.e. the frequency responses present the measured signal magnitude in the logarithmic scale without any additional normalization. However, to give a feeling about the experiments, the Sound Pressure Level of the noise in the room was between 75 dB

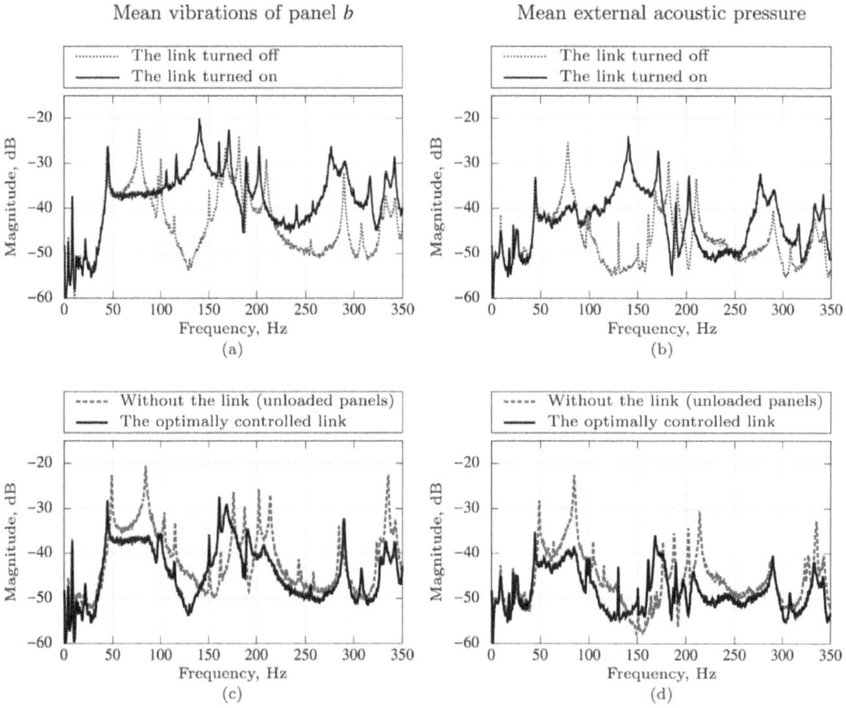

Figure 5.3 Frequency characteristics for the semi-active control experiments performed for the link mounted between the panels at $x_{L,1} = 0.5L_{a,x}$ and $y_{L,j} = 0.5L_{a,y}$. Plots (a) and (b) show a comparison of a double panel with the semi-active link turned off and on. Plots (c) and (d) show a comparison of a double panel without the link and with the optimally controlled semi-active link [175]. Reprinted from *Mechanical Systems and Signal Processing*, 154:107542, S. Wrona, M. Pawelczyk, and L. Cheng, "Semi-active links in double-panel noise barriers", Copyright (2022), with permission from Elsevier.

and 85 dB. Vibration measurements of the panel were taken point-by-point with the vibrometer over a uniform grid of 22 × 20 points, giving a total of 440 points, spaced at intervals of 0.02 m, hence covering the whole surface of panel *b*. After completing the measurements, the frequency analysis for all points was performed, and the mean vibration velocity was obtained by averaging all obtained frequency characteristics. The mean squared external acoustic pressure was calculated analogously by averaging frequency characteristics obtained with the microphones over the rectangular measurement grid, 1.00 m wide and 0.76 m high, 0.1 m away from the surface of panel *b*.

It follows from the analysis of Figs. 5.3(a) and (b) that the activation of the link strongly alters the frequency response of the barrier. What is most noticeable, is that the aforementioned mass-air-mass resonance is relocated from 78 Hz to 140 Hz. It is very beneficial from the point of view of semi-active control. For example, to minimize the acoustic pressure due to noise transmitted through the barrier, the link should be turned on for tonal noise frequencies below 90 Hz, and turned off for frequencies between 90 Hz and 160 Hz (with some exceptions for narrow resonance peaks at 97 Hz and 100 Hz). Assuming that the noise is mainly tonal, although,

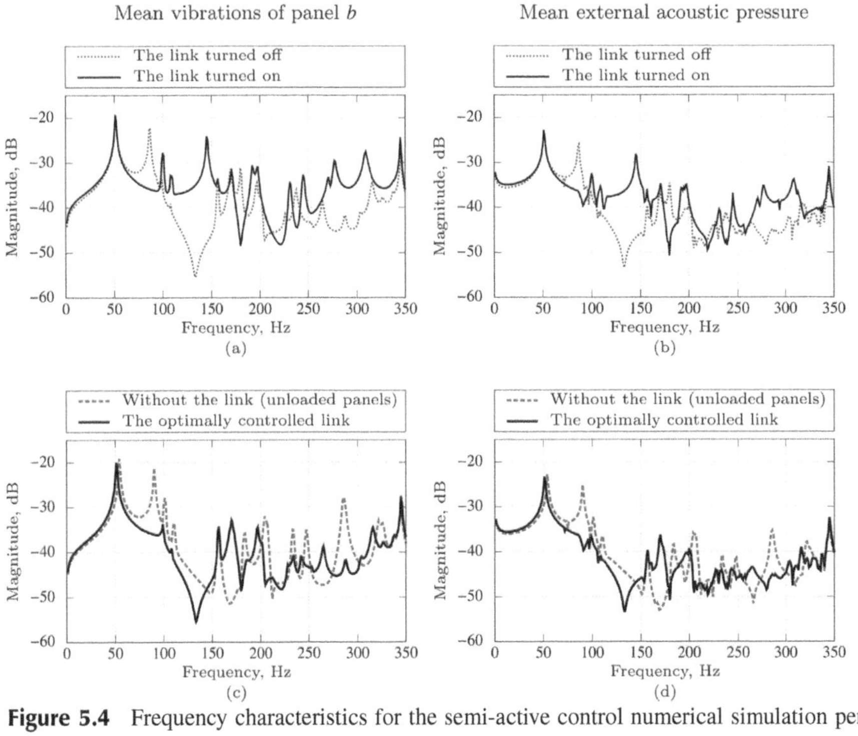

Figure 5.4 Frequency characteristics for the semi-active control numerical simulation performed for the link mounted between the panels at $x_{L,1} = 0.5L_{a,x}$ and $y_{L,j} = 0.5L_{a,y}$. Both panels are of the same thickness equal to 1 mm [175]. Reprinted from *Mechanical Systems and Signal Processing*, 154:107542, S. Wrona, M. Pawelczyk, and L. Cheng, "Semi-active links in double-panel noise barriers", Copyright (2022), with permission from Elsevier.

it can be non-stationary, such action can reduce the mean acoustic pressure due to transmitted noise by even more than 16 dB. Thus, the objective is to protect the structure against excitation of the low-order structural-acoustic resonances, not to truly attenuate them (it would require much more energy). Shifting the resonances "away from the noise spectrum" in the frequency domain with semi-active elements is a much more efficient approach, as long as the noise is tonal or narrow-band, thus there are frequency bands where the resonances can be safely shifted to. Otherwise, for a broadband noise this approach would generally be not suitable, however, such cases are out of the scope of the proposed method.

The frequency characteristics for the "optimally controlled link" are presented in Figs. 5.3(c) and (d). The "optimally controlled link" means that the semi-active link is turned on or off, depending on which frequency characteristic is better for a given frequency (the minimum of both "on" and "off" characteristics is taken). The binary output controlling the link should be generated automatically by a controller. The controller is responsible to choose the more beneficial frequency response of the barrier according to a predefined cost function and the continuously monitored noise spectrum. The presented results assume a scenario where the noise is purely tonal. It is a simplification and the real noise might be more complex, however, it is assumed that in the targeted applications a single tone (or a narrow-band noise) is dominating the noise spectrum, thus the presented behaviour of avoidance of resonances excitation should be achievable in practice.

Obtained "optimal" characteristics are compared with a double-panel barrier without the link (unloaded panels) for a better evaluation of the provided performance. The total mass of the semi-active link is 0.056 kg, while both panels of the considered barrier weight 2.57 kg. The semi-active link increases the overall mass of the barrier by only approximately 2%. Hence, the mass of the link has a negligible impact on the noise reduction and the obtained noise reduction is a results of switching the link, hence choosing a more beneficial frequency response of the barrier for particular noise frequency.

It follows from the analysis that for most frequencies a barrier with just one semi-active link can provide better noise reduction of sound than the unloaded barrier without the link. Hence, it is worth exploring what performance could be achieved with other arrangements of panels and links. This concept is undertaken in the numerical simulation studies presented in the following section, basing on the already derived and validated mathematical model.

5.2.3 CONTROL RESULTS

A semi-active link in the centre of the panels

First, a scenario already analysed experimentally in Subsection 5.2.2 is once again considered using the simulation environment (as presented in [175]). The frequency characteristics obtained are presented in Fig. 5.4; the results are presented in an analogous manner as in Fig. 5.3. Both Figs. 5.3 and 5.4 are very consistent, especially regarding the main points, once again confirming the model accuracy. First, the simulations presented also show that the semi-active link located in the center of the panels enables efficient mitigation of the mass-air-mass resonance originally located around 90 Hz (the excitation of the mass-air-mass resonance can be completely avoided due to switching of the semi-active link, what would not be possible with an addition of only a static mass). However, the fundamental mode (1,1) with panels vibrating in-phase is difficult to reduce in such a configuration when panels are of the same thickness. Moreover, the semi-active control enables the reduction of noise transmission for several other higher resonances. Some bands are slightly enhanced compared to the unloaded double-panel structure (e.g. around 170 Hz). However, the bands with successful noise reduction outweighs them.

A semi-active link with panels of different thickness

In this simulation, the thickness of panel a was increased from 1 mm to 2 mm. The frequency characteristics obtained are presented in Fig. 5.5. The first important

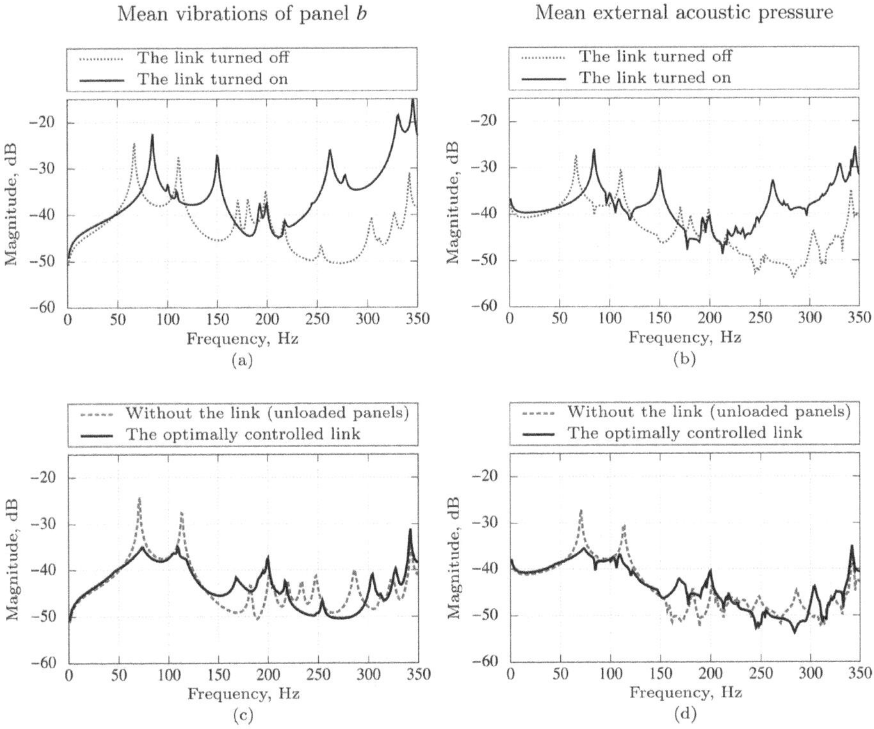

Figure 5.5 Frequency characteristics for the semi-active control numerical simulation performed for the link mounted between the panels at $x_{L,1} = 0.5L_{a,x}$ and $y_{L,j} = 0.5L_{a,y}$. Panel b thickness h_b = 1 mm, while panel a thickness h_a = 2 mm [175]. Reprinted from *Mechanical Systems and Signal Processing*, 154:107542, S. Wrona, M. Pawelczyk, and L. Cheng, "Semi-active links in double-panel noise barriers", Copyright (2022), with permission from Elsevier.

conclusion is that for a double-panel barrier with panels of different thicknesses, the semi-active link is able to change, thus reducing the fundamental frequency, i.e. when both panels vibrate in-phase with mode shape (1,1). The mass-air-mass resonance is mitigated in a similar manner to symmetric panel configuration. The noise transmission in the remaining part of the frequency band being considered remains rather similar to the symmetric configuration. Thus, it shows that asymmetric configuration is substantially more beneficial compared to the symmetric one.

A link with additional mass

If a symmetric double-panel barrier configuration has to be employed for any reason, e.g. the application of a semi-active control system to an already existing structure, an additional passive mass can be added to one of the panels to alter its natural frequencies and help mitigate the fundamental frequency with the semi-active link. The simulation results obtained for symmetric panels with a passive mass of 0.25 kg attached to panel b at its centre are given in Fig. 5.6. Although the asymmetric

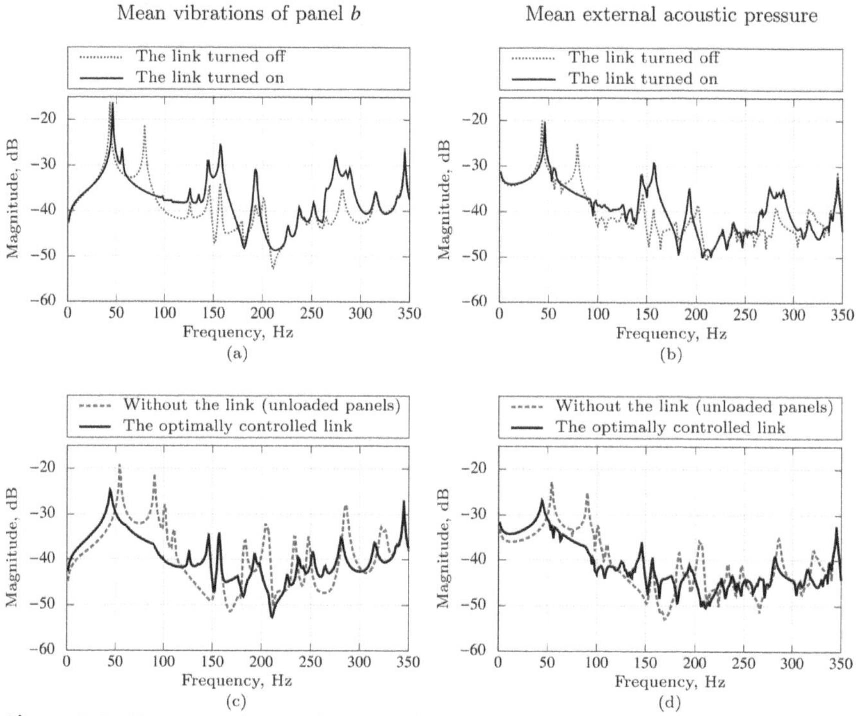

Figure 5.6 Frequency characteristics for the semi-active control numerical simulation performed for the link mounted between the panels at $x_{L,1} = 0.5L_{a,x}$ and $y_{L,j} = 0.5L_{a,y}$. Both panels are of the same thickness equal to 1 mm, but an additional mass of 0.25 kg is attached to panel b at its centre [175]. Reprinted from *Mechanical Systems and Signal Processing*, 154:107542, S. Wrona, M. Pawelczyk, and L. Cheng, "Semi-active links in double-panel noise barriers", Copyright (2022), with permission from Elsevier.

configuration, whenever possible, still seems to be a better choice, the addition of mass to one of the panels definitely helps in reduction of the fundamental mode; for the case considered, 7 dB more reduction of mean external acoustic pressure can be obtained compared to the case without the additional mass. An alternative could be to attach an additional stiffener to panel b as in [171, 176], instead of additional mass, in order to alter the fundamental mode frequency of one of the panels.

Three semi-active links

Finally, a scenario is considered with three semi-active links introduced into the gap cavity. The number of three links was arbitrarily chosen in order to explore the option of increasing the number of semi-active links, while maintaining still a reasonable number of them for practical applications. In the example presented, their locations have been arbitrarily chosen, however, a number of arrangements have been simulated to validate the conclusion presented in this subsection. The simulation results are given in Fig. 5.7. An asymmetric configuration with h_a = 2 mm and h_b = 1 mm has been used. The links were located at:

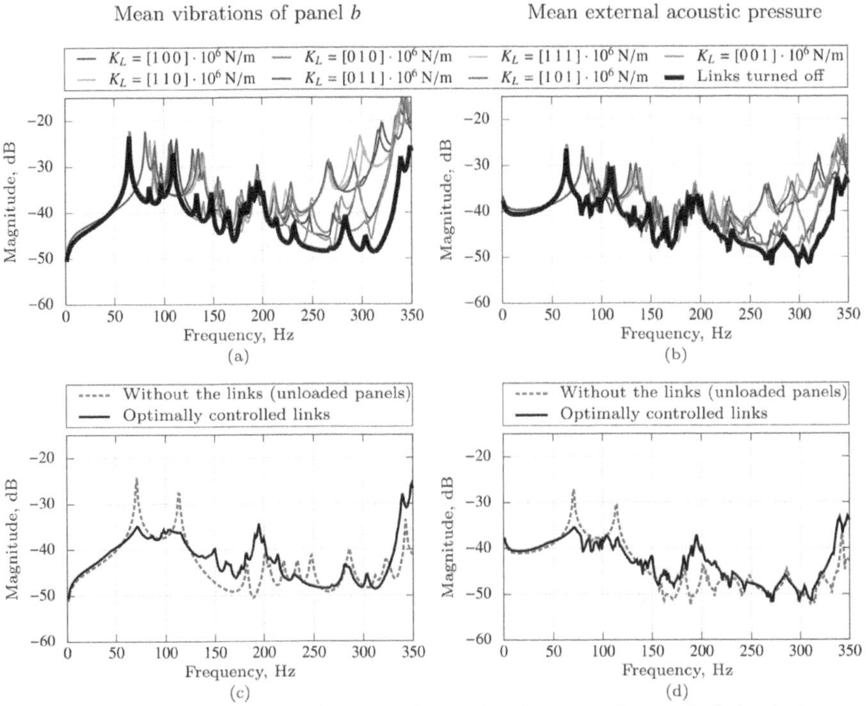

Figure 5.7 Frequency characteristics for the semi-active control numerical simulation performed for three links mounted between the panels. Panel b thickness h_b = 1 mm, while panel a thickness h_a = 2 mm [175]. Reprinted from *Mechanical Systems and Signal Processing*, 154:107542, S. Wrona, M. Pawelczyk, and L. Cheng, "Semi-active links in double-panel noise barriers", Copyright (2022), with permission from Elsevier.

$$x_{L,1} = 0.50L_{a,x}, \quad y_{L,1} = 0.50L_{a,y},$$
$$x_{L,2} = 0.83L_{a,x}, \quad y_{L,2} = 0.50L_{a,y},$$
$$x_{L,3} = 0.25L_{a,x}, \quad y_{L,3} = 0.20L_{a,y}.$$

To determine which links are turned on, a vector $K_L = [K_{L,1}\ K_{L,2}\ K_{L,3}]$ can be defined, assuming that $K_{L,i} = 0\,\mathrm{N/m}$ when the link is turned off, and $K_{L,i} = 10^6\,\mathrm{N/m}$ when the link is turned on.

The analysis of Fig. 5.7 leads to a conclusion that both fundamental and mass-air-mass resonances are successfully mitigated (as for a single link), but above a certain frequency (in the example considered, above 200 Hz), a configuration with all links turned off provides a continuously better performance; turning on any of the links can only worsen the transmission loss. It is due to a phenomenon that for higher frequencies the double-panel barrier itself provides good passive transmission loss, and adding a structural link at any location, although altering frequencies and mode shapes, enhances the overall energy transmission between the panels and generally worsens the transmission loss; a structural energy transmission path, in addition to the acoustic path, is added to the system.

In addition, comparing Figs. 5.5 and 5.7 leads to a conclusion, that the improvement due to an increased number of semi-active links is rather weak. A single link located at the centre of the panels is effective enough to mitigate low-order resonances (fundamental and mass-air-mass resonances, which are most responsible for weak transmission loss performance of the double-panel barrier). For higher resonances, the links considered are not useful, hence addition of more links is rather unjustified.

Summarizing the main conclusions, a single semi-active link located at the centre of both panels can successfully mitigate low-order resonances reducing the acoustic radiation of the barrier for narrow-band noise even by 16 dB. The external energy is only needed to switch states of the link. Moreover, there is clearly an optimal range of gap cavity depth $L_{g,z}$, when both acoustic and structural energy transmission paths are comparable, allowing the semi-active links to reach its best performance.

More links and different arrangements seem to be ineffective for improving higher frequency insulation due to physical phenomena occurring in double-panel structures. Nevertheless, the proposed semi-active link can be used jointly with other types of semi-active actuators that do not couple panels structurally, preserving the good transmission loss for higher frequencies provided by the structure itself. Such alternative actuators could semi-actively adapt stiffness or mass distribution of a panel. The semi-active link would then be responsible for the low-order resonances, while other semi-active actuators would alter higher resonances. Such combined technique would further develop the proposed semi-active approach.

5.3 COIL-BASED LINKS FOR DOUBLE PANELS

The coil-based link is a semi-active element, intended for damping of the transverse vibrations of the surface elements [136]. In line with idea of semi-active methods it does not require external energy to move the structure in a certain way, and the stability of the system is ensured. It is also possible to apply a switching law similar to piezoelectric elements. The provided energy is used only to supply the element to keep its properties—in this case, i.e. stiffness. The element consists of iron casing with rigidly mounted coil, and ferromagnetic rod moving freely and frictionless inside the coil. The coil is rigidly mounted to the vibrating surface (Fig. 5.8(a)) or to the supporting element, and the ferromagnetic rod is rigidly attached perpendicular (Fig. 5.8(b)) or parallel to the vibrating surface (Fig. 5.8(c)).

The lack of mechanical friction between the elements allows for a noiseless work. Moreover, it is possible to mount the element in various configurations, adapted to the application, including vibrations damping over a large area, dependently on the length of the ferromagnetic core and the used coil. The simple construction of the element allows to adjust maximal stroke of the ferromagnetic rod without impacting the mechanical structure. Such element can change the mode shapes of the vibrating structure, which influences the effectiveness of the structure's acoustic emission.

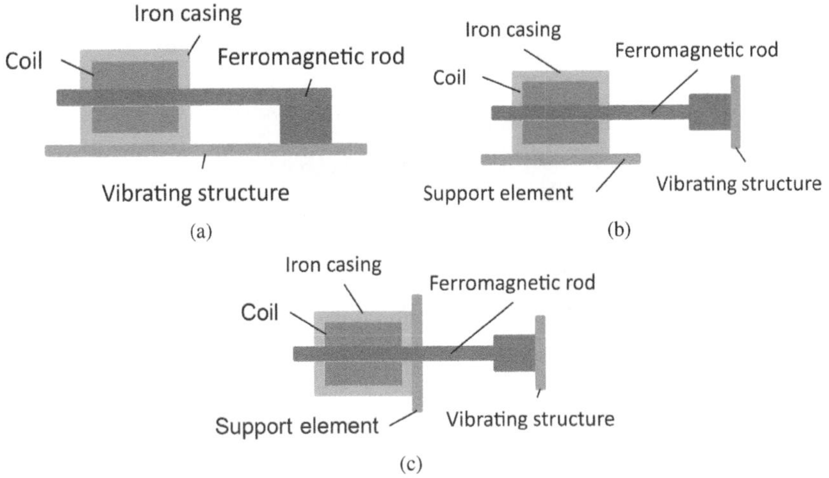

Figure 5.8 Coil-based link with coil mounted on the vibrating structure (a), on the perpendicular support element (b), and on the parallel support element (c).

5.3.1 HARDWARE PLATFORM

The idea of the coil-based link was implemented in the double-panel structure (DPS) constituting the front wall of the rigid device casing. The other casing walls are 3 mm thick plywood panels with additional bitumen layer to improve vibroacoustical insulation. The investigated element is a modified push-pull solenoid with a coil mounted on the incident (inner) panel of DPS by means of 3D-printed holder (Fig. 5.9(a)). The core (Fig. 5.9(b)) was shortened and rigidly connected with the radiating (outer) panel of DPS. The front wall of the rigid casing consists of 0.5 mm thick incident and 0.6 mm thick radiating steel panels. The size of each panel is 460 mm×460 mm, and the cavity gap between the panels is 50 mm. In this gap, five coil-based links are mounted: one at the centre of the cavity between the panels, and one in each corner to change a stiffness of the couplings between the panels.

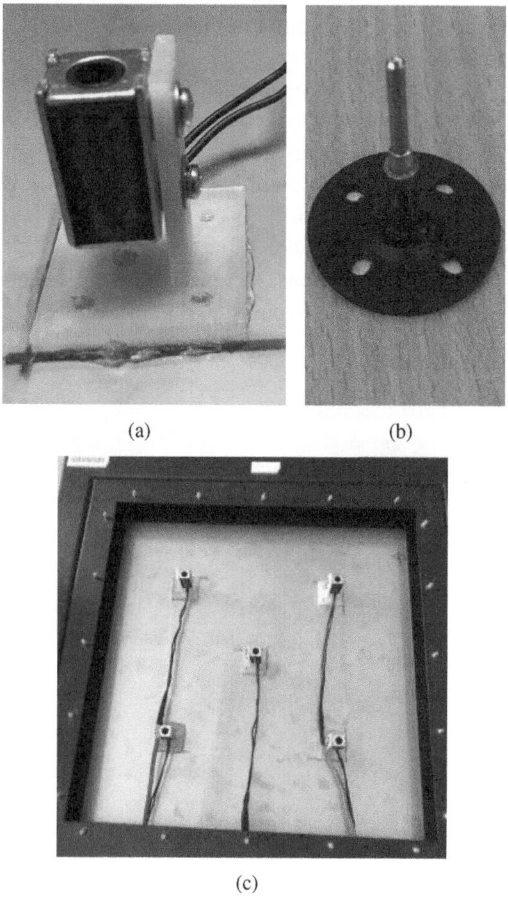

Figure 5.9 Single coil mounted on the panel (a), core (b), and five coils mounted on the panel (c).

The parameters of push-pull solenoid are listed in Table 5.1. The values come from manufacturer's documentation. However, after the modifications of the solenoid the parameters may have changed.

The stiffness of the element is changed by applying the voltage to the coil. Due to the high current flowing through the coil it was decided to use Pulse-Width Modulation (PWM) controllers, based on the Metal-Oxide Semiconductor Field-Effect Transistors (MOSFETs). Thus, the force generated by the element is indirectly set by change of the duty cycle of PWM signal. 0% of duty cycle corresponds to OFF state, when the element is not supplied, and core moves freely inside the coil, and

Table 5.1
Selected parameters of investigated coil-based links [151].

Parameter	Value, unit
Rated voltage	6 V
Rated current	0.3 A
Max. force	5 N
Max. voltage	12 V
Max. current	1.5 A
Max. stroke	10 mm

99% of duty cycle corresponds to ON state, when the element has the maximal stiffness and holds the core inside the coil in the same position. 100% of duty cycle was not used due to the possibility of overheating the coil.

The measurement and control system (Fig. 5.10) consists of previously described rigid frame casing with DPS. Inside the casing an active loudspeaker was placed to excite the structure to vibrate, by emitting tonal signals with the frequencies dependant on the type of experiment. The source of the signal was Direct Digital Synthesis (DDS) generator. Measurements of vibrations of the DPS were acquired using various types of the sensors: Macro-Fiber Composite (MFC) structures, vision or infrared (IR) cameras, and Laser Doppler Vibrometer (LDV). The data were stored, processed and presented in NI LabVIEW graphical programming environment dedicated to build the measurement and control systems, especially based on the NI devices. The PWM signal was provided to PWM controllers using NI myRIO automatically (in standalone mode) or manually via the Graphical User Interface (GUI) on the PC.

To ensure maximal efficiency of the coil-based links it was necessary to investigate how the mechanical properties of the element and parameters of the PWM signal influence the generated force. Therefore, the force was measured dependently on the following factors:

- duty cycle of PWM signal, varying from 0% to 100%,
- displacement between central points of the core and coil: 7 mm and 11 mm,
- frequency of the PWM signal: 500 Hz and 1 kHz,

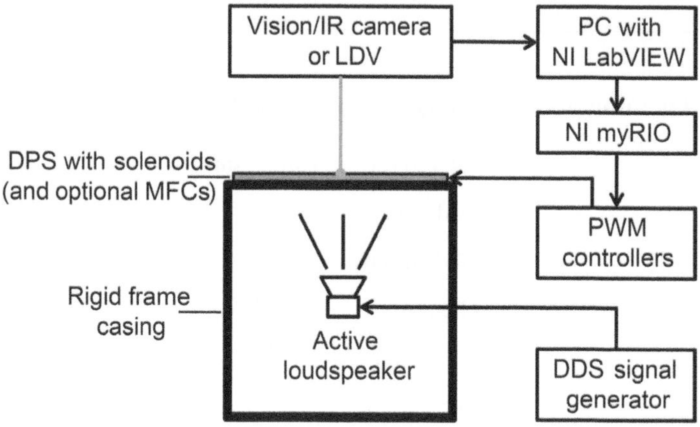

Figure 5.10 Scheme of the measurement and control system.

- coil's voltage, varying from 9 V to 11 V with 1 V step.

The parameters listed above were combined into five different scenarios. Characteristics of generated force obtained during the experiments are presented in Fig. 5.11. Following colours correspond with specific combinations of parameters:

- dotted—displacement: 11 mm, PWM frequency: 1 kHz, voltage: 9 V,
- dot-dashed—displacement: 7 mm, PWM frequency: 500 Hz, voltage: 9 V,
- dashed—displacement: 7 mm, PWM frequency: 1 kHz, voltage: 9 V,
- solid grey—displacement: 7 mm, PWM frequency: 1 kHz, voltage: 10 V,
- solid black—displacement: 7 mm, PWM frequency: 1 kHz, voltage: 11 V.

The most important observation is that the maximal force, generated by the element, is three times less than the value declared by the manufacturer (Table 5.1). On the one hand, as it was stated previously, the maximal voltage supplying the coil was reduced to 11 V to avoid overheating of the coil. On the other hand, the core was shortened to fit the element in cavity gap between the panels. The length of core before modifications was optimal to generate an expected force. Moreover, the displacement between central points of the core and coil has a crucial impact on the efficiency of the element. It was observed in comparison of the blue and red characteristics, where the only difference was the displacement. PWM frequency has insignificant impact, while the coil's voltage is important, however, not more than the displacement.

5.3.2 CONTROL ALGORITHM

As mentioned in Section 5.3.1, the basic version of the system controlling the solenoids allows to turn on or off the selected solenoids, as well as to modulate a

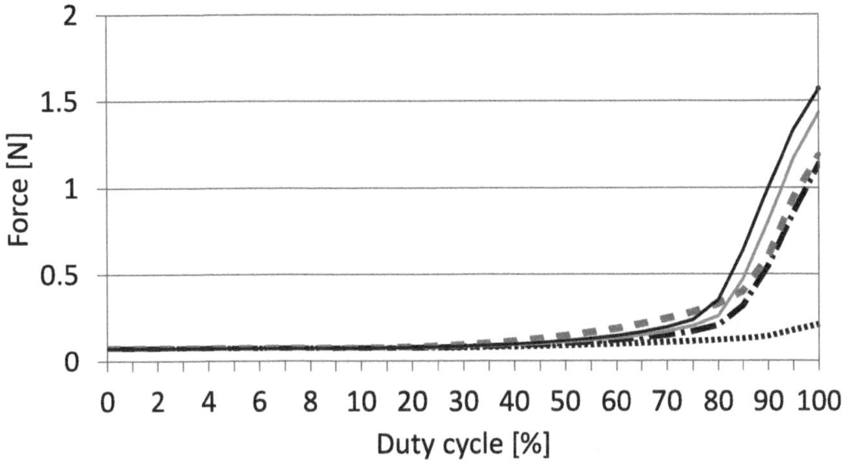

Figure 5.11 Force generated by the element dependently on duty cycle of PWM and selected parameters.

voltage signal supplying the solenoid's coils being the components of the coupling points. To indirectly modify the stiffness of the proposed electromagnetic couplings, Pulse Width Modulation (PWM) method is employed. Signal of a peak-to-peak voltage equal to 11 V is modulated by change of duty cycle, which as a percentage determines how much of each signal's period the signal is on. PWM voltage signal's duty cycle is modified and then the signal as a square wave supplies solenoid coils. Depending on the duty cycle value, the induced electromagnetic force varies, and thus the stiffness of double panel coupling changes. This in turn influences vibroacoustical properties of the double panel structure, what was observed during the research experiments.

Another aspect of control in the described system which may be developed and deployed as an improvement is the possibility to automatically change a number of activated couplings and duty cycle of PWM signal, based on the research outcome and experience gained from the previous experiments. Based on the obtained results, a heuristic lookup table algorithm may be introduced for a control of electromagnetic couplings to achieve the best results in terms of both noise and vibration reduction. The optimal setup for specific inputs can be looked up in case of their known values, which reduces computational burden.

For unknown values of the input features, one has to determine a way to estimate output. One of the solutions is to employ machine learning methods to estimate the best couplings' setup based on the dataset consisting from the previous research results. This can be achieved by means of open-source Python libraries such as scikit-learn which provide tools to program various algorithms as well as to visualize the predictions and the decision trees. Such approach depends heavily on a quality of prepared dataset and on a proper splitting process. Any bias in the input sets may result in a poor performance for real-time samples. There exists the risk of underfitting or overfitting the model. However, in case of many possible combinations of number of active couplings and duty cycle value, computational complexity of finding the optimal solution may be unacceptable in the real use cases. Hence, to simplify the problem, different algorithms may be employed to find acceptable output estimates for the presented problem with good outcome metrics.

For instance, if the best setup of number of activated couplings and duty cycle value is known for particular frequency and corresponding sound transmission loss value, then for actual measurements of these two quantities one may predict the best combination of activated couplings and duty cycle value in real time, taking into account a delay added to the system due to the use of multi-output regression algorithm. Such approach can be described in the form of scheme as in Fig. 5.12. This idea may be extended to a higher number of input features. If the dataset is extended in the meantime, flag in Fig. 5.12 is set to True and the metrics of the algorithm are checked to eventually tune it automatically by means of a selected method. Tuning process can also be done outside the loop.

Fig. 5.13 shows part of the decision tree which outputs number of activated couplings and duty cycle value. They should be set based on the measured sound transmission loss and noise frequency. Such tree may also be based on different kind of features, such as vibration, as well as temperature and/or other environmental conditions. Such approach may increase an accuracy of the proposed method with assumption that all added features are significant.

Another use case of such methods may be a prediction of sound transmission loss, power spectral density or other indicator of a vibroacoustical performance for a given sound frequency based on the experimental dataset. Such approach may be helpful in case of complex laboratory environments, which may be difficult to be modelled numerically.

Fig. 5.14 presents an example of a simple tree of a decision tree regression algorithm, where sound transmission loss value in dB may be estimated based on an input frequency of noise. Depending on the depth of the tree, the accuracy of the output value estimation changes. Such model, if tuned properly to find an optimal tree depth, may help to predict the behaviour of the laboratory system.

5.3.3 CONTROL RESULTS

The coil-based link can be used to change the mode shapes of vibrating structure. The experiments were performed for the fifth resonance. Four different scenarios were assumed: switched off links, single, central link activated, corner links activated, and five links activated. In Fig. 5.15 each scenario corresponds with each row. The results are compared with simulations. The Chladni method with image enhancement was originally adopted for this analysis using authors' laboratory setup [151].

The first observation is that the mode shape for coil-based links is not symmetrical. It may results from inequality of forces generated by the elements, because it was not possible to ensure reliable method of the assembly of the element between the panels, where the high precision is required. Despite such issue, it is observed that

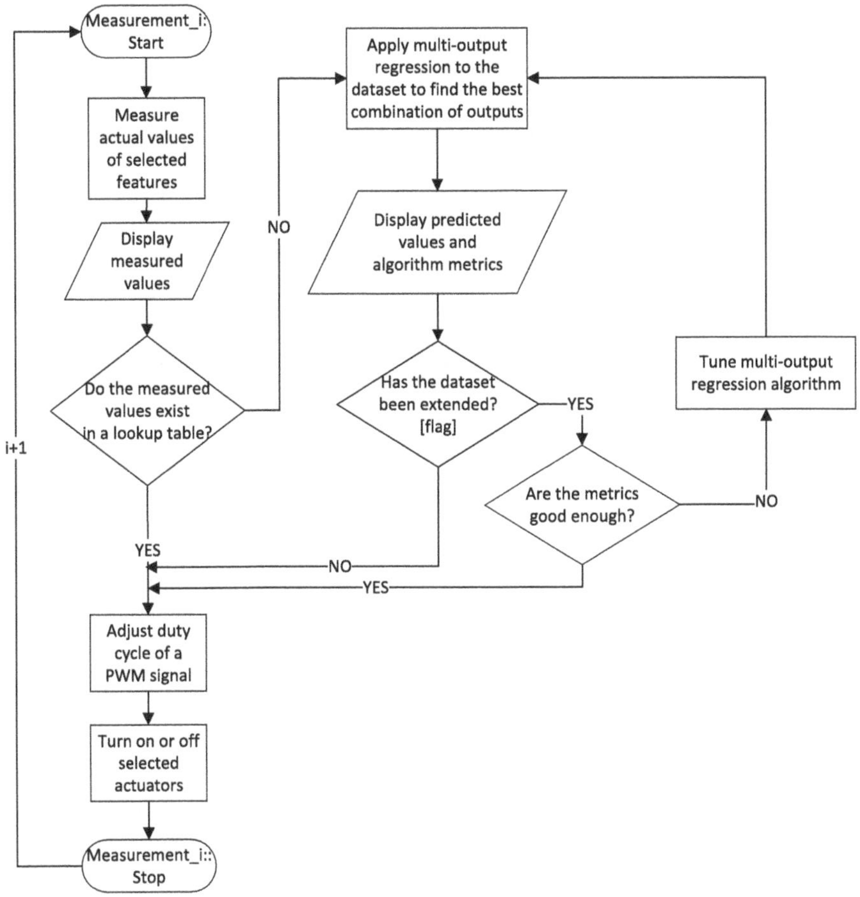

Figure 5.12 Algorithm scheme.

with activation of the single link the central area of the panel is more stiff (Fig. 5.15(e)), which corresponds to simulations results (Fig. 5.15(d)). However, for the four (Fig. 5.15(h)) and five (Fig. 5.15(k)) elements' scenarios the shapes are significantly deformed. The reason is the rapid temperature increase of the coil (Table 5.2), which significantly worsens the link efficiency.

Moreover, such fluctuations of the temperature may cause the thermal expansion of the steel plate in DPS and, in consequence, the resonance shifting to the lower part of frequency range. Due to this issue the experiments were repeated using links made from the solid magnets. It is observed, that asymmetry of the lower part of the panel (Figs. 5.15(c) and (f)) is similar as for the coil-based links (Figs. 5.15(b) and (e)). Thus, the main reason of such asymmetry is not inequality in generated forces, as it was supposed, but the properties of the panel itself. However, the activation of corner elements (Figs. 5.15(i) and (l)) caused the removal of asymmetry.

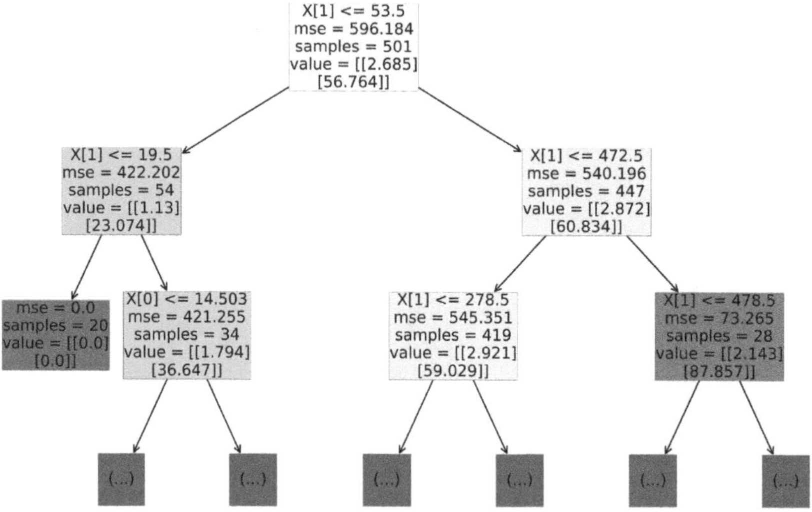

Figure 5.13 Part of the decision tree for multi-output regression (max. visualized depth = 3).

Another approach to research on the double-panel structure with introduced modifications was focused on the double panel's vibroacoustics. As the solenoids were mounted between the panels of the double-panel structure, the goal of the research experiments described below was to determine an influence of such modification on vibroacoustics of the double panel [141]. As the sensors of vibration, Macro-Fiber Composites (MFC) were used [140]. To examine acoustical performance of the double panel, microphones as the sensors were employed. The results showed that solenoids have a significant impact on the vibroacoustics of the double panel – both as the mass loadings and as the active couplings stiffening connection point between the panels. As part of the research experiments, measurement were acquired by means of MFC as the sensors. The signals were then processed to obtain Power Spectral Density curves for each channel in each of the examined setups.

Figs. 5.16–5.20 present the minimal Power Spectral Density curves built from the curves obtained for all scenarios available for the selected MFC channel. In each figure, there are three subplots. First subplot presents minimal PSD curve built of the curves obtained for five scenarios:

- zero activated couplings,
- 1 activated coupling supplied by PWM signal of duty cycle equal to 25%,
- 1 activated coupling supplied by PWM signal of duty cycle equal to 50%,
- 1 activated coupling supplied by PWM signal of duty cycle equal to 75%,
- and 1 activated coupling supplied by PWM signal of duty cycle equal to 99%.

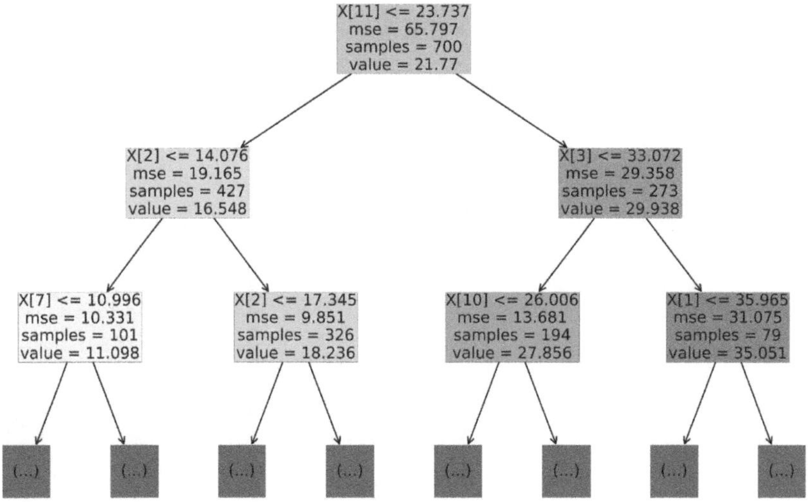

Figure 5.14 Part of the decision tree for decision tree regression to predict vibroacoustical indicators' levels (max. visualized depth = 3).

Second subplot presents minimal PSD curve built of the curves obtained for five scenarios:

- zero activated couplings,
- 4 activated couplings supplied by PWM signal of duty cycle equal to 25%,
- 4 activated couplings supplied by PWM signal of duty cycle equal to 50%,
- 4 activated couplings supplied by PWM signal of duty cycle equal to 75%,
- and 4 activated couplings supplied by PWM signal of duty cycle equal to 99%.

Third subplot presents minimal PSD curve built of the curves obtained for five scenarios:

- zero activated couplings,
- 5 activated couplings supplied by PWM signal of duty cycle equal to 25%,
- 5 activated couplings supplied by PWM signal of duty cycle equal to 50%,
- 5 activated couplings supplied by PWM signal of duty cycle equal to 75%,
- and 5 activated couplings supplied by PWM signal of duty cycle equal to 99%.

Fig. 5.16 presents such comparison for MFC no. 1 located in the left upper corner of the double panel. Minimal PSD estimate curve is of similar shape in all three subplots. Higher PSD levels are observed between 60 Hz and 300 Hz. Each curve is built

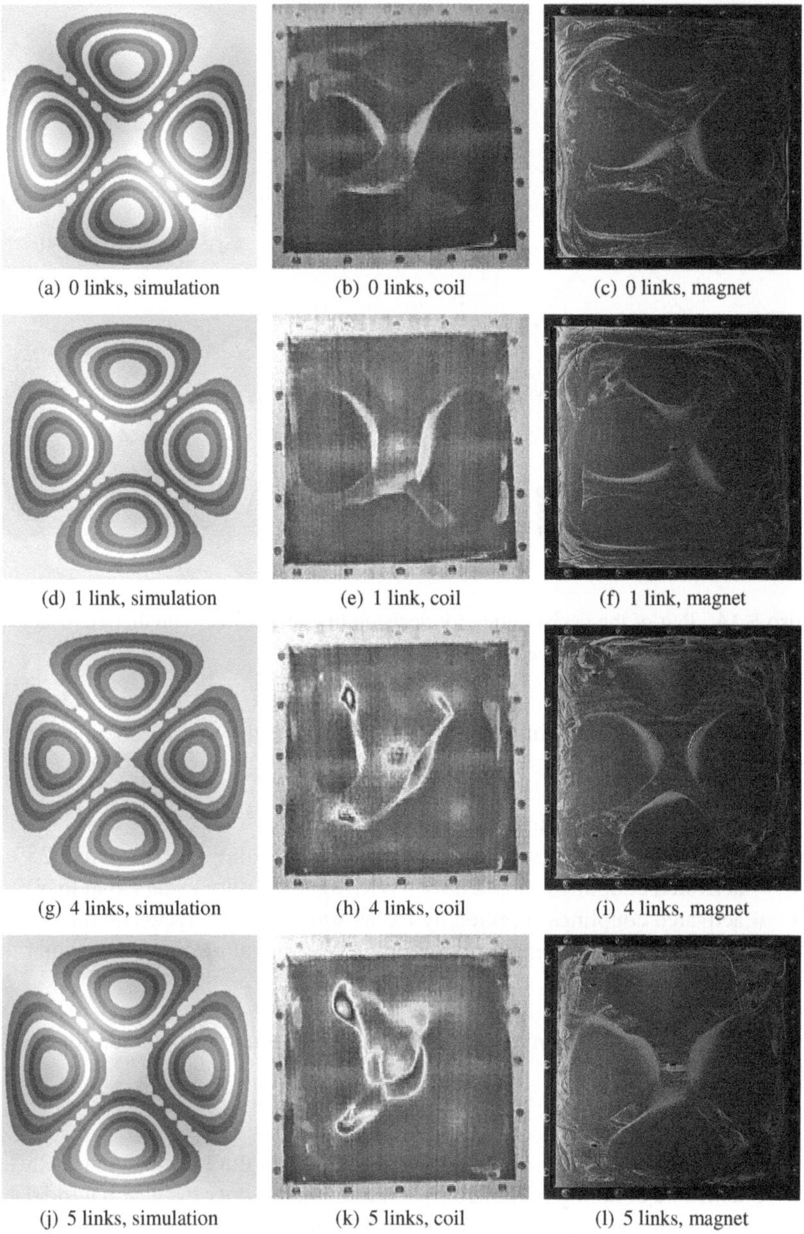
(a) 0 links, simulation
(b) 0 links, coil
(c) 0 links, magnet
(d) 1 link, simulation
(e) 1 link, coil
(f) 1 link, magnet
(g) 4 links, simulation
(h) 4 links, coil
(i) 4 links, magnet
(j) 5 links, simulation
(k) 5 links, coil
(l) 5 links, magnet

Figure 5.15 Mode shape for the fifth resonance: simulation (left column), coil-based links (middle column), and magnet-based links (right column). The patterns in the middle and right columns have been obtained with the Chladni method.

Table 5.2
Coil's temperature vs. time of element's operation. Coil's voltage: 11 V, coil's current: 0.85 A [151].

Time (s)	Temp. (°C)
0	22.1
5	25.9
10	29.6
15	32.6
20	35.9
25	39.8
30	41.5

Reprinted from *Sensors*, 20(15):4084, J. Rzepecki, et al., "Chladni figures in modal analysis of a double-panel structure", Copyright (2022), with permission from MDPI (STM).

of all five scenarios' curves, but for each subplot the order of the best setups at the specific frequencies is different. If the setup with 0 activated couplings is considered, its presence in the subplots is observed around the same frequency bands in most of the cases, e.g. around 50 Hz, 120 Hz, 230 Hz and 275 Hz. Around these frequencies, the most beneficial setup regarding vibration reduction around MFC no. 1 is not to activate any couplings.

Fig. 5.17 presents a comparison of minimal PSD estimates for MFC no. 2 located in the right upper corner of the double panel. As in Fig. 5.16, all curves are similar to each other. This means that minimal PSD estimate of similar levels can be achieved even if only 1 or 4 solenoids are activated. In each of the subplots, setup with duty cycle set to 99% is the best solution over significantly wide frequency bands, e.g. 265–305 Hz for 1 activated solenoid or 370–405 Hz for 5 activated solenoids. In Fig. 5.17 the similarities for setup with 0 activated couplings are not observed as in Fig. 5.16. The order of setups is different although MFC no. 2 is placed symmetrically to MFC no. 1 with respect to the vertical line at the panel's centre.

Fig. 5.18 presents a comparison of minimal PSD estimates for MFC no. 3 located at the centre of the double panel. In this case, minimal PSD estimates levels are in general higher than in Figs. 5.16 and 5.17 at the low frequencies up to 250 Hz, and around narrow 80 Hz frequency band they increase over −40 dB level. This is expected as the double panel's central area vibrates stronger than the areas located near the corners, as the panels' edges are fully clamped and in theory do not vibrate. At the low frequencies, central area of the panel covers with nodes, not anti-nodes,

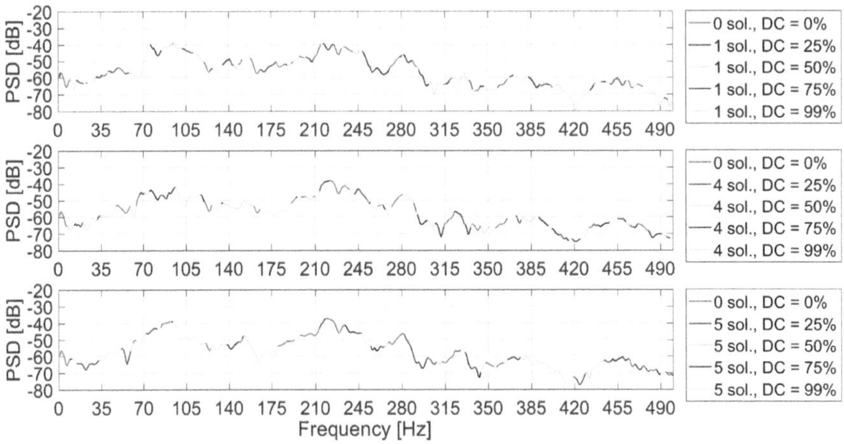

Figure 5.16 Minimal PSD estimates built of the selected scenarios for MFC channel no. 1.

for most of the mode shapes. However, the PSD levels do not increase significantly, which indicates a beneficial influence of the solenoids on vibration reduction. In all three subplots, a significant decrease of PSD levels is observed at the frequencies higher than 260 Hz. This is probably caused by the fact that at the higher frequencies mode shapes of the vibrating panels include high amplitude peaks around central area of the panels more frequently, and thus MFC no. 3 which is located at the panel's centre senses stronger vibration reduction. Most of the best scenarios at the higher frequencies are the ones with one or more activated couplings.

Fig. 5.19 presents a comparison of minimal PSD estimates for MFC no. 4 located in the left bottom corner of the double panel. PSD levels at the lower and higher frequencies (up to and above 260 Hz) are similar to those observed in Figs. 5.16 and 5.17. However, the shapes of PSD minimal estimates slightly differ from those obtained in cases of MFCs located in the upper corners of the double panel. The widest frequency bands of one most beneficial setup are observed for duty cycle set to 99% again.

Fig. 5.20 presents a comparison of minimal PSD estimates for MFC no. 5 located in the right bottom corner of the double panel. Significant PSD level decrease is observed over 300 Hz in each of three subplots. At most of the higher frequencies, PSD level does not exceed −70 dB, which is not observed in Figs. 5.16 and 5.19. Moreover, shapes of PSD estimates are different than for MFCs located in other corners of the double panel. Such difference may indicate undesirable panel's asymmetry, probably introduced during the manufacturing process. Such minor factors make the whole laboratory environment more complex and harder to model. However, as stated in Section 5.3.2, machine learning techniques may be applied so that the system behaviour may be accurately predicted based on the large sets of experimental data.

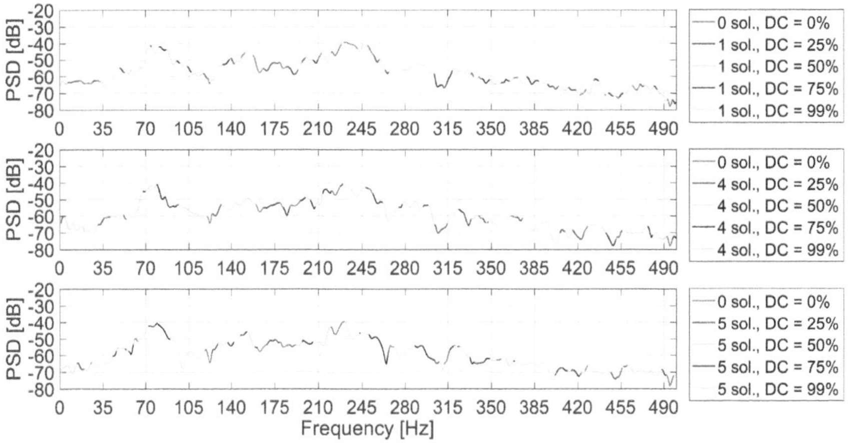

Figure 5.17 Minimal PSD estimates built of the selected scenarios for MFC channel no. 2.

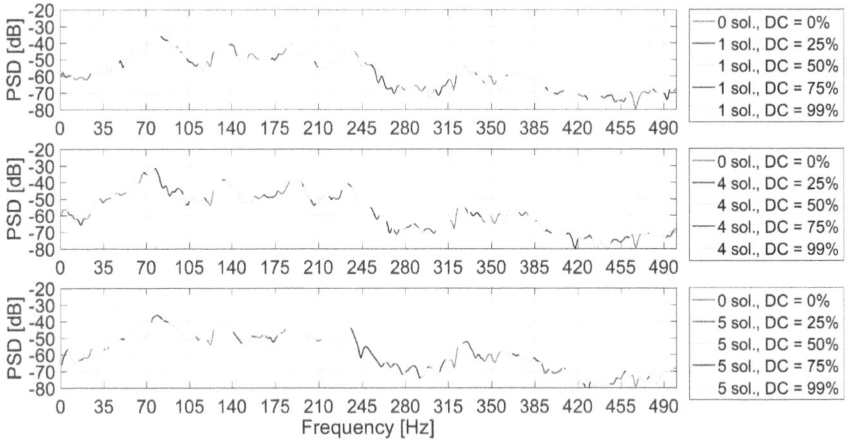

Figure 5.18 Minimal PSD estimates built of the selected scenarios for MFC channel no. 3.

Another approach—to determine acoustical properties of the double panel—may be to estimate its sound transmission loss based on the signals acquired by the microphones. Then, STL estimates can be compared for different scenarios. Fig. 5.21 presents three subplots. First subplot shows a difference between STL estimate obtained for a scenario with 1 activated coupling and duty cycle equal to 99%, and a scenario for zero activated couplings (a reference scenario). Second subplot shows a difference between STL estimate obtained for a scenario with 4 activated couplings and duty cycle equal to 99%, and a scenario for zero activated couplings.

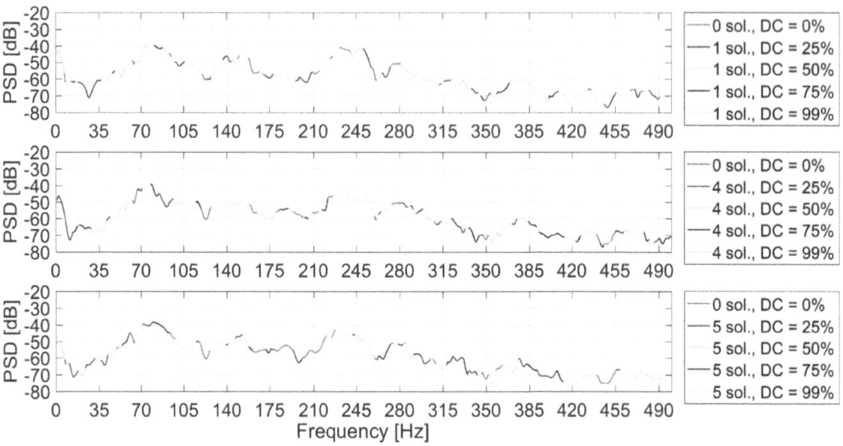

Figure 5.19 Minimal PSD estimates built of the selected scenarios for MFC channel no. 4.

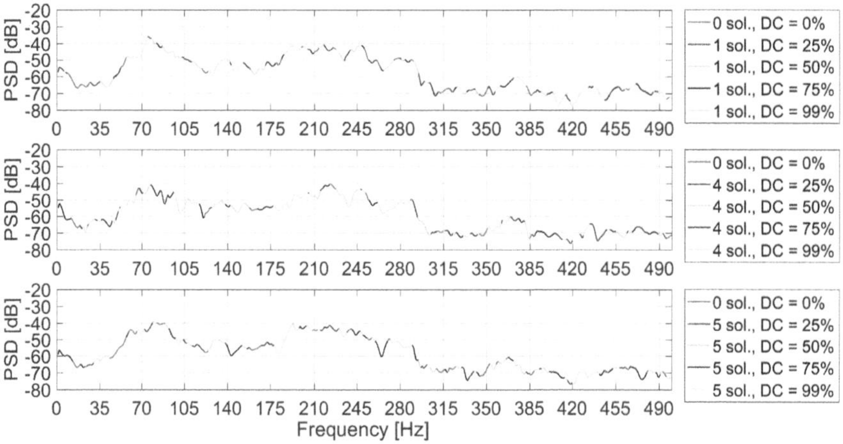

Figure 5.20 Minimal PSD estimates built of the selected scenarios for MFC channel no. 5.

Third subplot presents a difference between STL estimate obtained for a scenario with 5 activated couplings and duty cycle equal to 99%, and a scenario for zero activated couplings. All graphs are plotted within the same range of the axes. If selected scenario with any activated couplings performs better than the reference one, the difference—marked as green line—is above zero threshold level. In the opposite case, the line is marked as red.

Comparison to the reference scenario shows that the number of activated couplings is important if acoustical performance of the double panel is taken into account. Although all three subplots contain components above and under the threshold, STL improvement is observed within several frequency bands, especially in the case of 1 activated coupling. In the first subplot, improvement of STL for 1 activated coupling may exceed 20 dB. The most beneficial frequency band is 180–215 Hz. In theory, this is where mass-air-mass resonance frequency lies in case of the presented double-panel structure (c.a. 184 Hz). Hence, in this particular case, STL is improved around mass-air-mass resonance—which is perceived beneficial in terms of overall acoustic performance of the double-panel structure.

Common narrow frequency bands where STL is improved by means of solenoids' activation for all three subplots are i.a. around 70 Hz, 140 Hz, 330 Hz, 370 Hz and 430 Hz. However, there are also frequencies around which STL is worse by even 10 dB, e.g. 170 Hz, 220 Hz and 410 Hz.

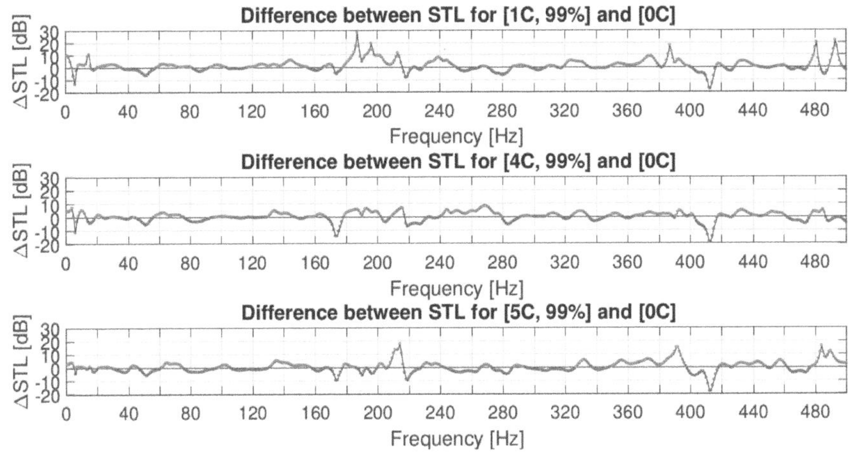

Figure 5.21 Differences between STL estimates for the selected scenarios.

Presented control results prove that coil-based links coupling the double-panel structure may be beneficial in terms of vibroacoustical properties improvement. Vibration may be significantly reduced by setting proper combination of activated couplings and duty cycle value. Acoustic performance of the double panel depends on the use of solenoids as their activation may lead to STL increase or decrease, depending on the selected setup and considered sound frequency.

As the presented laboratory environment is complex and many factors have to be taken into account, a machine learning approach can be applied to obtain multi-output prediction of vibroacoustic indicators, or suggested combination of number of activated couplings and duty cycle value – depending on the use case.

5.4 MASS MOMENT OF INERTIA ACTUATOR

The effective transmission loss of noise barriers can be increased by altering mode shapes and tuning resonances, shifting them away from the dominant part of the noise spectrum (as presented in [174]). Such adaptation has been performed using fixed passive elements in Section 4.2. However, it has rarely been approached in a semi-active manner, which allows a real-time tuning of panel response to the time-varying requirements. The authors believe that the ability to adapt to changing conditions would significantly enhance performance of the aforementioned noise barriers. In the literature, related semi-active approaches have been proposed for earthquake protection in civil structures [30, 29]. A sandwich plate with adjustable core layer thickness has also been proposed. This method employs a compressible open-cell foam core between panels and enables a compressible open-cell foam core between panels and enabling the adjustment of the structure's vibration behaviour by changing the core compression using different actuation pressures [44].

The method described in this section fills this gap and introduces a novel semi-active actuator with a tunable mass moment of inertia, capable of modulating the response of a panel during its operation [174]. The actuator is based on an additional mass mounted on a guide (the body of the actuator), which is attached perpendicularly to the surface of the panel. The movable mass can be shifted closer to or further away from the panel surface, tuning its effective mass moment of inertia. The proposed semi-active device only requires energy for shifting the movable mass, otherwise the mass is self-blocked. The actuator can shift the resonant frequencies and alter the mode shapes of the entire panel. These shapes determine the acoustic radiation efficiency of particular modes [176].

The proposed actuator enables an adaptation of the frequency-dependent transmission loss of a barrier to the current noise spectrum, e.g. the sound radiation of the barrier can be minimized in the targeted frequency bands. Such an approach is intended to the efficient reduction of transmission of non-stationary narrow-band noise, which is commonly encountered in real-life, both in the case of industrial devices and household appliances. The proposed solution can greatly enhance the effective transmission loss of a barrier, while being significantly lighter and requiring less space than vibration absorbers or Helmholtz resonators (which can be bulky when tuned for low frequencies). Simultaneously, it is less demanding in terms of system complexity and in energy required compared to a fully active approach.

When employing the proposed actuator, it is important to consider its placement on the panel surface. Two cost functions are proposed and evaluated for optimization of the location of the actuators. Depending on the objective of the considered application, different placements and configurations of the actuator allow for optimal performance. For this purpose, the concept of an equally weighted modal response is introduced into the optimization process in order to provide more general solutions.

5.4.1 HARDWARE PLATFORM

A rectangular steel plate with a prototype of the proposed semi-active actuator was used as a noise barrier for the method validation [174]. The plate was mounted to a heavy concrete box and excited with a loudspeaker placed inside the box. The loudspeaker was driven to generate a broadband, random noise (a band-limited white noise). The concrete walls of the box provided high noise attenuation, and hence

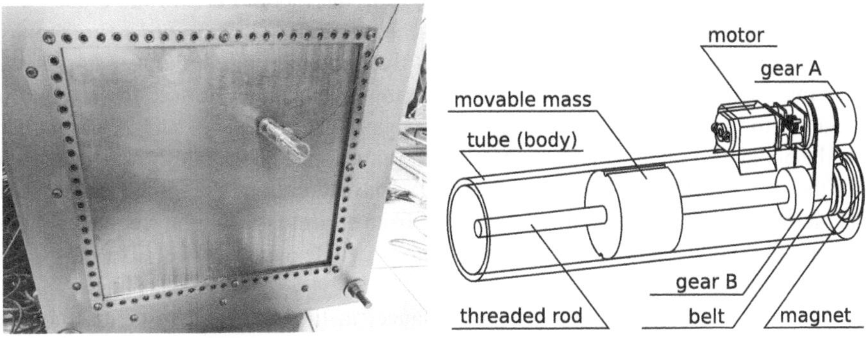

(a) (b)

Figure 5.22 (a) A photograph of the heavy concrete box with a panel and a mounted semi-active actuator. (b) A schematic of the semi-active actuator (own design) [174]. Reprinted from *Journal of Sound and Vibration*, 509:116244, S. Wrona, M. Pawelczyk, and L. Cheng, "A novel semi-active actuator with tunable mass moment of inertia for noise control applications", Copyright (2022), with permission from Elsevier.

most of the acoustic energy which was transmitted outside the box was transmitted through the steel plate. The acoustic modes of the cavity inside the box affected to some extent the acoustic excitation distribution over the panel, however, all of the vibration modes of the panel, theoretically expected in the considered frequency range, were excited enough to be captured with the laboratory equipment. A photograph of the laboratory setup is shown in Fig. 5.22(a). The dimensions of the panel area that was free to vibrate (i.e. the area inside the clamping frame) were 0.420 m $\times$ 0.390 m.

The schematic representation of the actuator prototype is presented in Fig. 5.22(b). The body of the actuator was made from a poly-methyl methacrylate (PMMA) tube, guiding the well-fitted movable mass inside it. The mass, with a threaded hole inside, is shifted using a threaded rod, thus adjusting the distance between the mass and the panel $z_{a,1}$. The threaded rod is rotated through a micro motor and a belt transmission. The threaded rod is mounted in a ball bearing at the base of the actuator. The actuator is attached to the panel surface using a neodymium magnet, also attached to the base of the actuator. The micro motor is equipped with an encoder allowing a precise determination of the movable mass current position. Although the mechanical design of this prototype could still be improved, it is a clear practical realization of the proposed type of semi-active actuator. The distance $z_{a,1}$ should be automatically adjusted by a dedicated controller, calculating the optimal distance $z_{a,1}$ based on continuously monitored frequency spectrum of the noise [174].

5.4.2 CONTROL ALGORITHM

The proposed actuator can alter natural frequencies and mode shapes of the panel by adjusting the distance $z_{a,1}$, as described in [174]. The underlying mechanisms behind this phenomena are based on increasing the modal mass of the modes with the increased mass moment of inertia of the actuator. However, in order for the mass moment of inertia of the actuator to affect the particular mode, the actuator should "swing" while the mode vibrates. It is determined by the location of the actuator in relation to the particular mode shape—the rotations of the panel surface are highest at the nodal lines, while they are absent at the anti-nodes (at the anti-nodes the motion of the panel surface is solely translational). The increase of the modal mass due to actuator is local (not uniform), hence apart from shifting the natural frequency of the mode, the mode shape is also altered in an irregular manner, what also effects the modal acoustic radiation efficiency.

To evaluate the best possible performance, in the presented research, the best transmission loss is selected a posteriori among all available $z_{a,1}$ values. In final application, the optimal distance $z_{a,1}$ should be selected automatically in real-time based on the noise spectrum and acquired frequency responses of the plate in function of $z_{a,1}$, however, at this stage it is considered out the scope.

5.4.3 CONTROL RESULTS

Simulations have been performed considering frequencies up to 300 Hz, thus including approximately the first ten eigenmodes of the panel. For the unloaded panel, frequency responses obtained from the model are shown in Fig. 5.23. The acoustic response shown (denoted hereafter as $A(\cdot)$) is the mean sound pressure amplitude obtained by averaging over a uniform measurement grid of 26×20 points, giving a total of 520 points, spaced at intervals of 0.04 m. The measurement grid was 1.00 m wide and 0.76 m high, and 0.1 m from the panel surface. This choice of

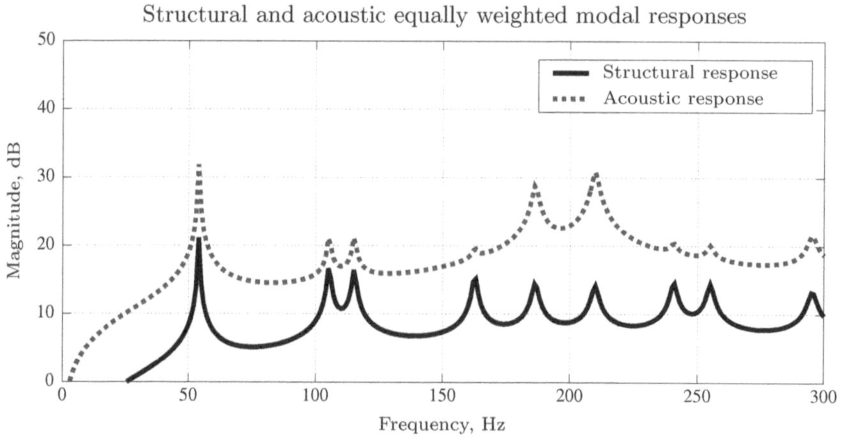

Figure 5.23 Structural (vibration) and acoustic equally weighted modal responses of the unloaded panel (without the semi-active actuator) [174]. Reprinted from *Journal of Sound and Vibration*, 509:116244, S. Wrona, M. Pawelczyk, and L. Cheng, "A novel semi-active actuator with tunable mass moment of inertia for noise control applications", Copyright (2022), with permission from Elsevier.

measurement grid follows from the experimental setup used for modelling validation, and is described in more detail in [176]. Such averaging provides a reliable estimate of the overall acoustic radiation generated by the panel. The structural (vibration) response shown is the mean vibration velocity obtained by averaging over a measurement grid of 22 × 20 points, giving a total of 440 points, spaced at intervals of 0.02 m, hence covering the whole surface of panel.

In all presented simulations, the excitation of the panel was obtained by directly applying an equal excitation to all loaded structural modes, instead of explicitly simulating a specific acoustic or structural excitation. Thus, the obtained responses have been named *equally weighted modal responses*, in order to better distinguish them from usual system responses. Such approach is motivated by the objective to design a barrier (or a sound source) dedicated for any kind of excitation (within the assumed frequency band), which is unknown and theoretically can excite any of the modes. Therefore, within the process of optimization of actuator's location, all of the modes should be well reflected in the simulated scenario in order to ensure that the actuator is able to sufficiently influence each mode of the panel. Moreover, it has been assessed that it is the best scenario to "prepare" the barrier and evaluate its performance for any kind of excitation, providing a more general solution as the result of the optimization process.

The movable mass is a replaceable component of the actuator and it can be changed during the preparation phase to achieve a desired mass. In the optimization process, the total mass of the semi-active actuator (including the body of the device) is limited to a maximum value of 0.2 kg (15% of the panel weight, which is equal to 1.3 kg). The real actuator used in the experimental setup may have some imperfections compared to simulations, e.g. the total mass may slightly differ, the distance of the centre of the mass of the actuator from the panel mid-plane $z_{a,1}$ may differ to some extent, also the mounting may be not perfectly rigid, etc. However, the achieved consistency between the simulations and experiments is high, hence, the authors are confident that any potential inaccuracies should be small enough to not affect the overall performance of the proposed solution. Nevertheless, for the semi-active control purpose, experimental frequency responses should be used instead of theoretically predicted ones to take into account any potential inaccuracies.

Having in mind potential practical applications, the control system should employ as few actuators as possible to achieve the imposed objective. The simpler solution is often more attractive. Hence, in the simulation studies, optimization is performed for a single actuator, which proved to be sufficient for the considered objectives. Nevertheless, more actuators can be considered as well to enhance the performance further.

The distance $z_{a,1}$ is assumed to be adjustable in a range between 0.01 m and 0.10 m (in practice these limits will depend on the chosen configuration of the actuator and the setup as a whole). For the purpose of optimization, the continuous domain of $z_{a,1}$ was discretized with a step size of 0.01 m, thus considering 10 possible settings of $z_{a,1}$ for the semi-active actuator. The aim of the optimization is to find the optimal location for the actuator on the panel surface, considering all available settings of $z_{a,1}$. It is also assumed for optimization purposes that only a tonal noise will be attenuated by the barrier, or that only a tonal sound should be emitted by the panel. Thus, during the cost function evaluation, each frequency in the considered range is evaluated individually (incremented by steps of 1 Hz and ranging up to 300 Hz) in choosing the most suitable setting of $z_{a,1}$.

A population-based memetic algorithm was used to carry out the optimization. For each optimization process, the population consisted of 300 individuals (considered solutions), the maximum number of generations was set to 15, and the probabilities of crossover, mutation and individual learning were 0.20, 0.30 and 0.06, respectively. For a detailed introduction to memetic algorithms, please refer to [130].

SIMULATION STUDIES

The adopted scenario considers the panel used as a semi-active noise barrier [174], which when excited should radiate noise to the environment on its other side as little as possible (considering excitation originating from both air-borne sound and structural vibrations). To this end, the acoustic radiation of the panel should be minimized over a wide frequency range. This objective is encapsulated in the following cost function

$$J_1 = \sum_{f=1}^{f=f_{max}} \left[\min_{z_{a,1}} A(f, z_{a,1}) \right] , \tag{5.2}$$

where f_{max} is a maximum frequency limiting the frequency range of interest; and $A(f, z_{a,1})$ is the mean sound pressure amplitude as a function of both frequency and the parameter $z_{a,1}$ (calculated according to the approach presented in Chapter 3). The cost function J_1 takes into account the overall acoustic radiation of the panel. To evaluate the cost function, the maximum frequency was set to f_{max} = 300 Hz.

The memetic algorithm achieved a solution with the actuator placed at $x_{a,1}$ = 0.287 m, $y_{a,1}$ = 0.098 m, with mass $m_{a,1}$ = 0.081 kg. The results of this optimization are shown in Figs. 5.24 and 5.25. To analyse the obtained configuration, it is instructive to first consider Fig. 5.24, which presents all ten structural and acoustic responses obtained for different values of $z_{a,1}$. The fundamental frequency gradually decreases when $z_{a,1}$ increases. For the following modes it is less clear due to mode superposition and changes in the order of peaks, however, similar phenomena occur. In addition to shifting the natural frequencies, changing $z_{a,1}$ can also strongly affect the amplitude of individual modes in the acoustic response due to the alteration of the modal acoustic radiation efficiency. The modal acoustic power P_i of first ten eigenmodes of the panel in relation to the distance $z_{a,1}$ has been compared in Table 5.3. Also note that a shift in the distance $z_{a,1}$ has a stronger effect on modes, which have nodal lines of their mode shapes near the actuator (the rotation amplitude of the panel surface is highest at the nodal lines).

Having all of the acoustic responses stored, an optimal value of $z_{a,1}$ can be determined for each frequency. The optimal solution (obtained by taking a minimum value of all responses for each frequency) is presented in Fig. 5.25 with a black solid line. The individual responses for different $z_{a,1}$ are presented with a light grey colour for reference, while the acoustic response of the unloaded panel (without the semi-active mass) is shown as a dashed line. Comparing the optimal solution with the response of the unloaded panel, it can be seen that all of the strongest modes can be mitigated by more than 10 dB, assuming a single tonal or narrowband excitation at a time. The acoustic response of the panel can effectively be flattened by modifying the location of the mass, and only residual peaks are left. In nearly the entire frequency range,

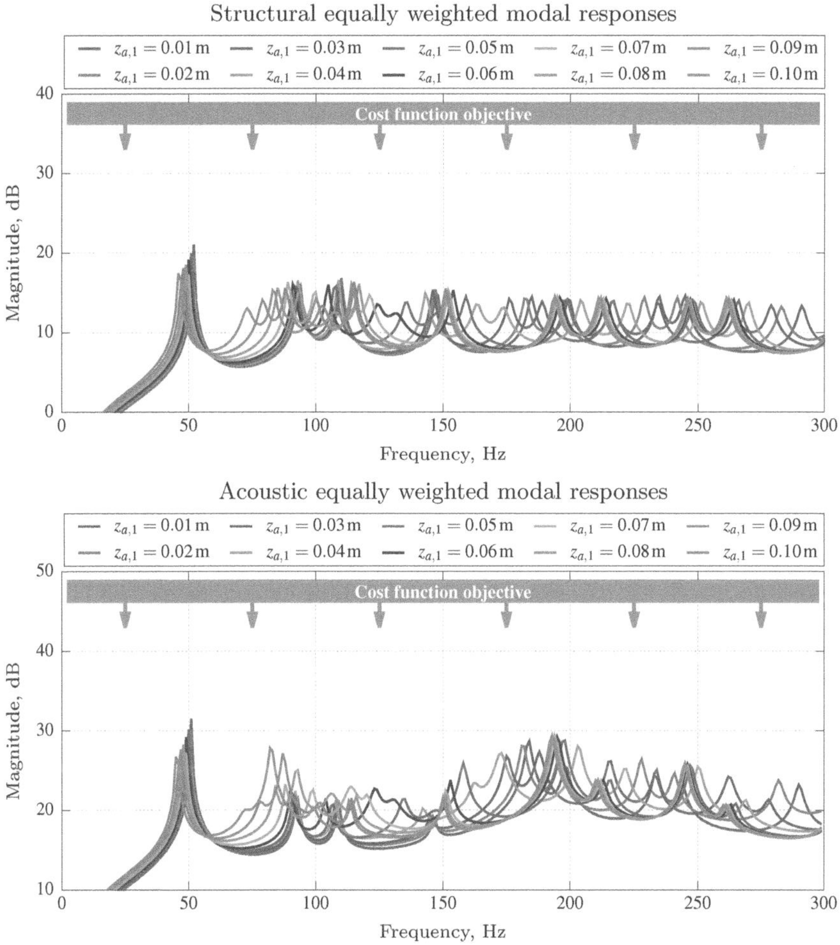

Figure 5.24 Individual structural (vibration) and acoustic equally weighted modal responses of the panel, obtained for optimization index J_1, with the actuator set for different $z_{a,1}$, in the range from 0.01 m to 0.10 m, incremented by 0.01 m. The actuator was attached at $x_{a,1} = 0.287$ m, $y_{a,1} = 0.098$ m, with $m_{a,1} = 0.081$ kg [174]. Reprinted from *Journal of Sound and Vibration*, 509:116244, S. Wrona, M. Pawelczyk, and L. Cheng, "A novel semi-active actuator with tunable mass moment of inertia for noise control applications", Copyright (2022), with permission from Elsevier.

Table 5.3

Comparison of modal acoustic power P_i in [dB] of first ten eigenmodes of the panel in relation to the distance $z_{a,1}$ [174].

$z_{a,1}$	Mode number									
	1	2	3	4	5	6	7	8	9	10
0.01	20.4	15.1	13.8	11.3	24.8	21.2	20.8	18.6	19.9	9.9
0.02	20.4	15.2	13.8	11.1	24.4	22.1	22.3	19.9	20.5	16.5
0.03	20.4	15.3	13.8	11.0	22.9	24.1	22.6	20.5	21.7	16.5
0.04	20.3	15.5	13.9	12.6	18.2	25.2	24.6	21.5	22.0	15.8
0.05	20.2	15.8	14.0	17.2	13.9	22.0	25.4	19.1	22.0	15.1
0.06	20.1	16.2	14.5	19.4	18.8	18.2	25.5	18.7	22.0	14.6
0.07	20.0	16.6	17.1	21.2	15.4	16.5	25.6	18.6	22.1	14.4
0.08	19.8	16.3	20.4	20.2	14.2	15.7	25.6	18.6	22.1	14.3
0.09	19.6	15.1	22.3	16.5	14.0	15.3	25.6	18.6	22.1	14.2
0.10	19.3	13.6	23.2	13.9	14.0	15.0	25.6	18.6	22.1	14.1

the optimal response of the loaded panel is lower than the response of the unloaded panel. Thus the conclusion can be drawn that this semi-active control approach is fully capable of avoiding the excitation of resonant frequencies of the whole panel by using only a single semi-active actuator.

Slightly different optimization results can be obtained by using a modified cost function J_2

$$J_2 = \max_{f} \left[\min_{z_{a,1}} A(f, z_{a,1}) \right] . \tag{5.3}$$

The cost function J_2 results in the minimization of the acoustic radiation of the most radiating modes. While the cost function J_1 seeks for a wider trade-off and may allow slight increase of the acoustic radiation in one band, in exchange for a bigger reduction in another band, the cost function J_2 always takes into account only the highest peak in the acoustic response. It leads to as flatten acoustic response as possible. The algorithm achieved a solution with the actuator placed at $x_{a,1} = 0.261$ m, $y_{a,1} = 0.121$ m, with $m_{a,1} = 0.080$ kg. Results of this optimization are shown in Fig. 5.26. It can

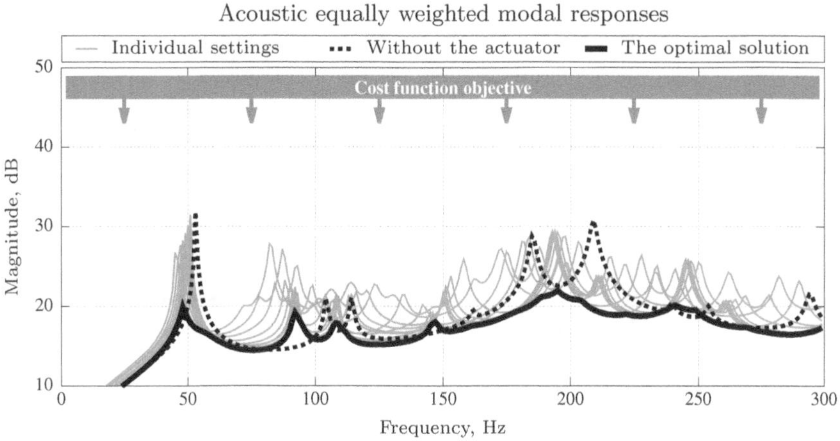

Figure 5.25 Acoustic equally weighted modal response of the panel, obtained for optimization index J_1 (solid grey lines—panel with the actuator set for different $z_{a,1}$, in the range from 0.01 m to 0.10 m, incremented by 0.01 m; dashed line—the unloaded panel; solid black line—panel with optimally controlled semi-active mass). The actuator was attached at $x_{a,1}$ = 0.287 m, $y_{a,1}$ = 0.098 m, with $m_{a,1}$ = 0.081 kg [174]. Reprinted from *Journal of Sound and Vibration*, 509:116244, S. Wrona, M. Pawelczyk, and L. Cheng, "A novel semi-active actuator with tunable mass moment of inertia for noise control applications", Copyright (2022), with permission from Elsevier.

been seen in Fig. 5.26 that the response is quite similar to the previous one, although flattened even more at the expense of slight enhancements of the amplitude at some specific frequencies. However, these enhancements are practically negligible, hence the cost function J_2 can potentially provide even better solutions depending on the adopted objective.

It is worth mentioning that during practical operations, the individual responses of the barrier for different distances $z_{a,1}$ can be stored in a look-up table and, if needed, periodically experimentally recaptured in order to update the lookup table used for choosing the optimal configuration of $z_{a,1}$.

REAL CONTROL EXPERIMENT

In this section additional verification experiments were performed for the optimal configuration followed from cost function J_1 (as presented in [174]). The laboratory setup presented in Section 5.4.1 was used. The cost function J_1, as shown in earlier in this section, resulted in $x_{a,1}$ = 0.287 m, $y_{a,1}$ = 0.098 m, and $m_{a,1}$ = 0.081 kg. Due to relatively low total mass, the moving mass was comparable with the mass of the actuator body. Hence, the range of $z_{a,1}$ was limited in practice to a range between 0.03 m and 0.10 m.

First, experimentally measured individual acoustic responses of the panel with the actuator set for different $z_{a,1}$ are given in Fig. 5.27. The figure corresponds to

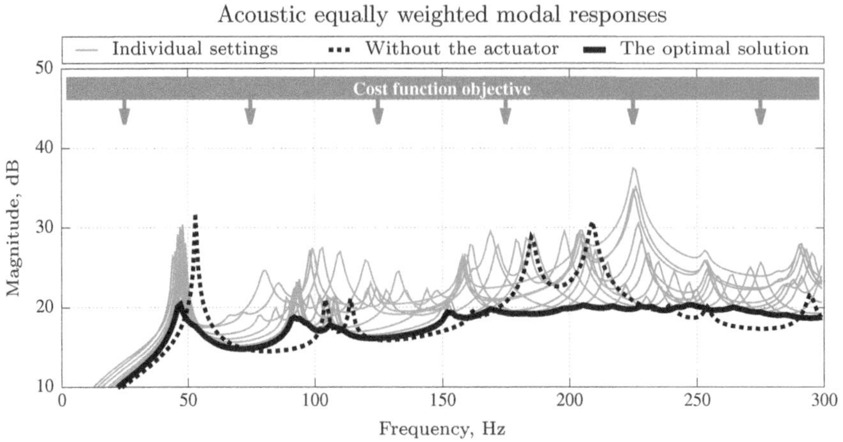

Figure 5.26 Acoustic equally weighted modal response of the panel, obtained for optimization index J_2 (solid grey lines—panel with the actuator set for different $z_{a,1}$, in the range from 0.01 m to 0.10 m, incremented by 0.01 m; dashed line—the unloaded panel; solid black line—panel with optimally controlled semi-active mass). The actuator was attached at $x_{a,1}$ = 0.261 m, $y_{a,1}$ = 0.121 m, with $m_{a,1}$ = 0.080 kg [174]. Reprinted from *Journal of Sound and Vibration*, 509:116244, S. Wrona, M. Pawelczyk, and L. Cheng, "A novel semi-active actuator with tunable mass moment of inertia for noise control applications", Copyright (2022), with permission from Elsevier.

simulation results given in Fig. 5.24. Although there are discrepancies, partially due to uneven acoustic excitation, the predicted influence of the semi-active actuator actions is coherent with experimental results. Individual acoustic responses of the panel without the actuator and with the actuator set for $z_{a,1}$ = 0.05 m are also compared with theoretical predictions in Fig. 5.28.

It is noteworthy that the impact of changing $z_{a,1}$ decreases for increasing frequencies. It is probably due to imperfect magnetic mounting (not perfectly rigid). A more rigid design should mitigate this issue. Nevertheless, in the frequency range up to 300 Hz a single semi-active actuator was able to achieve considerable reduction, often exceeding 10 dB (cf. Fig. 5.29). Results presented in Fig. 5.29 correspond to simulation results given in Fig. 5.25. The predicted results of the semi-active control are coherent with the experiments. The authors believe that these results clearly present the potential of the proposed semi-active control approach.

This study shows that employment of even a single actuator provides substantial benefits. However, the utilization of multiple actuators would introduce more degrees of freedom and dimensions into the space of possible configurations of semi-active actuators, thus further extending the capabilities of the proposed approach. In addition, the proposed approach can be used to support active noise barriers [107], e.g. to enhance the controllability of inertial actuators for targeted frequency bands, thus forming a hybrid active/semi-active control system.

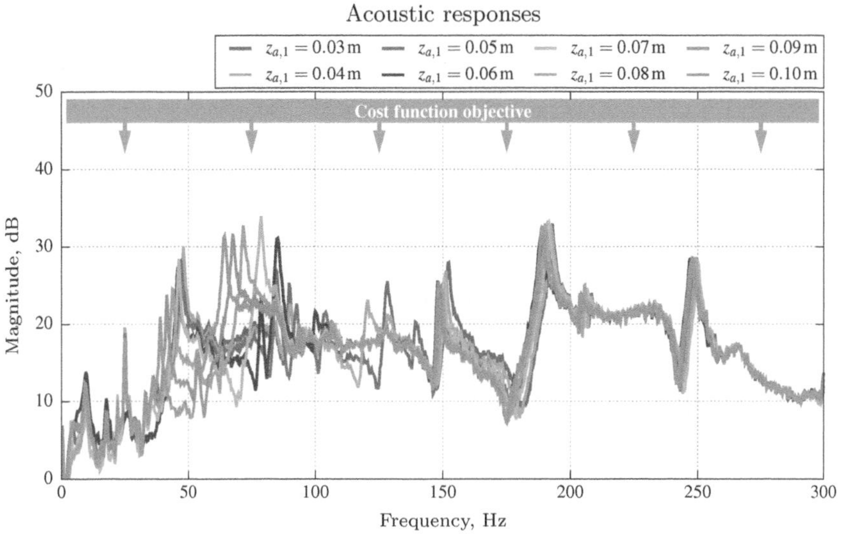

Figure 5.27 Experimentally measured individual acoustic responses of the panel, obtained for optimization index J_1, with the actuator set for different $z_{a,1}$, in the range from 0.03 m to 0.10 m, incremented by 0.01 m. The actuator was attached at $x_{a,1}$ = 0.287 m, $y_{a,1}$ = 0.098 m, with $m_{a,1}$ = 0.081 kg. The figure corresponds to simulation results given in Fig. 5.24 [174]. Reprinted from *Journal of Sound and Vibration*, 509:116244, S. Wrona, M. Pawelczyk, and L. Cheng, "A novel semi-active actuator with tunable mass moment of inertia for noise control applications", Copyright (2022), with permission from Elsevier.

5.5 SEMI-ACTIVE SHUNT SYSTEMS

In passive shunt systems it is difficult to achieve the full potential of such systems. Some power obtained from mechanical circuit must be lost to drive switches and by the circuit used to implement control algorithm. Complex control algorithms that theoretically should perform better in practice may perform worse because more power is lost in the control circuit. In semi-active systems, the control circuit and switches are powered from the external power source. In such systems more complex control algorithms may be used. The performance is additionally increased because the energy from extracted from piezoelectric element is used only for conversion to mechanical energy. Unfortunately, some energy must be lost due to non-ideal elements. When compared to classical active control systems, in semi-active shunt systems high-voltage amplifiers are not needed to drive piezoelectric elements. Simple transistor keys are sufficient.

5.5.1 HARDWARE PLATFORM

The basic approach for control is the same as in passive circuits described in Section 4.3.

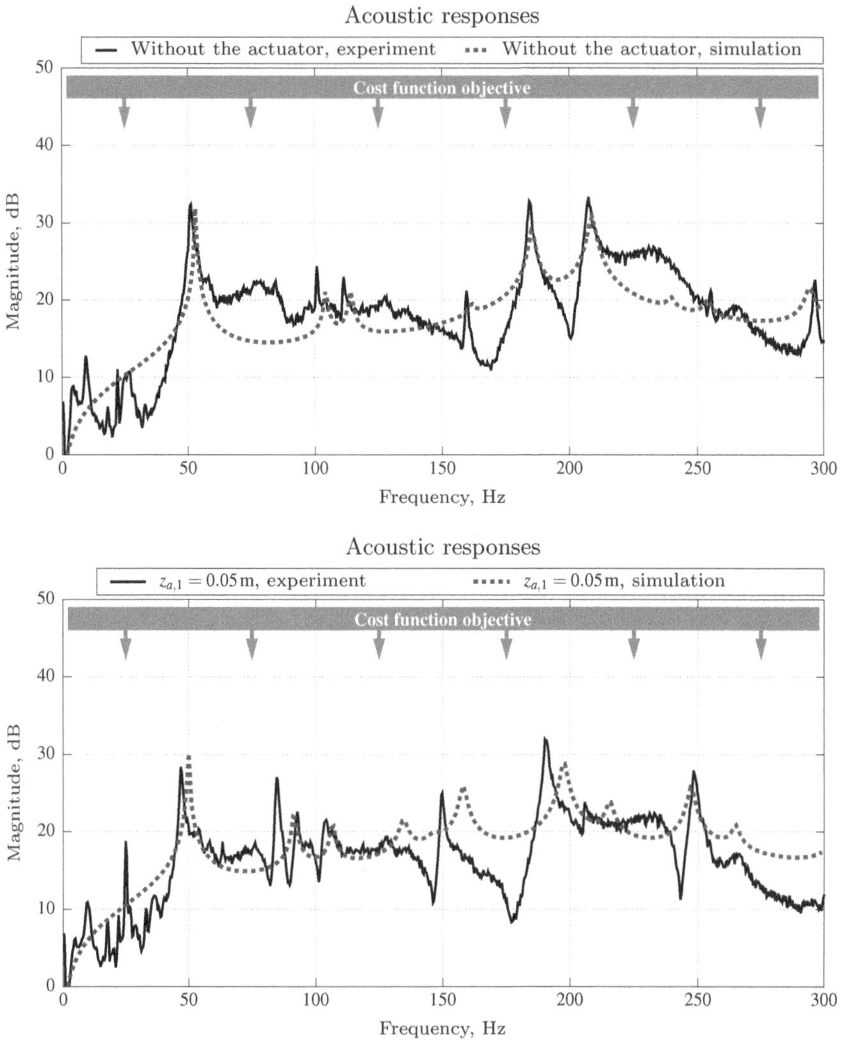

Figure 5.28 Experimentally measured and simulated individual acoustic responses of the panel, obtained for optimization index J_1, without the actuator (top) and with the actuator set for $z_{a,1}$ = 0.05 m (bottom). The actuator was attached at $x_{a,1}$ = 0.287 m, $y_{a,1}$ = 0.098 m, with $m_{a,1}$ = 0.081 kg [174]. Reprinted from *Journal of Sound and Vibration*, 509:116244, S. Wrona, M. Pawelczyk, and L. Cheng, "A novel semi-active actuator with tunable mass moment of inertia for noise control applications", Copyright (2022), with permission from Elsevier.

In opposite to passive implementation a switching between open and short states in semi-active SSDI is based on MOSFETs (Fig. 5.30). Bipolar Junction Transistors (BJT) are used to provide +/− 15 V MOSFET drive signal, converted from

3.3 V source coming from microcontroller. The oscillations are damped using resistors connected with gates, and D1/D2 diodes block reverse currents.

The `V_SWITCH` is voltage measured by microprocessor on switch circuit. In Fig. 5.31 an implementation of this function is presented. To eliminate rectified

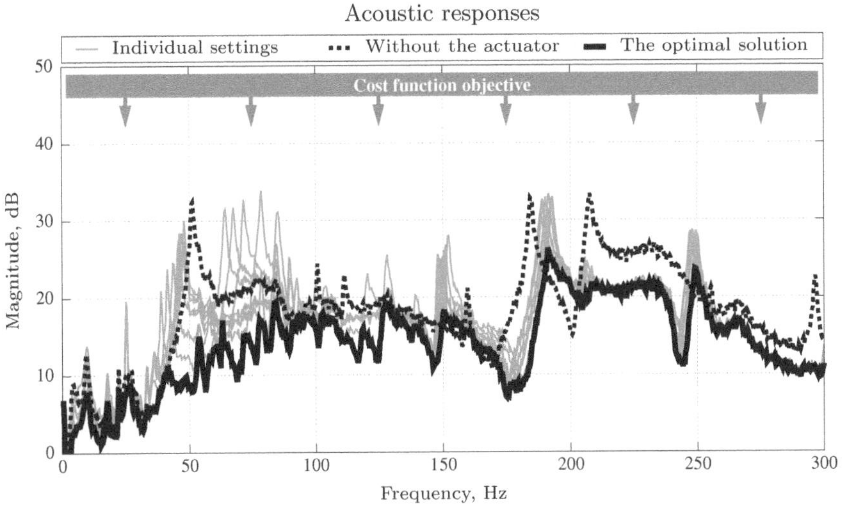

Figure 5.29 Experimentally measured acoustic response of the panel, obtained for optimization index J_1 (solid grey lines—panel with the actuator set for different $z_{a,1}$, in the range from 0.03 m to 0.10 m, incremented by 0.01 m; dashed line—the unloaded panel; solid black line—panel with optimally controlled semi-active mass). The actuator was attached at $x_{a,1} = 0.287$ m, $y_{a,1} = 0.098$ m, with $m_{a,1} = 0.081$ kg. The figure corresponds to simulation results given in Fig. 5.25 [174]. Reprinted from *Journal of Sound and Vibration*, 509:116244, S. Wrona, M. Pawelczyk, and L. Cheng, "A novel semi-active actuator with tunable mass moment of inertia for noise control applications", Copyright (2022), with permission from Elsevier.

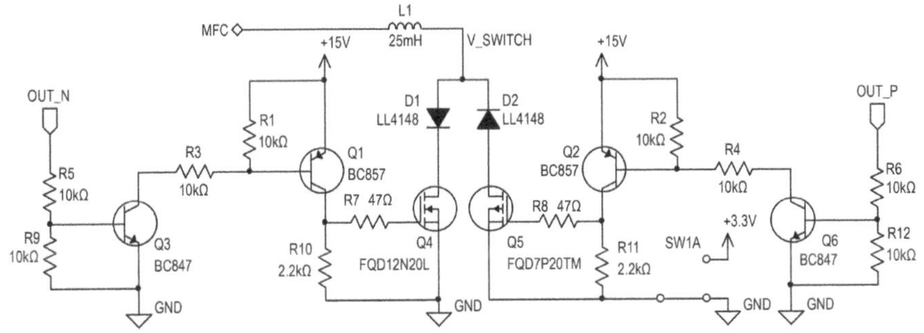

Figure 5.30 Scheme of semi-active SSDI electrical circuit [151]. Reprinted from *Sensors*, 21:2517, K. Mazur, et al., "Vibroacoustical performance analysis of a rigid device casing with piezoelectric shunt damping", Copyright (2022), with permission from MDPI (STM).

voltage R1 and C1 are used as Radio Frequency Interference (RFI) filter. Both the elements are also parts of an antialiasing filter. To extend measurement range to ±110 V optionally, a 1/11 divider based on R4 and R6 is used. D1 and D3 diodes protect ADC from too low or too high voltage, and D2 and D4 diodes have the same function with respect to the amplifier. The bipolar input signal come do the ADC, through the U1A amplifier which convert it to unipolar. In this configuration, the U1B amplifier is used as buffer [104].

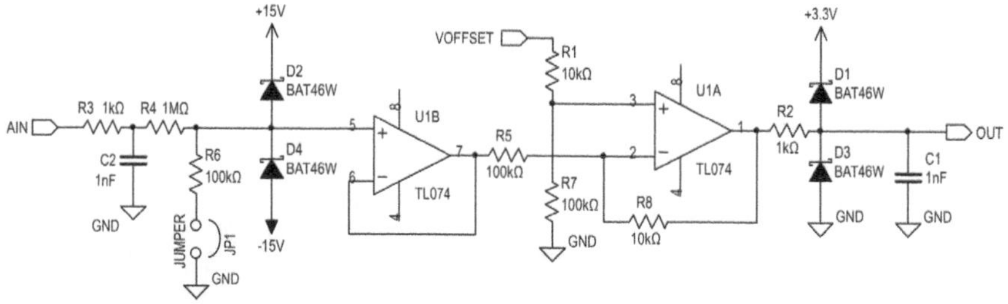

Figure 5.31 Scheme of MFC voltage measurement electrical circuit [151]. Reprinted from *Sensors*, 21:2517, K. Mazur, et al., "Vibroacoustical performance analysis of a rigid device casing with piezoelectric shunt damping", Copyright (2022), with permission from MDPI (STM).

The laboratory setup is similar to the measurement and control system used in experiment with passive shunt system. The aluminium panel with nine MFCs was mounted on rigid frame and was excited to vibrations using active loudspeaker placed inside the casing. The tonal signal used as an excitation was generated by the dedicated circuit, placed on the same PCB as semi-active SSDI implementation. Two-way communication between the PCB and PC was implemented, thus, it was possible to control the whole system remotely. The measurement system is based on the microphones and LDV, placed identically as in experiment with passive system. The signals were pre-filtered using anti-aliasing filters, and acquired using dSpace platform.

In the experiment, similarly as for the passive case, three types of circuits were compared: open, short and SSDI. The MFC elements were connected in series and the (1,3) and (3,1) modes were investigated.

5.5.2 CONTROL ALGORITHM

The SSDI switching law is described in subsection 4.3.2. When the MFC voltage is measured by the microcontroller the following variant of SSDI can be directly used (Fig. 5.32):

If $\frac{du_{MFC,i}(t)}{dt} > 0$ Then

    SW1 $\leftarrow$ Open, SW2 $\leftarrow$ Short;

Else

    SW1 $\leftarrow$ Short, SW2 $\leftarrow$ Open;

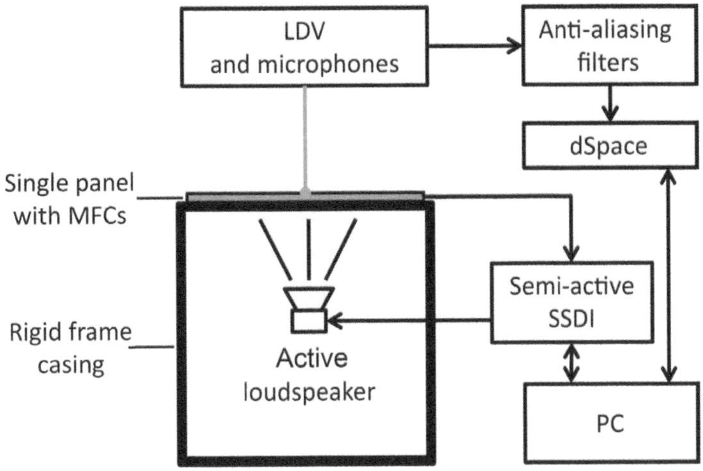

Figure 5.32 Scheme of measurement and control system for the experiment.

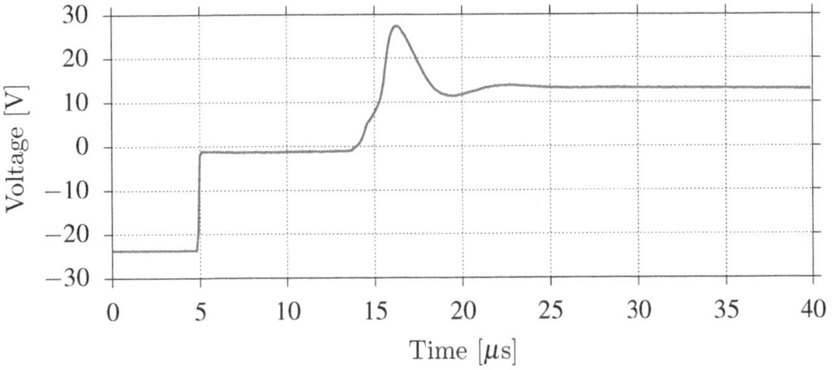

Figure 5.33 Voltage on the switch (`V_SWITCH`) during negative to positive voltage transition.

An alternative is to use extrema detection like in passive implementation. In a microcontrolled-based implementation it is easier to solve some issues related non-ideal operation of the circuit. Fig. 5.33 shows the voltage on the switch (`V_SWITCH`) during negative-to-positive voltage transition. A simple maximum detector (or system that use $\frac{\mathrm{d}u_{MFC,i}(t)}{\mathrm{d}t}$) may detect a false maximum at 16 μs. Additionally, when the switch is performed too early the voltage may still decrease, but switching voltage back will introduce significant energy loss. Thus, disabling switching for some time after each switch is beneficial. When this time is smaller than half-period of the sine disturbance it will not cause any performance loss. In the semi-active implementation of SSDI used on the casing this dead-time is equal to 1.5 ms.

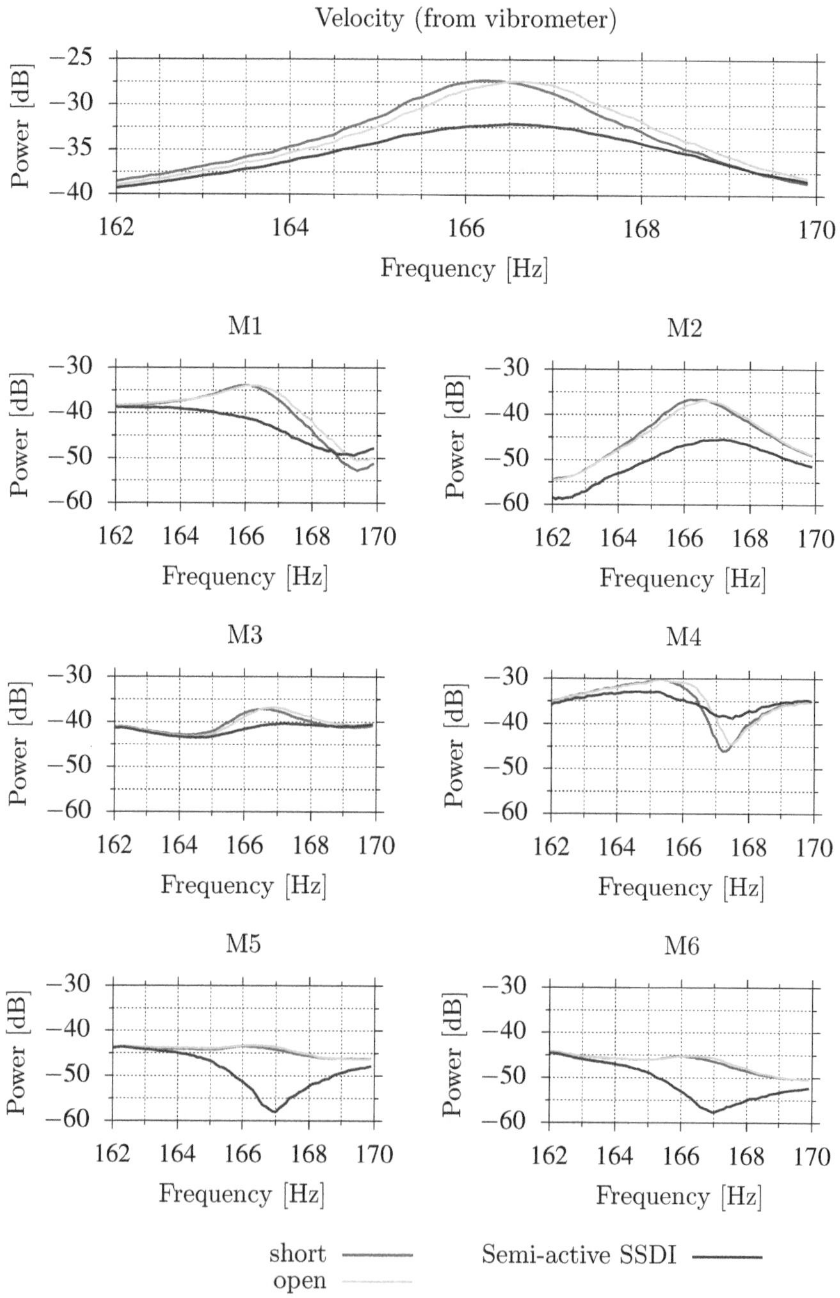

Figure 5.34 Measured signal powers for a different excitation frequencies.

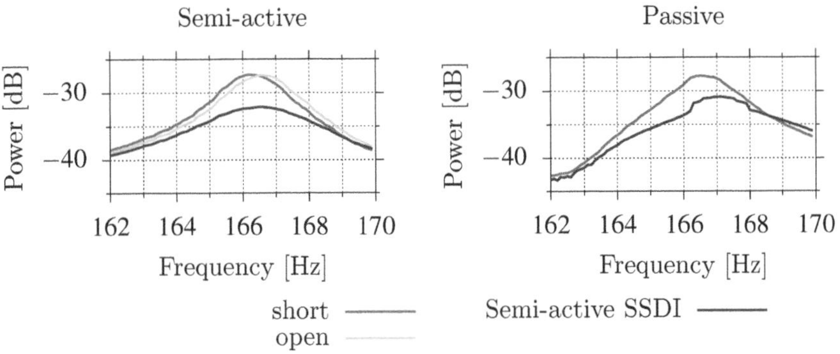

Figure 5.35 Measured velocity signal powers for a different excitation frequencies.

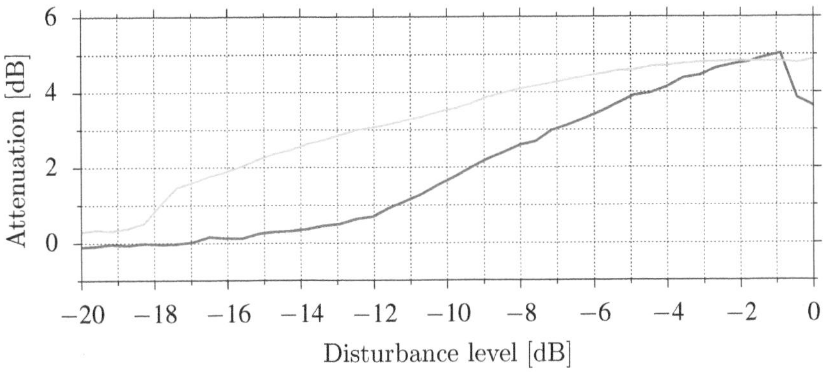

Figure 5.36 Vibration velocity attenuation level for a different excitation level at open circuit resonance.

In implementation on the microprocessor it is easier to set an arbitrary threshold used by the extrema detector. The passive implementation has this threshold equal to transistor base-emitter voltage, ab. 0.7 V. In the semi-active implementation the ADC quantum is equal to 89 mV ($\pm$110 V measurement range). However, due to oversampling the effective quantum is equal to 28 mV. The threshold has been set to 316 mV.

5.5.3 CONTROL RESULTS

Figure 5.34 shows comparison of the performance of passive and semi-active SSDI system for a different excitation frequencies with 0.1 Hz step. For each frequency the power of all signals was measured for three systems: the system with MFC with open terminals, the system with MFC with shorted terminals, and the semi-active

SSDI system. Semi-active SSDI circuit provides vibration reduction up to 5.0 dB compared to the short circuit (piezoelectric does not generate any force) and up to 4.7 dB compared to the open circuit. The noise reduction at frequencies close to resonance is observed on all microphones. On some microphones, however, the resonance is not visible significantly. But, even for that microphones noise reduction is visible and the sound power drops to levels lower than observed for frequencies far from resonance.

Fig. 5.35 shows a comparison of the performance of passive and semi-active SSDI systems for a different excitation frequencies. The semi-active system achieves better vibration reduction and avoids vibration amplification that occurs at higher frequencies for the semi-passive system. The semi-active system uses the same control law and similar algorithm to detect maximum voltage on MFC. However, the microprocessor-based maxima detector used in semi-active system is slightly better, especially more robust on noise and different events that occur after switching. The microprocessor-based detector waits some time after switching and avoids undesired switching caused by potential false maximas.

The performance of SSDI systems strongly depends on the excitation level. Fig. 5.36 shows the measured velocity signal power for a different excitation levels at open circuit resonance.

At lower excitation levels the voltage on piezoelectric element is too low for the SSDI circuit to work correctly. Both the passive and semi-active systems have some voltage drops on nonlinear elements on the side connected to MFC, and both use diodes. The passive system uses the energy from MFC element also to supply the switch implemented on bipolar transistors. The semi-active system uses external energy source to drive MOSFET switches and thus provides superior vibration reduction at lower excitation levels. At high levels the vibration reduction is comparable because both the systems use the same control law. However, the performance of semi-active system can be improved by implementing better maxima detection, with lower delay. The semi-active system has no performance drop at very high excitation levels.

5.6 SUMMARY

As it was shown in this chapter, there is a wide variety of semi-active control approaches that can be adopted for noise barriers, including especially noise-reducing casings.

First, original solutions designed by the authors for double-panel walls are presented: switched mechanical and coil-based links. Both of them take advantage of the cavity between panels and attempt to beneficially managing energy transfer between panels. Hence, the transmission loss of the noise barrier can be enhanced with a low energy effort.

Second, solutions for individual panels are considered: mass moment of inertia actuator and semi-active shunt system. They alter properties of the panel, reducing the noise transmission through it. They offer a considerable improvement, while maintaining a low energy demand.

6 Active control systems

6.1 INTRODUCTION

The third and last category of noise reduction methods considered in this monograph are active solutions [129, 58, 138, 137]. They have a greater energy demand, but achieve the highest levels of noise reduction. They most often consist of actuators, sensors and a controller which, basing on the measurement signals, calculates the control signals in accordance with the selected algorithm. The actuators and sensors may be both acoustic (e.g. speakers and microphones) and structural (e.g. electrodynamic exciters and accelerometers). These systems introduce additional energy directly into the system, but they do not have to completely eliminate system vibration. Often, it is enough to alter the vibration distribution appropriately to enhance the effective insulation of the acoustic barrier.

6.2 HARDWARE PLATFORM

6.2.1 OPTIMIZATION OF ACTUATORS ARRANGEMENT

Active noise barriers exhibit many advantages over their passive counterparts, but they have to be carefully implemented in order to operate efficiently and achieve a high level of performance [135, 124, 50, 82, 173]. One of the critical aspects in the design of an active barrier is the arrangement of the actuators, such that they are able to effectively control the vibration of the plate that forms the noise barrier. It is noteworthy that the optimization of the actuator arrangement is also an important step in the design of systems where plates are intentionally designed to emit sound [79].

Different techniques have been proposed over the years to optimize the arrangement of actuators for control applications (as given in [173]). One approach primarily focuses on selecting a control strategy and defining a performance index, and then simultaneously optimizing the locations of the actuators and the controller parameters. Liu et al. [92] used a genetic algorithm and the spatial H_2 norm of the closed-loop system as the performance index. Arabyan and Chemishkian [25] presented a computational method to design an H_∞ controller and the corresponding optimal actuator locations. Kumar et al. [81] considered the performance of an LQR controller as an objective. Chhabra et al. [26] used the modified control matrix and the singular value decomposition approach for optimal placement of piezoelectric actuators. However, in such approaches, optimality of the obtained solution is dependent on the choice of the control strategy.

Another approach concentrates on an open-loop system analysis, which is independent of the controller choice. The controllability Gramian was used in the optimization criterion by Leleu et al. [87]. Hale and Daraji [56] presented a modified H_∞

DOI: 10.1201/9781003273806-6

norm based method for the optimal placement of piezoelectric sensor/actuator pairs mounted on a cantilever plate. The optimal placement of piezoelectric actuators for active vibration control of a membrane structure using the controllability Gramian and the particle swarm optimization algorithm was studied by Liu et al. [93].

The aforementioned studies provide methods for the optimization of actuator locations mainly for the Active Vibration Control (AVC) of plates. Although the same actuator configuration can also be used for Active Structural Acoustic Control (ASAC) as employed in active noise barriers [125, 126, 168, 28, 122], it is not necessarily the optimum arrangement for this purpose. The optimization of the actuator arrangement for vibration control entails a search for a solution that generally reaches a trade-off between controlling numerous modes of vibration. Some of these structural modes may radiate sound efficiently, whilst others may vibrate considerably without contributing strongly to the noise transmission or radiation; as a result, these modes do not need to be controlled in the context of a noise barrier. Therefore, in the context of noise-reducing casings, an acoustic radiation-based approach to the optimization of the arrangement of actuators is proposed. A model of acoustic radiation is introduced into the optimization process and new cost functions are formulated to focus on modes that are truly relevant to the overarching goal of the barrier, which is to block the transmission of noise. The main contribution is thus providing new insight into the optimization process that should be adopted for the positioning of actuators in active noise barriers.

OPTIMIZATION PROCESS

In this section an optimization process is presented that aims to find the optimal placement of a number of actuators mounted to a vibrating plate for the purpose of active control. The objective of the control system is to reduce the noise radiated from the acoustic enclosure via the Active Structural Acoustic Control (ASAC) approach [106, 107]. In order to reach this goal, the control system should be able to control the vibration modes of the plate in the frequency range of interest. The ability to control the ith mode can be described by an element on the diagonal of the controllability Gramian matrix, $\lambda_{c,i}$, as derived in Section 3.2.2. However, some of the vibrational modes are more important as they more strongly transmit or radiate noise when excited; while other modes behave in the exactly opposite manner and can be neglected, since they vibrate without strongly contributing to the radiated acoustic field [173]. In order to reflect this behaviour, the modal acoustic power corresponding to ith vibration mode of the plate, P_i, can be used (cf. Section 3.2.2). Taking this into account, an optimization problem defined by an appropriate cost function will be presented, which will enable an optimal solution to be found for the arrangement of the given actuators.

Optimization problem

The optimization variables defined for the considered problem are the coordinates of a predefined number of actuators, N_a. A flat rectangular plate is considered, hence two coordinates per ith actuator, $x_{a,i}$ and $y_{a,i}$, are sufficient to unambiguously describe its location. Hence, the optimization algorithm is required to find a solution in an $2N_a$-dimensional space.

Due to physical dimensions of the actuators, certain constraints have to be defined in order to maintain the practicability of the solution. Namely, margins from the plate edges and between the actuators should be maintained, with the assumption that the actuators can be attached only from one side of the plate. Inertial actuators are considered in this paper, which are most commonly manufactured with a round foot print, although the method could be extended to more complex geometries as required. The first resulting constraint ensures that the actuators are placed within the boundaries of the plate; the dimensions of the considered rectangular plate are $a \times b$, hence, the coordinates of ith actuator $x_{a,i} \in (\frac{1}{2}d_{a,i}, a - \frac{1}{2}d_{a,i})$ and $y_{a,i} \in (\frac{1}{2}d_{a,i}, b - \frac{1}{2}d_{a,i})$, where $d_{a,i}$ is the diameter of the ith actuator. The second constraint ensures that actuators do not overlap; for $i \neq j$, the ith and jth actuators should not be closer than a distance of $\frac{1}{2}d_{a,i} + \frac{1}{2}d_{a,j}$, which is represented by the following constraint: $(x_{a,i} - x_{a,j})^2 + (y_{a,i} - y_{a,j})^2 \geq (\frac{1}{2}d_{a,i} + \frac{1}{2}d_{a,j})^2$.

Cost functions

The cost functions for the described problem can be formulated in a number of ways [173]. In this research, six cost functions will be evaluated and analysed. First, three cost functions that do not take into account the acoustic radiation, J_1–J_3, are formulated as follows,

$$J_1 = \min_i \lambda_{c,i} \,, \tag{6.1a}$$

$$J_2 = N_J^{-1} \left(\sum_i \lambda_{c,i} \right) , \tag{6.1b}$$

$$J_3 = \left(\prod_i \lambda_{c,i} \right)^{N_J^{-1}} , \tag{6.1c}$$

for $i \in \{1, 2, ..., N_J\}$, where N_J is the number of modes considered in the cost function. The same range of i is also considered for the other cost functions. All three cost functions J_1–J_3 focus on maximizing the controllability of the system, however, they result in a different balance between the N_J controllability measures, $\lambda_{c,i}$, corresponding to the N_J considered modes. Cost function J_1 represents only the least controllable mode, and thus ensures that there are no uncontrollable resonances within the frequency range of interest. Cost function J_2, which represents the mean controllability of the modes within the frequency range of interest, may increase the controllability of certain modes, even if this happens at the expense of reducing the controllability of other modes. Finally, cost function J_3 should lead to solutions that provide a trade-off between J_1 and J_2, making sure that the smallest of the factors is maximized, whilst also benefiting to some extent an increase in the controllability of the other modes in the frequency range of interest.

Subsequently, three additional cost functions are defined, J_4–J_6, which are analogous to the initial three cost functions, but take into account the acoustic radiation. These cost functions are defined as

$$J_4 = \min_i \left(\frac{\lambda_{c,i}}{P_i} \right) , \tag{6.2a}$$

$$J_5 = N_J{}^{-1} \left(\sum_i \frac{\lambda_{c,i}}{P_i} \right) , \tag{6.2b}$$

$$J_6 = \left(\prod_i \frac{\lambda_{c,i}}{P_i} \right)^{N_J{}^{-1}} . \tag{6.2c}$$

In each case, the division of $\lambda_{c,i}$ by P_i forces the optimization algorithm to seek solutions with better controllability (more energy efficient) for the ith mode, if the ith mode acoustic radiation measure P_i is higher. That is, the cost functions are weighted to focus the effort into the controllability of the strongly radiating structural modes.

Optimization algorithm

The search space that follows from the optimization problem described in the previous subsections is very complicated and contains numerous local maxima [173]. Therefore, an efficient algorithm must be employed in order to find a solution that satisfies the defined requirements. A Memetic Algorithm (MA) can be utilized for such a task, which is a hybrid form of a population-based approach coupled with separate individual learning [130]. The MA combines advantages of a global search, as offered by evolutionary algorithms, and local refinement procedures, which enhance convergence to the local maxima [130, 128]. Due to these complementary properties, MA are particularly suitable for solving complex multi-parameter optimization problems, such as the placement of sensors and actuators [165, 172].

ANALYSIS OF OPTIMIZATION RESULTS

In this section, an analysis of the optimization results obtained for the arrangement of real actuators is presented [173]. Dayton Audio DAEX32EP-4 are considered as actuators in this paper. They have a circular form factor, with a mass $m_{a,i} = 0.115$ kg and a diameter $d_{a,i} = 0.060$ m. The dimensions of the considered plate are $a = 0.420$ m and $b = 0.420$ m, hence, based on the constraints defined in Subsection 6.2.1, the coordinates of the ith actuator are given as $x_{a,i} \in (0.030, 0.390)$, $y_{a,i} \in (0.030, 0.390)$ and $(x_{a,i} - x_{a,j})^2 + (y_{a,i} - y_{a,j})^2 \geq (0.060)^2$ for $i \neq j$.

The configurations for three, six and nine actuators have been optimized using the six cost functions, J_1–J_6, defined in the previous section. The objective was to maximize the controllability of the plate used as an active acoustic barrier. The low frequency range was considered, hence the first $N_J = 12$ vibration modes of the plate were considered in the optimization process. The obtained results are summarized in Table 6.1.

Table 6.1

Results of the optimization for cost functions J_1–J_6 with $N_a = 3$ and $N_J = 12$. The natural frequencies ω_i are given in [Hz], while values of the cost functions J_1–J_6, $\lambda_{c,i}$, P_i and $\lambda_{c,i}/P_i$ are given in [dB]. Resulting values of the cost functions used as the optimization index are marked with bold font. Individual modes of high acoustic radiation ($P_i \geq 30$ dB) are highlighted with a grey background. The actuators placement is also given [173].

		Cost functions used in the optimization					
		Neglecting the acoustic radiation			Taking into account the acoustic radiation		
		J1	J2	J3	J4	J5	J6
Obtained values	J1	**50**	11	49	40	29	47
	J2	55	**62**	56	58	61	55
	J3	53	33	**54**	51	46	53
	J4	17	-12	17	**21**	8	13
	J5	33	41	32	34	**43**	35
	J6	29	9	28	25	22	**32**
		ω_i $\lambda_{c,i}$ P_i $\frac{\lambda_{c,i}}{P_i}$	ω_i $\lambda_{c,i}$ P_i $\frac{\lambda_{c,i}}{P_i}$	ω_i $\lambda_{c,i}$ P_i $\frac{\lambda_{c,i}}{P_i}$	ω_i $\lambda_{c,i}$ P_i $\frac{\lambda_{c,i}}{P_i}$	ω_i $\lambda_{c,i}$ P_i $\frac{\lambda_{c,i}}{P_i}$	ω_i $\lambda_{c,i}$ P_i $\frac{\lambda_{c,i}}{P_i}$
Modes	1	43 55 28 27	20 73 21 52	41 60 28 32	29 68 24 44	26 69 23 46	44 54 28 26
	2	75 58 22 36	69 41 22 19	65 60 26 34	82 47 23 24	45 65 18 47	72 61 21 39
	3	77 59 21 38	72 36 20 16	81 60 21 39	85 43 21 22	67 63 11 52	76 59 21 38
	4	102 52 21 31	109 49 37 13	100 54 22 32	113 56 35 21	83 36 20 16	103 51 21 30
	5	119 54 22 33	133 34 24 11	113 54 20 33	116 55 34 21	117 53 34 19	110 56 22 33
	6	136 56 19 37	142 23 25 -2	129 56 23 33	132 56 35 21	129 29 22 8	130 58 20 37
	7	146 50 22 28	155 37 28 9	151 51 29 22	149 52 17 35	137 46 34 12	155 51 16 35
	8	178 50 32 19	175 22 21 1	176 49 28 21	166 40 19 21	160 33 23 10	168 48 19 29
	9	184 50 33 17	220 13 24 -12	183 51 22 29	186 48 27 21	167 43 31 12	179 48 34 13
	10	194 50 19 31	225 11 15 -5	184 51 33 17	189 47 25 22	194 33 26 8	197 47 14 33
	11	203 50 30 20	247 38 30 8	201 50 27 23	217 51 23 27	212 42 27 15	199 53 14 38
	12	212 52 22 30	301 14 22 -8	212 50 27 22	223 48 22 26	236 42 23 20	203 50 17 34
Actuators placement	i	$x_{a,i}$ (m) $y_{a,i}$ (m)	$x_{a,i}$ (m) $y_{a,i}$ (m)	$x_{a,i}$ (m) $y_{a,i}$ (m)	$x_{a,i}$ (m) $y_{a,i}$ (m)	$x_{a,i}$ (m) $y_{a,i}$ (m)	$x_{a,i}$ (m) $y_{a,i}$ (m)
	1	0.379 0.317	0.194 0.254	0.380 0.384	0.377 0.385	0.187 0.239	0.097 0.373
	2	0.382 0.383	0.255 0.223	0.358 0.311	0.044 0.390	0.120 0.309	0.308 0.376
	3	0.174 0.381	0.194 0.194	0.244 0.376	0.202 0.200	0.293 0.134	0.382 0.372
		An overview:	An overview:	An overview:	An overview:	An overview:	An overview:

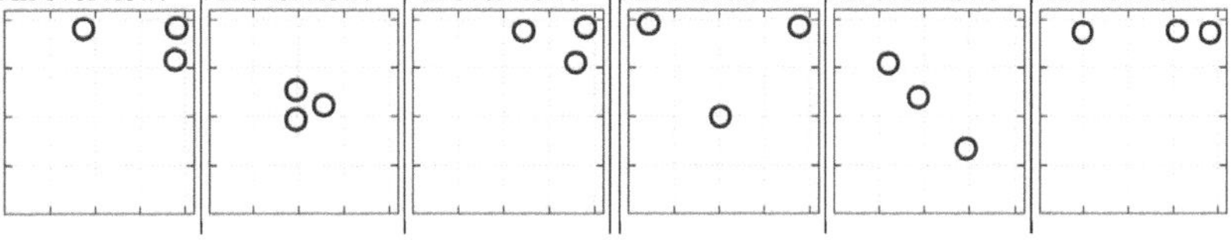

Reprinted from *Mechanical Systems and Signal Processing*, 147:107009, S. Wrona, M. Pawelczyk, and J. Cheer, "Acoustic radiation-based optimization of the placement of actuators for active control of noise transmitted through plates", Copyright (2022), with permission from Elsevier.

Table 6.2

Results of the optimization for the cost functions J_1 and J_4 with $N_J = 12$ and N_a equal to 3, 6 or 9. The placement of the actuators is also given [173].

		Cost functions used in the optimization: J1, $N_a = 3$	J1, $N_a = 6$	J1, $N_a = 9$	J4, $N_a = 3$
Obtained values	J1	**50**	**53**	**56**	40
	J2	55	59	63	58
	J3	53	56	59	51
	J4	17	19	21	**21**
	J5	33	38	44	34
	J6	29	31	37	25

		J1, $N_a=3$: ω_i	$\lambda_{c,i}$	P_i	$\frac{\lambda_{c,i}}{P_i}$	J1, $N_a=6$: ω_i	$\lambda_{c,i}$	P_i	$\frac{\lambda_{c,i}}{P_i}$	J1, $N_a=9$: ω_i	$\lambda_{c,i}$	P_i	$\frac{\lambda_{c,i}}{P_i}$	J4, $N_a=3$: ω_i	$\lambda_{c,i}$	P_i	$\frac{\lambda_{c,i}}{P_i}$
Modes	1	43	55	28	27	32	66	25	41	22	71	22	48	29	68	24	44
	2	75	58	22	36	46	65	19	47	40	66	14	52	82	47	23	24
	3	77	59	21	38	79	56	28	29	61	60	26	34	85	43	21	22
	4	102	52	21	31	86	53	23	30	71	60	21	39	113	56	35	21
	5	119	54	22	33	104	57	24	33	82	60	14	47	116	55	34	21
	6	136	56	19	37	114	56	28	28	90	57	18	38	132	56	35	21
	7	146	50	22	28	124	55	31	25	104	58	17	41	149	52	17	35
	8	178	50	32	19	132	55	24	31	112	56	31	25	166	40	19	21
	9	184	50	33	17	144	53	15	38	121	56	27	29	186	48	27	21
	10	194	50	19	31	152	53	34	19	127	56	29	27	189	47	25	22
	11	203	50	30	20	174	53	29	24	141	56	20	37	217	51	23	27
	12	212	52	22	30	186	53	25	28	147	56	36	21	223	48	22	26

	i	J1, $N_a=3$: $x_{a,i}$ (m)	$y_{a,i}$ (m)	J1, $N_a=6$: $x_{a,i}$ (m)	$y_{a,i}$ (m)	J1, $N_a=9$: $x_{a,i}$ (m)	$y_{a,i}$ (m)	J4, $N_a=3$: $x_{a,i}$ (m)	$y_{a,i}$ (m)
Actuators placement	1	0.379	0.317	0.380	0.272	0.214	0.118	0.377	0.385
	2	0.382	0.383	0.037	0.385	0.249	0.050	0.044	0.390
	3	0.174	0.381	0.313	0.312	0.381	0.369	0.202	0.200
	4			0.035	0.176	0.063	0.256		
	5			0.385	0.032	0.369	0.033		
	6			0.123	0.164	0.135	0.248		
	7					0.380	0.301		
	8					0.186	0.180		
	9					0.205	0.382		
		An overview: 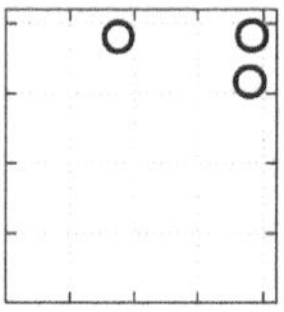		An overview: 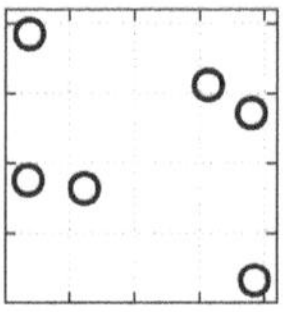		An overview: 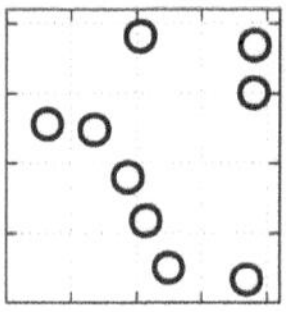		An overview: 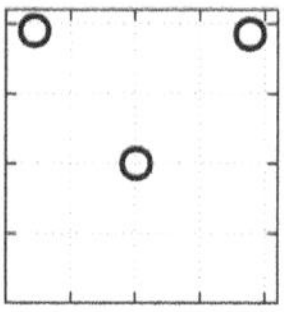	

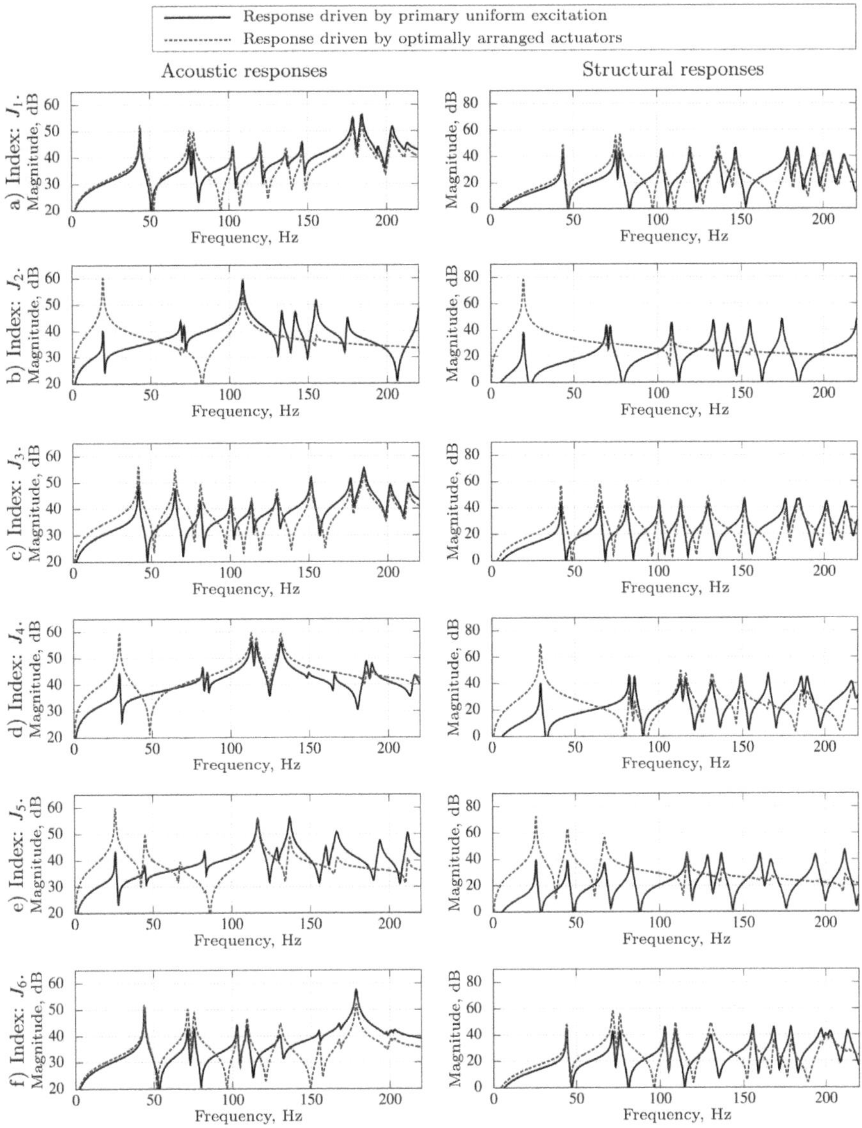

Figure 6.1 Acoustic and structural vibration responses of the plate, obtained for different optimization indices J_1–J_6 as summarized in Table 6.1. Responses are shown for both the primary uniform excitation and when excited using the optimally arranged actuators [173]. Reprinted from *Mechanical Systems and Signal Processing*, 147:107009, S. Wrona, M. Pawelczyk, and J. Cheer, "Acoustic radiation-based optimization of the placement of actuators for active control of noise transmitted through plates", Copyright (2022), with permission from Elsevier.

It follows from an analysis of the results presented in Table 6.1 that the introduction of an acoustic radiation estimate into the cost function J_1, obtaining J_4, enables an increase in the controllability measure $\lambda_{c,i}$ of more than 5 dB for acoustically-relevant modes (where $P_i \geq 30$ dB; in Table 6.1 they are highlighted with a grey background). An increase in the controllability measure $\lambda_{c,i}$ means that the ith mode is more excited with the same control effort (e.g. an increase of $\lambda_{c,i}$ by 5 dB means that the modal velocity of the ith mode is by 5 dB greater with the same control effort). The price that is paid for this increase is a smaller controllability for modes that are less responsible for acoustic radiation or transmission. The minimal controllability measure $\lambda_{c,i}$ for J_1 is 50 dB, while for J_4 it is 40 dB. However, the least controllable mode for the solution obtained with J_4 has $P_i = 19$ dB, which means that its role in acoustic radiation or transmission will be minor compared to the other modes that have a P_i that is more than 15 dB higher.

It is also interesting to highlight that modifying cost function J_1 to give J_4 provides a similar increase in the controllability of the acoustically-relevant modes to that achieved by employing additional actuators, as shown by the results presented in Table 6.2. By optimizing the actuator locations using J_4, a similar controllability can be reached as achieved when the number of actuators are doubled and optimized using J_1. In other words, by using J_4 the number of actuators, N_a, could be reduced, e.g. from 6 to 3 per plate, whilst maintaining a similar level of controllability in terms of the acoustically-relevant modes. In practical noise control applications, this reduction in the required number of actuators offers a significant reduction in the cost and control system complexity, which is a considerable advantage.

Referring again to the results presented in Table 6.1, it can be seen that the results obtained for both J_2 and J_5 are in general inferior to the results obtained with J_1 and J_4. The reason for this is that if a sum is employed in the cost function (J_2 and J_5), it can be beneficial to maximize only one component of the sum and neglect the others. In the considered optimization problem, the controllability of the first mode was maximized, but the remaining modes were neglected and, as a result, these cost functions do not meet the objective.

It is interesting to note from the results presented in Table 6.1 that both J_1 and J_3 result in similar cost function values. The controllability of all considered modes has been maximized using these cost functions. However, introduction of acoustic radiation measure into cost function J_3, which gives J_6, provides unsatisfactory results. It turns out that it is beneficial for J_6 to have a single mode of high acoustic radiation and lower controllability, while maximizing controllability of the other less acoustically-relevant modes. This is, therefore, an unacceptable solution for the considered application.

Acoustic and structural vibration responses of the plate, obtained for the different optimization indices J_1–J_6, are presented in Fig. 6.1. These responses are calculated for the solutions summarized in Table 6.1, hence in all cases the number of actuators $N_a = 3$. Both responses, driven with the primary uniform excitation and by the optimally arranged actuators, are presented. The responses of the plate due to primary uniform excitation are obtained by applying an equal excitation to all structural modes, instead of simulating an external acoustic excitation. These responses correspond to the result of a uniform wide-band external excitation that can be produced by many types of common noise sources. The responses due to excitation by the optimally arranged actuators are obtained by simulating actuator action as forces acting at the optimized actuator locations $(x_{a,i}, y_{a,i})$. A wideband signal again was used as the input to the actuators. The magnitude of the input signals to the actuators was arbitrarily chosen and was the same in all evaluated cases and for all actuators. The larger the response due to the actuators (shown by the dashed line) compared to the response when driven by the primary uniform excitation (shown by the black line), the easier it will be for the control system to reduce the noise transmission or radiation in the considered frequency range.

It follows from analysis of Fig. 6.1 that in the case of the structural responses obtained for J_1 and J_3, the responses due to optimally arranged actuators nearly match for all considered peaks in the responses due to primary uniform excitation. However, in the case of the acoustic responses obtained for J_1 and J_3, the highest peaks in the responses due to primary uniform excitation are higher than the corresponding responses due to the actuators, while the response due to the optimally arranged actuators obtained for J_4 dominates the highest peaks in the acoustic response of the plate due to primary uniform excitation. Still, an enhanced performance in acoustic response obtained for J_4 is traded for a weaker structural response for the modes that are less responsible for acoustic radiation or transmission. These remarks are consistent with the previous conclusions drawn from the analysis of Table 6.1.

To summarize, introduction of the acoustic radiation measure into the cost function in the form of J_4 offers best performance and enables an increase in the controllability measure of more than 5 dB for acoustically-relevant modes. The increase in controllability for these modes is comparable to that achieved by employing additional actuators (at least doubling the number of actuators for the considered system). From a different point of view, this method could also be used to rearrange the actuators in order to try to reduce their number, while maintaining the same level of controllability. Such reduction in practical noise control applications entails a significant reduction in the cost and the control system complexity.

These advantages are traded for a reduction in the controllability of the modes that are less responsible for acoustic radiation or transmission. However, such modes have been shown to have a modal acoustic power that is at least 15 dB lower than that due to the dominant modes, which means that their contribution to the noise transmission and radiation is negligible.

6.2.2 OPTIMIZATION OF ERROR MICROPHONES

The goal of the active noise-cancelling casing is to reduce noise emitted to the environment. In case of feedback systems or adaptive feedforward systems that try to control noise emission directly, the noise emission must be measured or estimated. The noise emission can be estimated by measuring acoustic pressure on a sphere around the casing [119]. In practice the acoustic pressure can be measured only in a finite number of locations at the same time, and the average noise emission, $J(n)$, needs to be approximated by using a numerical quadrature:

$$J(n) \approx \frac{1}{N_M} J_1(n) = \sum_{i=0}^{M_M} a_i m_i^2(n), \tag{6.3}$$

where m_i is the acoustic pressure at ith location, a_i is the weight related to ith location, and $M_M = N_M - 1$ and N_M is the number of positions in which acoustic pressure is measured. Sampling of the sphere introduces a spatial aliasing. The spatial aliasing is unavoidable, but generally it can be reduced by increasing number of samples. The spatial aliasing also depends on locations at which acoustic pressure is sampled. The acoustic pressure can be measured by microphones. Weighting needs to be introduced for cases with nonuniform microphone arrangement. When all microphones have the same measurement uncertainty, the uncertainty of the weighted sum is the lowest when all weights are the same. This can be obtained by using uniform microphone arrangement. When uniform microphone arrangement is used, the approximation can be simplified to:

$$J(n) \approx \frac{1}{N_M} J_2(n) = \sum_{i=0}^{M_M} m_i^2(n). \tag{6.4}$$

Uniform microphone arrangement can be obtained if the microphones are positioned in vertices of Platonic solids. Unfortunately, only 5 Platonic solids exists, with: 4, 6, 8, 12 and 20 vertices. For other number of microphones only approximations exists [59]. The placement, however, can be uniform enough to avoid weighting.

The number of needed microphones is related to spatial aliasing. In the simplest one-dimensional case with two microphones, the unique pressure distribution between microphones can be obtained only when the distance between microphones is smaller than a half of the wavelength. In extreme case, for standing wave with the wavelength equal to distance between microphones, both microphones may measure the same sinusoidally changing acoustic pressure, but the reported amplitude may be from zero, if microphones are located in nodes of a standing wave, up to a maximal amplitude when microphones are located in antinodes of a standing wave. Unfortunately, the acoustic pressure profile can be arbitrary, the Fourier decomposition may contain any frequencies. However, if we assume that the active noise cancelling casing is the only sound source, the maximal possible frequency can be estimated. If we assume that nonlinearities in the casing are not significant, the maximal frequency generated by the casing is the maximal frequency generated by the primary source. Additionally, the casing provides passive reduction of high frequencies. The passive noise reduction of the casing can be included by observing the noise outside of the casing.

When uniform spacing between microphones on a sphere is used, the average area per microphone is equal to:

$$S_m = \frac{4\pi r^2}{N_M}, \tag{6.5}$$

where r is the radius of a sphere, thus, each microphone approximately covers a disc with a radius:

$$r_m \approx \sqrt{\frac{S_m}{\pi}} = \frac{2r}{\sqrt{N_M}}. \tag{6.6}$$

The distance between microphones is approximately equal to $2r_m$. The distance must be shorter than a half of a wavelength, so the number of required microphones is equal to:

$$N_M > 64r\frac{f_{\max}}{c}, \tag{6.7}$$

where $f_{\max}$ is the maximal frequency, and c is the speed of sound. Alternatively, we can estimate the upper frequency limit for a given number of microphones:

$$f_{\max} < \frac{1}{8}\frac{c}{r}\sqrt{N_M}. \tag{6.8}$$

This limit is only a rough approximation. In the simplest case a catastrophic failure, when all microphones are places in nodes of a standing wave, generated by destructive interference controlled by active noise control, may occur at $2f_{\max}$. In such case only a local noise reduction will be achieved, not global.

For non-uniform microphone arrangements the maximal distance between microphones for the same number of microphones must be larger than the distance in uniform arrangement. Thus, the upper frequency limit for non-uniform arrangements is lower than for uniform arrangements. So, uniform arrangement not only avoids weighting in quadrature, and provides the lowest possible uncertainty of emission, but also provides the best possible upper frequency range.

The casing is rarely in free space. However, casing placed on a frame lattice, sometimes found in industry, can be a good approximation of such conditions. In most cases the casing is placed on a sound reflecting floor. In such case microphones can be located only on half sphere. The number of microphones required to cover that space with the same distance between microphones, to cover the same frequency range, is halved. Placing a casing near walls may reduce the number of needed microphones even further. When the casing is placed near a wall only a quarter of microphones is needed. When the casing is placed in the corner only $\frac{1}{8}$ of microphones are needed. Unfortunately, such locations will provide respectively 3 dB, 6 dB and 9 dB amplification of low frequencies, due to reduced number of sound propagation directions [57]. For higher frequencies interference between primary and reflected sound may provide more complex behaviour, for instance destructive interference. The corner may also work as corner reflector [23].

Fig. 6.2 shows exemplary comparison of sound pressure level around a washing machine with active noise/vibration control and two microphone arrangements. The washing machine was placed on a floor near wall. Remaining four walls were actively controlled. The results are shown for spinning at 1200 rpm. To verify global noise reduction the sound pressure level was measured at 17 locations. Two microphone arrangements were tested, octahedral with 0.9 m radius and dodecahedral with 1 m radius. In case of dodecahedral arrangement a larger radius was required to provide

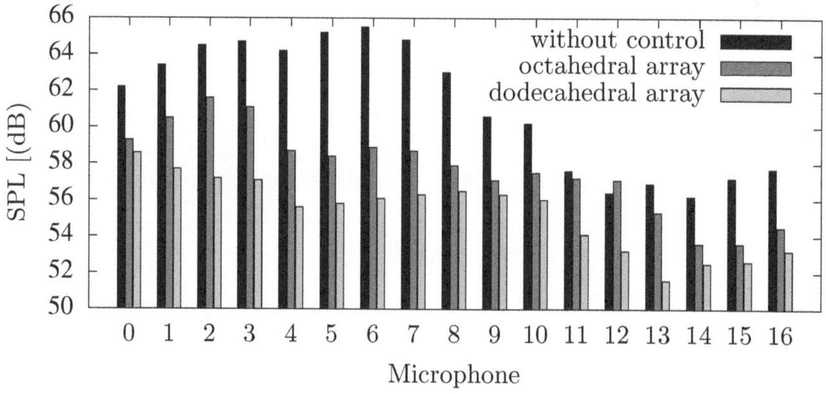

Figure 6.2 Sound pressure level at monitoring microphones.

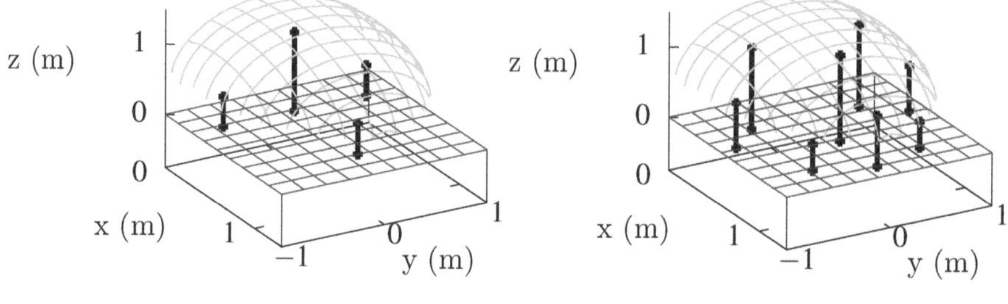

Figure 6.3 Locations of microphones.

larger distance between the casing and microphones. A spherical octahedral arrangement has 6 microphones and should be suitable up to 116 Hz, assuming $c = 343\ \frac{\text{m}}{\text{s}}$. The dodecahedral arrangement should be suitable up to 192 Hz.

The locations of microphones, in 3-dimensional Euclidean space $[x_i, y_i, z_i]^T$, are shown in Table 6.3 and Fig. 6.3. Due to floor and wall, only microphones in a part of a sphere were present: 4 for octahedral arrangement and 8 for dodecahedral arrangement. The floor corresponds to a $z = 0$ plane, the wall corresponds to a $x = 0$ plane. The centre of the casing is located at $[0.31, 0, 0.43]^T$. In case of octahedral arrangement each error microphone was placed 0.6 m in front of each actively controlled wall.

Without control the maximal observed SPL (*Sound Pressure Level*) is equal to 65.5 dB. With octahedral array the maximal sound pressure level drops to 61.6 dB, a 3.9 dB improvement. The SPL distribution is, however, different. At location were

Table 6.3
Locations of microphones.

	Octahedral arrangement			Dodecahedral arrangement		
	x_i (m)	y_i (m)	z_i (m)	x_i (m)	y_i (m)	z_i (m)
M0	1.01	0	0.43	1.04	0.52	0.37
M1	0.31	0.7	0.43	1.04	−0.52	0.37
M2	0.31	−0.7	0.43	0.56	0.84	0.64
M3	0.31	0.0	1.13	0.56	−0.84	0.64
M4	–	–	–	0.37	0.52	1.16
M5	–	–	–	0.37	−0.52	1.16
M6	–	–	–	0.75	0.00	1.22
M7	–	–	–	1.16	0.00	0.72

the SPL was the highest without control, the SPL drops to 58.9 dB, 6.6 dB improvement. There is also location 12, where the SPL increases from 56.4 dB to 57.1 dB, 0.7 dB higher than without control. This behaviour is related to no observable reduction of 225 Hz to 235 Hz band at monitoring microphones. This frequency band is, however, reduced at error microphones. Fig. 6.4 shows the average PSD (*Power Spectral Density*) of signals from monitoring microphones. So, for that frequency the active control system does not reduce that band globally, but only shifts sound emission in a way in which at locations of error microphones noise is reduced. This is, however, expected, because that arrangement should be acceptable only up to 116 Hz. The problem occurs at ab. 230 Hz, twice the maximal frequency. It actually occurs at frequency for which the wavelength is equal to the distance between microphones and such failure was expected.

The dodecahedral arrangement provides better SPL reduction on all monitoring microphones. The maximal SPL is equal to 58.6 dB, a 6.9 dB improvement compared to system without active control and a 3.9 dB improvement over a system with octahedral microphone arrangement. The system with dodecahedral arrangement provides average sound pressure reduction for all important frequencies, both on monitoring microphones (Fig. 6.4) and on error microphones (Figs. 6.5–6.6). In that case the local noise reduction on error microphones provides global noise reduction in a whole room.

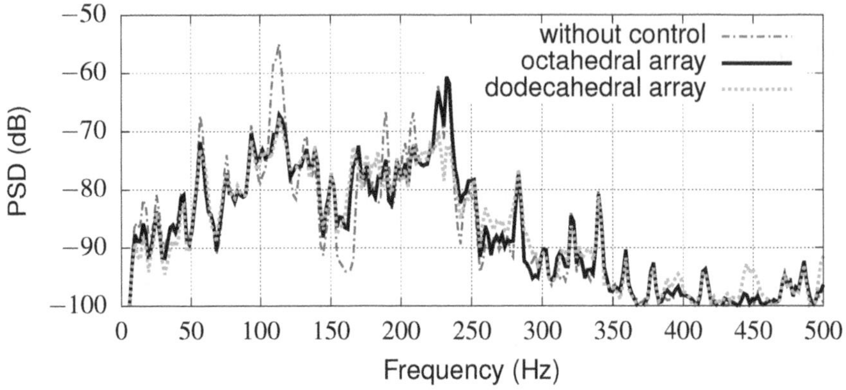

Figure 6.4 Average PSD of signals from monitoring microphones.

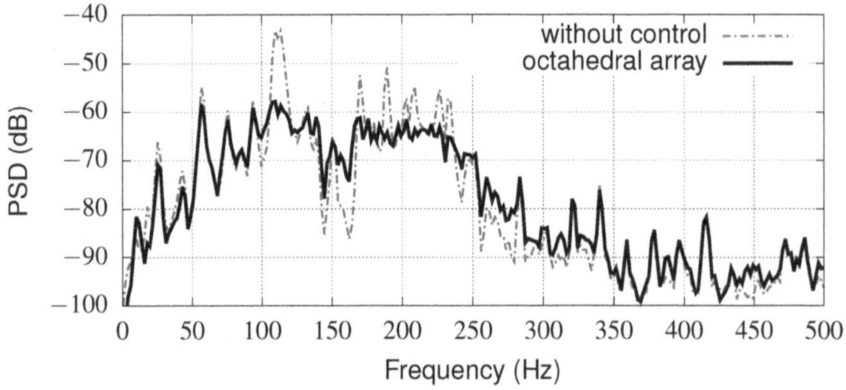

Figure 6.5 Average PSD of signals from error microphones for an octahedral arrangement.

6.2.3 CONTROL UNIT

Modern Active Control systems use digital signal processing. Analogue signals from error sensors are filtered to avoid aliasing, converted to digital signals, processed by microprocessors or programmable logic systems, and digital controller outputs are converted to analogue signals, filtered by reconstruction filters, amplified and send to actuators. Smallest systems may use only one error sensor, one actuator and require computing power provided by even cheapest microcontrollers. Bigger, multichannel Active Control systems needed for active noise cancelling casings, may require tenths of sensors and actuators and very high computing power. The required computing power scales badly with increasing number of channels, the number of secondary paths is equal to the number of actuators multiplied by a number of error signals [118]. The number of control filters is equal to the number of actuators multi-

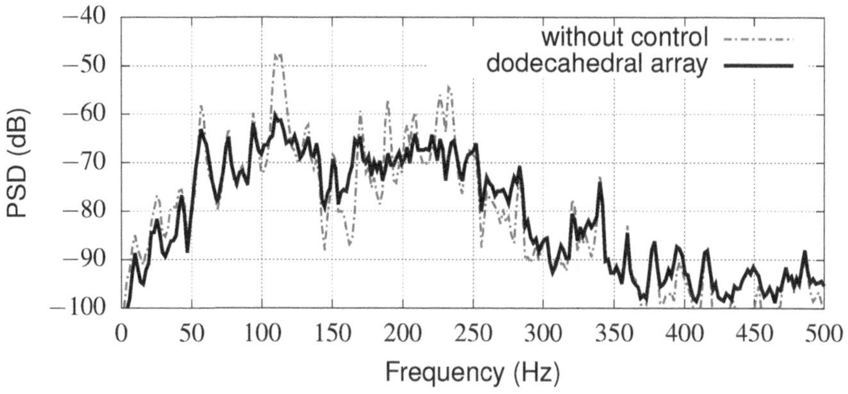

Figure 6.6 Average PSD of signals from error microphones for a dodecahedral arrangement.

plied by the number of reference sensors, and each adaptive control filter, generally, requires corrections from each error. Additionally increase of sampling frequency not only requires a higher number of control algorithm iterations per second, but also the length of FIR (*Finite Impulse Filter*) filters, commonly used as control filters and for secondary path modelling, must be increased, by the same factor. The needed computational power can easily exceed the computational power of a single processor. Fortunately, most Active Control algorithms have a small number of data dependencies and can be executed in parallel on multiple processors, even on distributed systems. Each node of the system may have a small amount of input and output channels and all necessary data is exchanged between nodes. An important advantage of such systems is scalability, the number of nodes can be chosen based on the number of channels of Active Control system. However, in such system the processing power scales linearly with the number of nodes and channels, but in many algorithms the computational requirements scale quadratically or even cubically. For such systems some simplifications of control algorithms, which will reduce dependence on the number of channels to linear, are valuable. Some of them will be presented in Section 6.4.4.

Fig. 6.7 shows the control unit used for all experiments in this chapter. The unit employs 5 dSpace DS1104 R&D boards placed inside a single system, a 19" rack $\times$ 86 server running GNU/Linux. Each DS1104 board has 8 analogue inputs and 8 analogue outputs, the whole control unit provides 40 analogue input and 40 analogue outputs. Each board has a Motorola MPC8240 PowerPC processor with 32 MiB local RAM (*Random Access Memory*). All boards are connected to the host with a PCI (*Peripheral Component Interconnect*) bus. Each board acts as a bus master on a PCI bus, when an access to a PCI bus is granted by the arbiter it may initiate bus transactions [139]. The PCI bus provides up to 133 MB/s bandwidth with low latency. In this setup the guaranteed latency is smaller than 5 μs. This allows for multiple data transfers between boards in each sampling period.

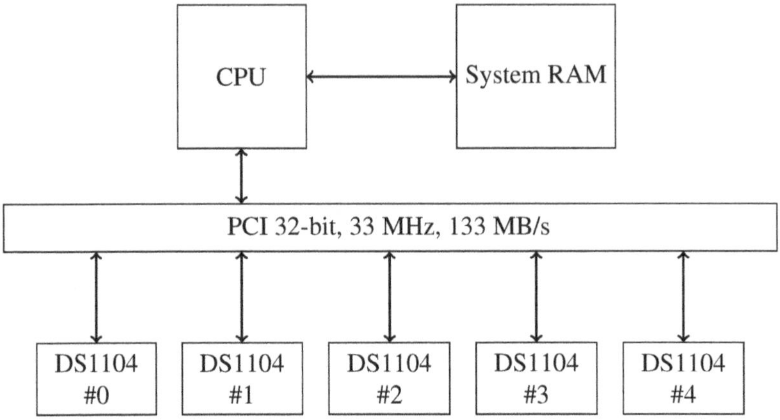

Figure 6.7 Control unit.

The Linux kernel support for DS1104 boards is provided by the `ds1104` driver [101]. The driver provides access to board's local memory from userspace, provides an interface needed to execute code on board, and provides 4 MiB shared memory region located in the System RAM that can be accessed from userspace and from code running on all DS1104 boards. Higher level support for DS1104 boards is provided by `ds1104-utils` [101]. The one of important utilities is the `ds1104-boot` program, which loads the program to DS1104 boards. The programs for DS1104 boards are compiled using a GNU toolchain [155]. A `newlib` [181] is used as a standard C library. The basic DS1104 hardware support is provided by the `ds1104lib` [101] library.

Due to practical constraints on the number of microphones the active noise-cancelling casings can provide global noise reduction to about 300 Hz. The antialiasing and reconstruction filters were designed to provide flat amplitude response up to 400 Hz. To provide wide transition band the sampling frequency was set to 2 kHz. All ADCs and DACs are synchronized using an external trigger signal.

6.3 CONTROL STRUCTURES

6.3.1 FEEDFORWARD CONTROL

The goal of all considered Active Control systems is to reduce noise outside of the casing. Without control the noise is equal to $\mathbf{z}(n) = [z_0(n), z_1(n), \ldots, z_{M_E}(n)]^T$ (Fig. 6.8), where $M_E = N_E - 1$ and N_E is the number of error signals. This noise is called the "primary noise". Active systems try to reduce noise by adding additional sound component, "secondary noise", $\mathbf{y}(n) = [y_0(n), y_1(n), \ldots, y_{M_E}(n)]^T$, which interferes with primary noise, and produces:

$$\mathbf{e}(n) = \mathbf{z}(n) + \mathbf{y}(n), \tag{6.9}$$

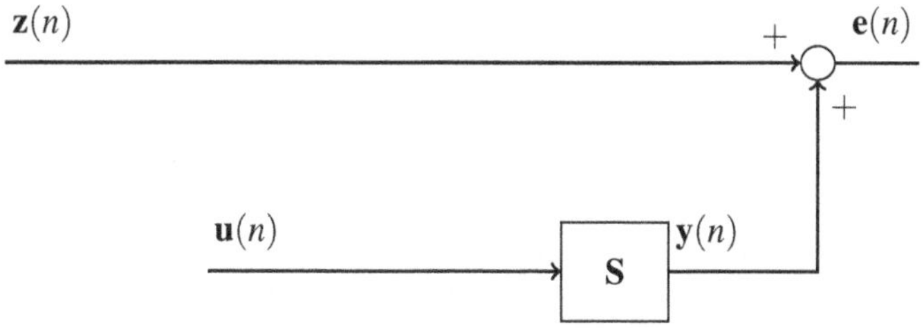

Figure 6.8 Active control system simplified block diagram.

where $\mathbf{e}(n) = [e_0(n), e_1(n), \ldots, e_{M_E}(n)]^T$. The resulting sound is treated as the error signal.

The "secondary noise" is not generated directly at points of interest, but propagates through secondary paths, $\mathbf{S}$:

$$\mathbf{y}(n) = \mathbf{Su}(n), \tag{6.10}$$

where:

$$\mathbf{S} = \begin{bmatrix} S_{0,0} & S_{0,1} & \ldots & S_{0,M_U} \\ S_{1,0} & S_{1,1} & \ldots & S_{1,M_U} \\ \vdots & \vdots & \ddots & \vdots \\ S_{M_E,0} & S_{M_E,1} & \ldots & S_{M_E,M_U} \end{bmatrix}, \tag{6.11}$$

where $S_{c,l}$ are some generic, causal operators related to cth control signal and lth error signal, and $M_U = N_U - 1$ and N_U is the number of control signals. Usually, linear filters are used for modelling secondary paths. In that case secondary paths operators can be expressed as polynomials of one-step delay operator q (definition: $qf(n) = f(n-1)$):

$$S_{c,l} = \frac{\sum_{i=0}^{M_S} s_{c,l,i} q_i}{1 + \sum_{i=1}^{A_S - 1} s_{c,l,-i} q_i}, \tag{6.12}$$

where $M_S = N_S - 1$ and N_S is the number of coefficients of numerator and A_S is the number of coefficients of denominator. To avoid overparametrization, the first coefficient of denominator is always equal to 1.

Acoustic secondary paths are usually very complex due to multiple reflections and for them FIR models ($A_S = 1$):

$$S_{c,l} = \sum_{i=0}^{M_S} s_{c,l,i} q_i, \tag{6.13}$$

require comparable number of coefficients, when compared to IIR (*Infinite Impulse Filter*) models, with $A_S > 1$.

Alternatively, discrete transfer functions $S_{c,l}(z^{-1})$ can be used instead. Both approaches are equivalent, due to relationship between one-step delay operator and z^{-1} operator: $q = \mathscr{Z}^{-1}\{z^{-1}\}, z^{-1} = \mathscr{Z}\{q\}$.

In classical ANC (*Active Noise Control*) systems the secondary paths include acoustic path between secondary sound source and error microphones. This acoustics path introduces large delays, at least few milliseconds in typical systems (ab. 3 milliseconds per meter). This significantly limits the possible noise reduction bandwidth of feedback systems. The acoustics path delay is comparable to periods of spectral components of the noise. For instance, the period of 200 Hz tone is equal to 5 ms and a 1 m acoustic path delay, 3 ms, is higher than a half period leading to instability of a simple control loop with proportional controller (if the gain is large enough). More complex controllers may compensate for this delay. However, it is very difficult to design a stable system with high gain, needed for high noise reduction, over a wide frequency range. Additionally, in real systems the response of acoustic path is more complex due to reflections. This path is also frequently non-stationary because of changes in acoustic environment: moving objects, temperature changes, etc. Due to those difficulties Active Noise Control systems frequently use feedforward control. The acoustic path is also a problem for ANVC (*Active Noise/Vibration Control*) systems, but not for AVC (*Active Vibration Control*) and ASAC (*Active Structural Acoustic Control*) systems. However, in AVC and ASAC systems, the secondary path might be also complex due to multiple resonances of vibrating plates and also non-stationary, for instance due to plate temperature changes [109].

The feedforward approach assumes that the primary noise at points of interest, $\mathbf{z}$, is generated by remote location(s), and propagates through primary paths $\mathbf{P}$ (Fig. 6.9):

$$\mathbf{z}(n) = \mathbf{P}\mathbf{d}(n), \tag{6.14}$$

where:

$$\mathbf{P} = \begin{bmatrix} P_{0,0} & P_{0,1} & \dots & P_{0,M_D} \\ P_{1,0} & P_{1,1} & \dots & P_{1,M_D} \\ \vdots & \vdots & \ddots & \vdots \\ P_{M_E,0} & P_{M_E,1} & \dots & P_{M_E,M_D} \end{bmatrix}, \tag{6.15}$$

where $P_{m,l}$ is the operator associated with mth disturbance and lth error, $M_D = N_D - 1$, N_D is the number of primary sources, and the primary noise can be independently measured by primary noise measurement block $\mathbf{X}$, producing reference signals $\mathbf{x}$:

$$\mathbf{x}(n) = \mathbf{X}\mathbf{d}(n), \tag{6.16}$$

where:

$$\mathbf{X} = \begin{bmatrix} X_{0,0} & X_{0,1} & \dots & X_{0,M_D} \\ X_{1,0} & X_{1,1} & \dots & X_{1,M_D} \\ \vdots & \vdots & \ddots & \vdots \\ X_{M_X,0} & X_{M_X,1} & \dots & X_{M_X,M_D} \end{bmatrix}, \tag{6.17}$$

where $X_{m,k}$ is the operator associated with mth disturbance and kth reference signal, where $M_X = N_X - 1$ and N_X is a number of reference signals.

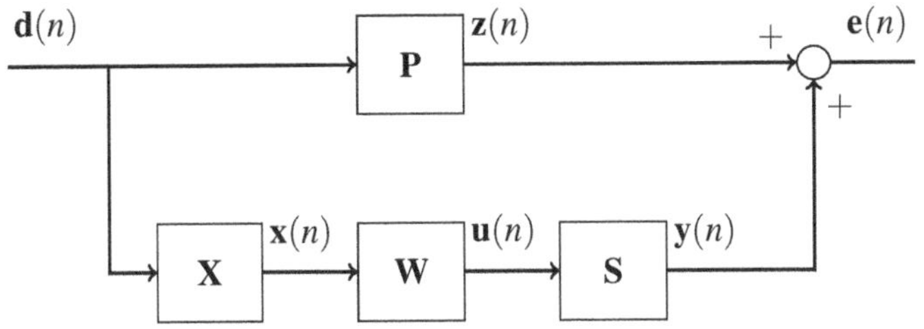

Figure 6.9 Feedforward ANC system block diagram.

The reference signals are processed by control filters $\mathbf{W}$:

$$\mathbf{u}(n) = \mathbf{W}\mathbf{x}(n), \tag{6.18}$$

where:

$$\mathbf{W} = \begin{bmatrix} W_{0,0} & W_{0,1} & \cdots & W_{0,M_X} \\ W_{1,0} & W_{1,1} & \cdots & W_{1,M_X} \\ \vdots & \vdots & \ddots & \vdots \\ W_{M_U,0} & W_{M_U,1} & \cdots & W_{M_U,M_X} \end{bmatrix}, \tag{6.19}$$

where $W_{c,l}$ is a path between cth control signal and lth error.

Because the feedforward system has no feedback, if the all subsystems are stable, the whole active control system is stable. In practice $\mathbf{P}$, $\mathbf{X}$ and $\mathbf{S}$ subsystems are always stable and potential instability is in the control filters $\mathbf{W}$. Because the stability of control filters does not depend on other subsystems it can be easily verified. In many applications the a finite impulse response filters, which are inherently stable, are used as control filters. This makes the whole system unconditionally stable.

When FIR filters are used the cth output is equal to:

$$u_c(n) = \sum_{k=0}^{M_X}\sum_{l=0}^{M_W} w_{k,c,l}q^l x_k(n) = \sum_{k=0}^{M_X}\sum_{l=0}^{M_W} w_{k,c,l}x_k(n-l), \tag{6.20}$$

where $M_W = N_W - 1$ and N_W is a number of coefficients. Because each output depends only on filter weights and reference signals, this step can be efficiently implemented on multiple processor systems or even distributed systems. In direct approach, where all calculations related to cth control signal are performed on a selected node of distributed system, each node needs only a full set of reference signals, with M_W previous samples, and control filter coefficients for channels handled by that node. The reference signals can be connected separately to each node, or connected only to one node and then sent digitally to other nodes. In this approach each node must have a local copy of reference signals regressors array. Alternatively, if data exchange is possible between nodes, to avoid duplication of reference signals regressor array, this array can be distributed to multiple nodes and each node will generate partial sums for each control output. Then, those partial sums can be aggregated.

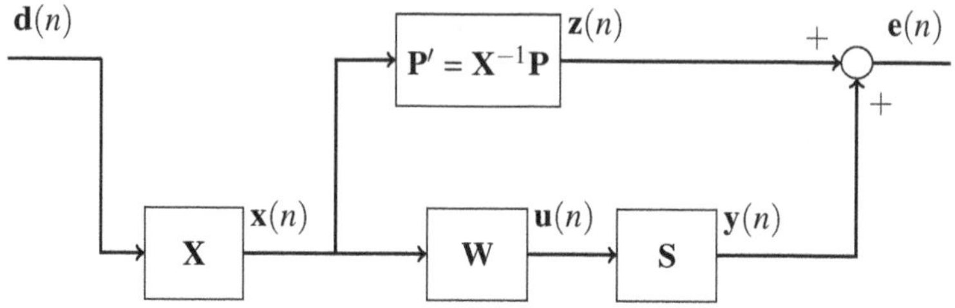

Figure 6.10 Alternative feedforward ANC system block diagram.

If

$$\mathbf{XWS} = \mathbf{P}, \tag{6.21}$$

the primary noise is completely eliminated at points of interest. However, to satisfy this condition the sum of delays in the reference paths, control filters and secondary paths must be smaller than delays in primary paths.

Because the $\mathbf{d}(n)$ signal cannot be measured, and thus, the primary path model $\mathbf{P}$ cannot be identified, sometimes the reference signal is used as the input for "primary path" (Fig. 6.10). However, in this approach the "primary path" model $\mathbf{P}'$ can be noncausal. Because now everything depends on the reference signal $\mathbf{x}$, $\mathbf{d}(n)$ and $\mathbf{X}$ can be removed. With this approach the Eq. 6.21 can be rewritten as:

$$\mathbf{WS} = \mathbf{X}^{-1}\mathbf{P} = \mathbf{P}'. \tag{6.22}$$

Now both secondary path $\mathbf{S}$ and new "primary path" $\mathbf{P}'$ models can be identified. In practice, due to causality issues, this equation cannot be satisfied and only an approximate solution can be used.

Wiener filter

For single-channel linear systems the optimal solution can be obtained using a Wiener filter. The Wiener approach requires that the control filter output is added to primary noise directly. However, if $\mathbf{W} = [W_{1,1}]$ and $\mathbf{S} = [S_{1,1}]$ operators commute, the order of operators can be virtually swapped for control filter optimization (Fig. 6.11). Because such system does not exist in reality, the secondary path $\mathbf{S}$ was replaced by its model $\hat{\mathbf{S}}$. Operators in a form of polynomials of q (or z^{-1}), due to cumulative property of multiplication of polynomials, commute with each other: $A(q)B(q)f(n) = B(q)A(q)f(n)$.

Now, the optimal control filter weight can be obtained by solving Wiener-Hopf equations [159]:

$$\mathbf{R_a w} = \mathbf{R_{az}}, \tag{6.23}$$

where $\mathbf{w} = [w_0, w_1, \ldots, w_{M_W}]^T$ is a vector of control filter coefficients, $\mathbf{R_{az}} = [R_{az}(0), R_{az}(1), \ldots, R_{az}(M_W)]^T$ is a vector with cross-correlation between $\mathbf{r}$ and $\mathbf{z}$,

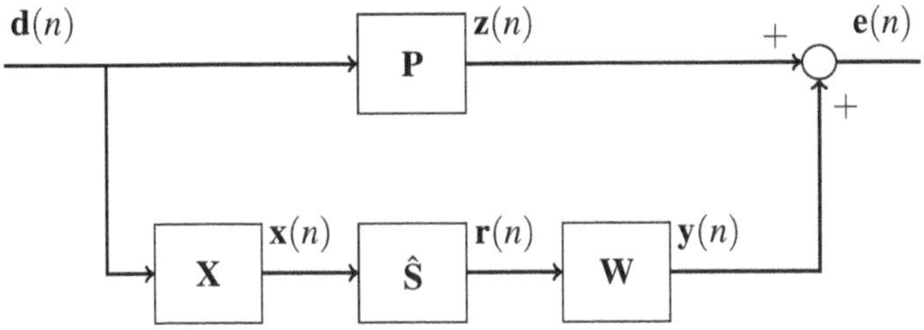

Figure 6.11 Modified Feedforward ANC system block diagram for Wiener solution.

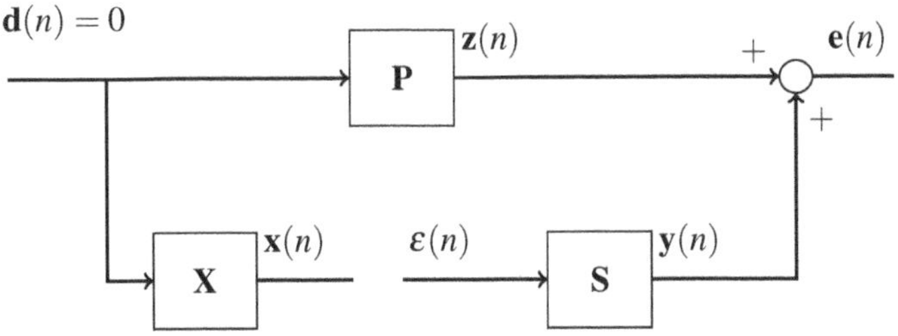

Figure 6.12 Feedforward ANC system block diagram used for secondary path identification.

and:

$$\mathbf{R_a} = \begin{bmatrix} R_a(0) & R_a(1) & \dots & R_a(M_W) \\ R_a(1) & R_a(2) & \dots & R_a(M_W-1) \\ \vdots & \vdots & \ddots & \vdots \\ R_a(M_W) & R_a(M_W-1) & \dots & R_a(0) \end{bmatrix} \tag{6.24}$$

is a **r** signal autocorrelation matrix. The required **z** and **r** signals, needed for cross-correlation and autocorrelation can be measured on real Active Control system, usually in two phases. In the first phase secondary path model $\hat{\mathbf{S}}$ must be identified (Fig. 6.12). The excitation signal $\varepsilon(n)$ is applied to secondary path. The secondary path model, $\hat{\mathbf{S}} = \sum_i^{M_W} \hat{s}_i q^i$, coefficients can be obtained using correlation:

$$\hat{\mathbf{s}} = \left[\hat{s}_0, \hat{s}_1, \dots, \hat{s}_{M_W}\right]^T = (R_\varepsilon)^{-1} R_{\varepsilon e}. \tag{6.25}$$

If possible, for easier estimation the primary noise source can be turned off in that phase. The same method for secondary path identification can be used for adaptive algorithms.

In the second phase (Fig. 6.13) the required signals can be measured with enabled primary noise source and disabled active control. In that case the primary noise **z**

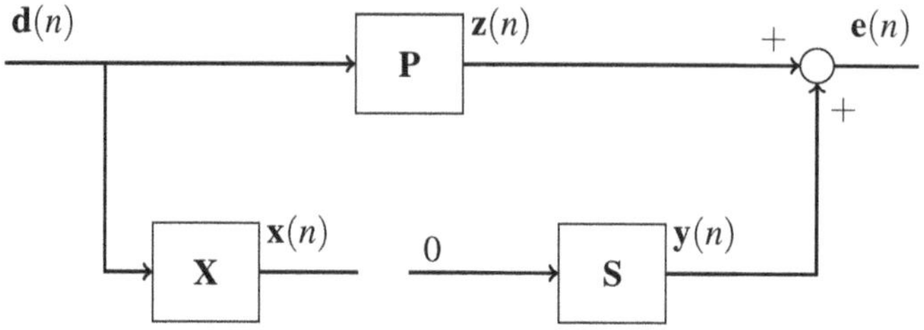

Figure 6.13 Feedforward ANC system block diagram used for primary path identification.

can be measured directly using error sensor, due to $\mathbf{z}(n) = \mathbf{e}(n) + S0 = \mathbf{e}(n)$. The second needed signal is reference signal **x**. The Wiener solution, however, requires the reference signal filtered by virtual secondary path: $\mathbf{r}(n) = \mathbf{S}\mathbf{x}(n)$. The secondary path model was estimated in the first phase.

6.3.2 ACOUSTIC FEEDBACK

In practice, both primary and secondary source have effect on the reference signal and an additional "acoustic feedback path" **F** must be added to the model (Fig. 6.14)

$$\mathbf{x} = \mathbf{X}\mathbf{d} + \mathbf{F}\mathbf{u}, \tag{6.26}$$

where:

$$\mathbf{F} = \begin{bmatrix} F_{0,0} & F_{0,1} & \cdots & F_{0,M_X} \\ F_{1,0} & F_{1,1} & \cdots & F_{1,M_X} \\ \vdots & \vdots & \ddots & \vdots \\ F_{M_U,0} & F_{M_U,1} & \cdots & F_{M_U,M_X} \end{bmatrix}, \tag{6.27}$$

where $F_{c,k}$ represents a path between cth control signal and kth reference signal.

This additional feedback changes the effective control filter, $\mathbf{W}' = (1 - \mathbf{W}\mathbf{F})^{-1}\mathbf{W}$, and adds a positive feedback loop, which can destabilize the system. The effect of this feedback loop must be taken into account when designing the control filter or alternatively, the effect on the feedback path can be removed, by adding additional feedback compensation $\hat{\mathbf{F}}$ term: (Fig. 6.15):

$$\mathbf{x} = \mathbf{X}\mathbf{d} + \mathbf{F}\mathbf{u} - \hat{\mathbf{F}}\mathbf{u}, \tag{6.28}$$

where:

$$\hat{\mathbf{F}} = \begin{bmatrix} \hat{F}_{0,0} & \hat{F}_{0,1} & \cdots & \hat{F}_{0,M_X} \\ \hat{F}_{1,0} & \hat{F}_{1,1} & \cdots & \hat{F}_{1,M_X} \\ \vdots & \vdots & \ddots & \vdots \\ \hat{F}_{M_U,0} & \hat{F}_{M_U,1} & \cdots & \hat{F}_{M_U,M_X} \end{bmatrix}, \tag{6.29}$$

where $\hat{F}_{c,k}$ is a model of $F_{c,k}$ path.

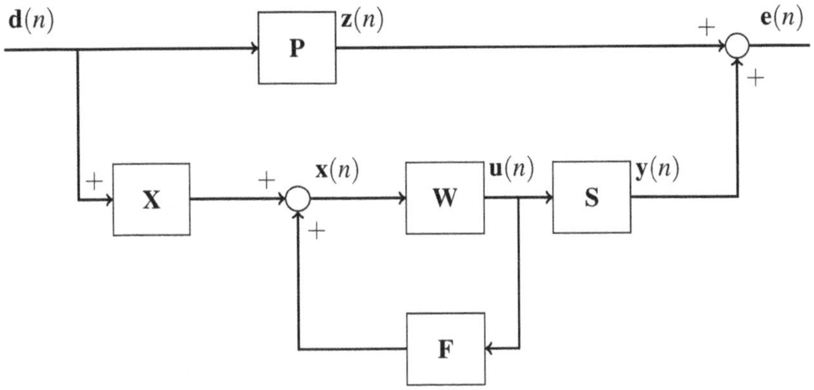

Figure 6.14 Feedforward ANC system block diagram with additional acoustic feedback path.

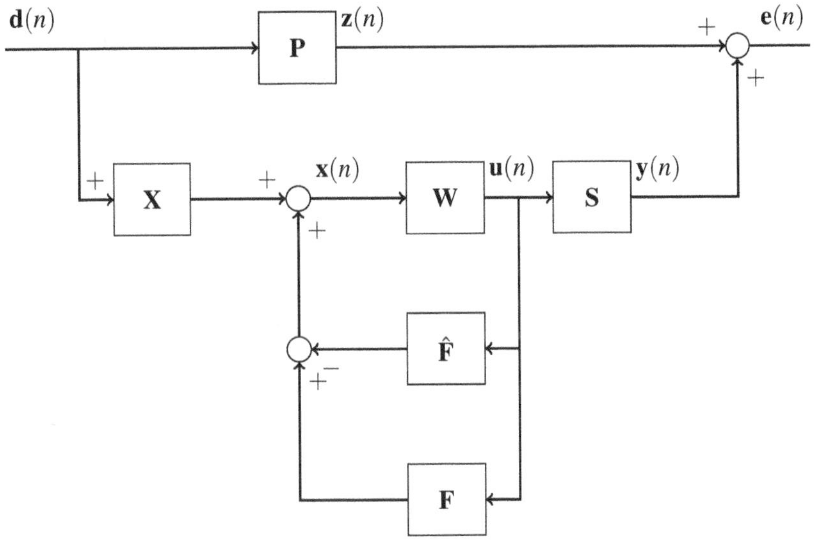

Figure 6.15 Feedforward ANC system block diagram with additional acoustic feedback path and its compensation.

In case of active noise-cancelling casings with the reference microphone located in the casing, the acoustic feedback is usually significant and must be eliminated. Because the acoustic feedback can be easily compensated in later parts of this chapter it will be assumed that the acoustic feedback does not exist or has been eliminated.

6.3.3 FEEDBACK CONTROL

Due to high delays in secondary paths feedback control in Active Noise Control systems is more difficult and less popular. However, there are applications for which the secondary path acoustic delay is small, for instance headphones or earplugs with active noise control. In case of Active Vibration Control acoustic path does not exist, the sound propagates through structures, with higher velocity and also in many applications actuators and sensors are collocated. When actuators and sensors are collocated the secondary path is simple and it is possible to use very simple feedback controllers, even proportional controller. When actuators and sensors are not collocated the secondary path is more complex and includes some delay.

A feedback control with very large delays in the secondary path is difficult, but possible. The first commonly known solution is to use Smith predictor [154] (Fig. 6.16). An explicit delay D is extracted from secondary path $\mathbf{S}$ so the secondary path becomes $\mathbf{S}q^D$, and then $\hat{\mathbf{S}} - \hat{\mathbf{S}}q^D$, where $\hat{\mathbf{S}}q^D$ is the secondary path model, is added to the error signal and feed back to controller $\mathbf{W}$. It is assumed that the controller will provide necessary negative gain. The regulator sees $\mathbf{S}q + \hat{\mathbf{S}} - \hat{\mathbf{S}}q^D$ object, and if the model of the secondary path is perfect the controller sees $\hat{\mathbf{S}}$, the secondary path without delay.

The error is equal to:

$$\mathbf{e}(n) = \frac{1 - \hat{\mathbf{S}}(1 - q^D)\mathbf{W}}{1 - \hat{\mathbf{S}}(1 - q^D)\mathbf{W} - \mathbf{S}q^D\mathbf{W}}\mathbf{z}(n). \tag{6.30}$$

With perfect model $\hat{\mathbf{S}} = \mathbf{S}$, this equation can be simplified to:

$$\mathbf{e}(n) = \frac{1 - \hat{\mathbf{S}}(1 - q^D)\mathbf{W}}{1 - \hat{\mathbf{S}}\mathbf{W}}\mathbf{z}(n) \tag{6.31}$$

Another possibility is to use IMC (*Internal Model Control*) (Fig. 6.17).

The error is equal to:

$$\mathbf{e}(n) = \frac{1 - \hat{\mathbf{S}}\mathbf{W}}{1 - \mathbf{S}\mathbf{W} + \hat{\mathbf{S}}\mathbf{W}}\mathbf{z}(n). \tag{6.32}$$

With perfect model $\hat{\mathbf{S}} = \mathbf{S}$, this equation can be simplified to:

$$\mathbf{e}(n) = \left(1 - \hat{\mathbf{S}}\mathbf{W}\right)\mathbf{z}(n). \tag{6.33}$$

The IMC system is equivalent to a feedback system with acoustic feedback path compensation and error sensors used as reference sensors. In the IMC system the

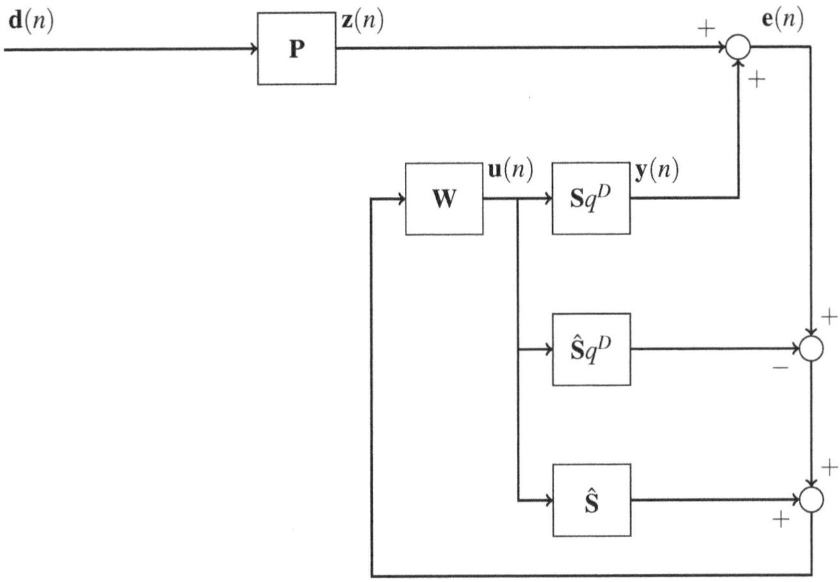

Figure 6.16 ANC system with Smith predictor.

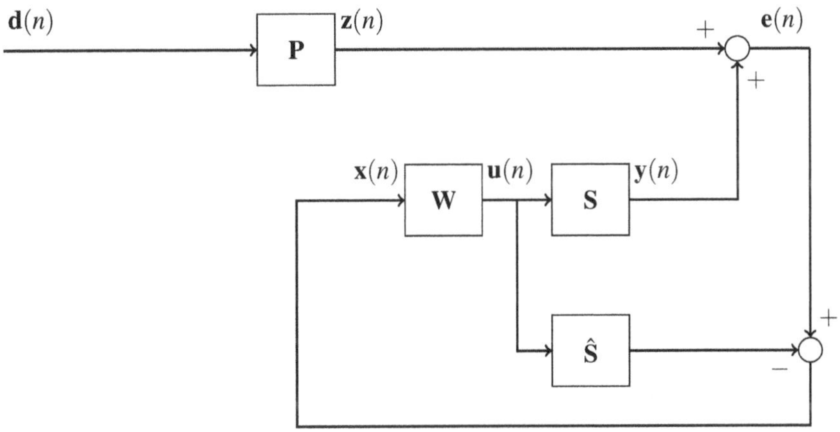

Figure 6.17 ANC system with IMC structure.

Table 6.4
Error signal levels for a different disturbances for different control strategies.

	Front (dB)	Right (dB)	Left (dB)	Top (dB)
113 Hz tonal disturbance				
without control	78.0	84.9	85.0	71.6
feed-forward	55.7	63.2	55.4	53.0
IMC	53.9	62.5	55.0	53.6
Simulated 1200 rpm spinning				
without control	73.2	79.1	79.7	72.9
feed-forward	63.8	69.8	67.0	65.1
IMC	66.4	71.9	72.1	68.5

reference signal is equal to $\mathbf{x}(n) = \mathbf{Pd}(n) + \mathbf{Su}(n) - \hat{\mathbf{S}}\mathbf{u}(n)$, while in a feedforward system with acoustic feedback path compensation the reference signal is equal to $\mathbf{x}(n) = \mathbf{Xd}(n) + \mathbf{Fu}(n) - \hat{\mathbf{F}}\mathbf{u}(n)$. If $\mathbf{X} = \mathbf{P}$, $\mathbf{F} = \mathbf{S}$ and $\hat{\mathbf{F}} = \hat{\mathbf{F}}$ both equations are exactly the same.

The IMC system, when compared to feed-forward control, in case of active casings usually provides better attenuation for simple, deterministic disturbances, like tonal noise [114]. The noise reduction results for a washing machine are shown in Table 6.4. The IMC system provides slightly better noise reduction for a 113 Hz tonal noise, but provides worse results for simulated 1200 rpm spinning noise, from a loudspeaker inside the washing machine. The spectrum of both signals on all microphones is shown in Figs. 6.18 and 6.19. The increased noise reduction for tonal disturbances is related to small nonlinearities that are usually present in the system. Linear feed-forward systems are not able to cope with harmonic frequencies caused by them and when the primary tone is significantly reduced, harmonic frequencies may dominate. Nonlinear feed-forward systems have no such problems [110, 112], but they have higher computational requirements. In vibroacoustic systems that use multiple actuators to cover wider frequency range it is possible to combine multiple channels into one, and use only one nonlinear filter [111].

6.3.4 VIRTUAL MICROPHONE CONTROL

Adaptive feedforward systems and feedback systems provide reduction of error signals. When the goal of the system is to reduce noise emission, the noise emission

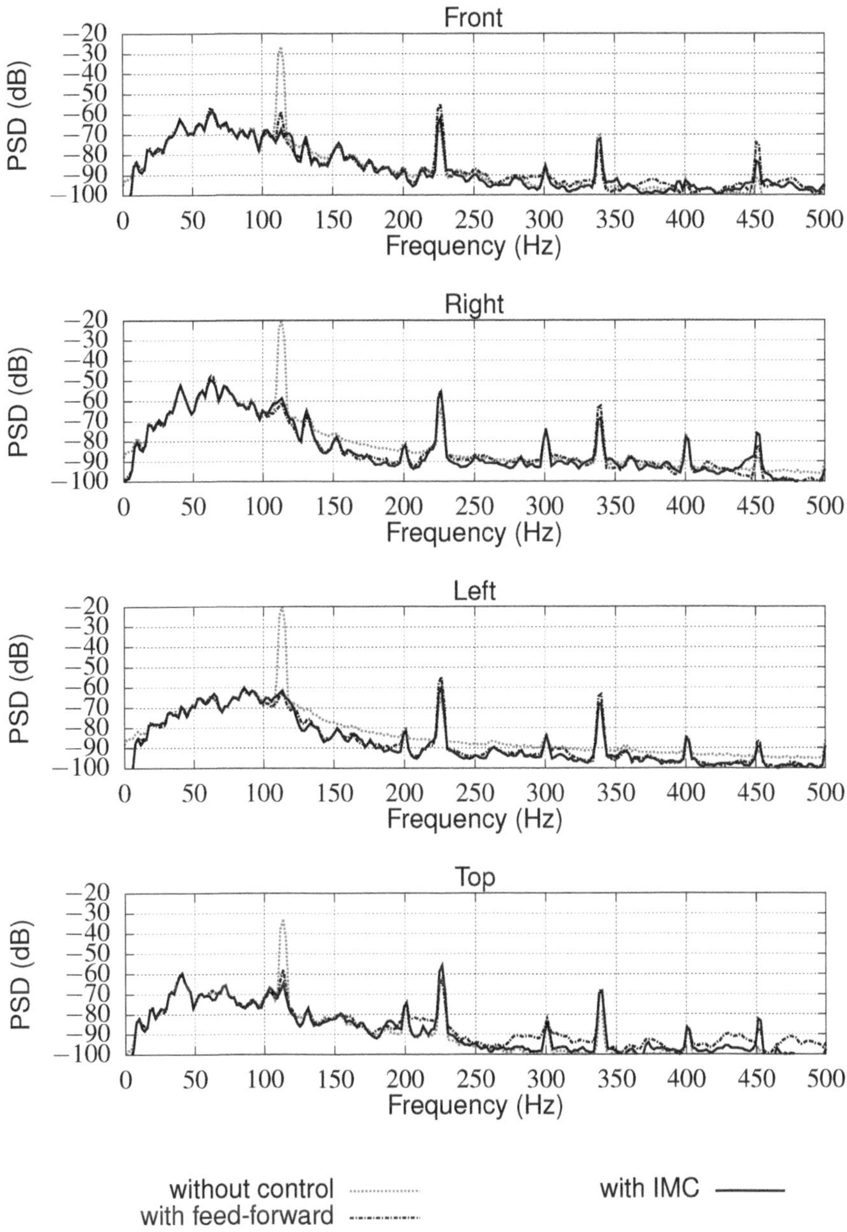

Figure 6.18 Error signal spectrums for a 113 Hz tone (washing machine).

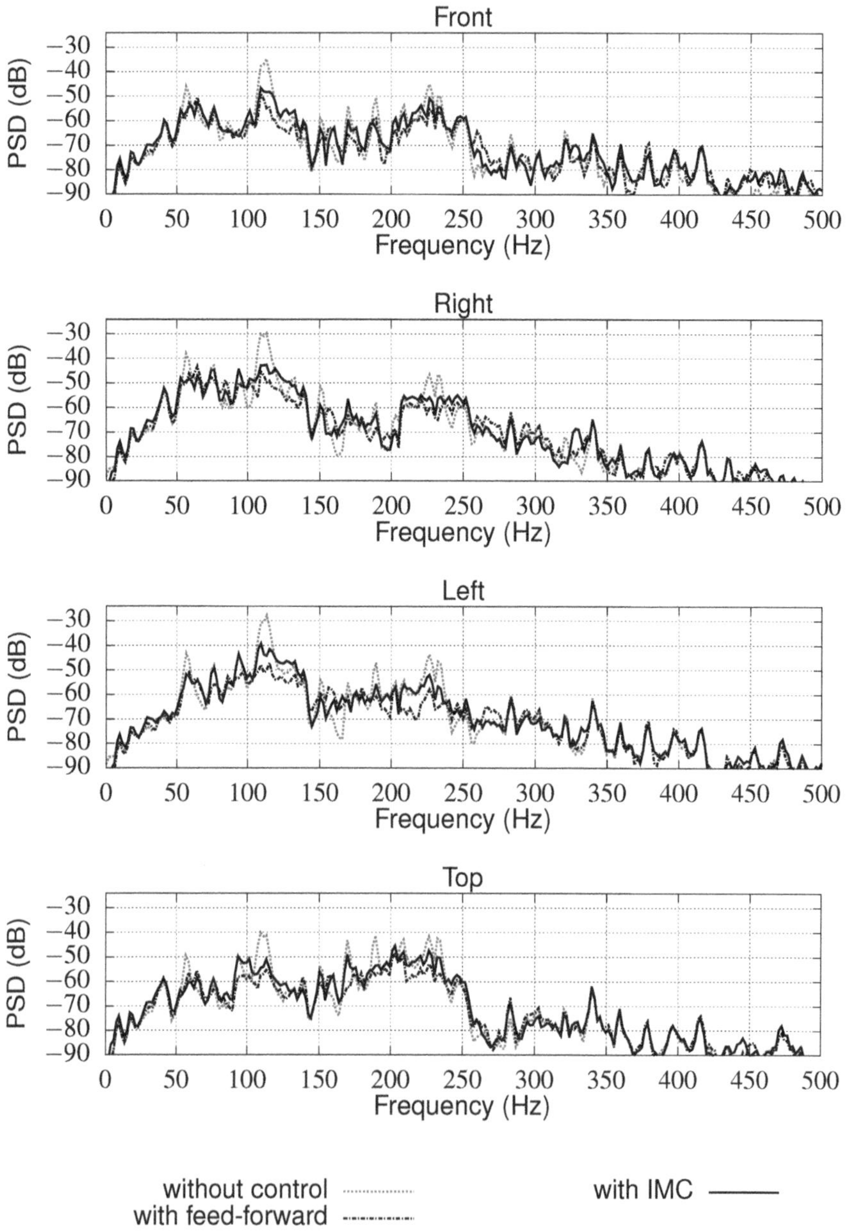

Figure 6.19 Error signal spectrums for a reproduced spinning noise at 1200 rpm (washing machine).

measured by error microphones or its estimate, must be available for controller. The most accurate solution is to measure noise emission directly by microphones. However, such approach requires potentially large number of microphones at specific locations. In many cases it is not acceptable. Two alternatives exits: do not use error sensors or use a different type of error sensors, for instance vibration sensors mounted on the casing [115, 116].

Without error sensors only a limited adaptation can be used. The simplest solution is to not use adaptation at all. Such solution have two important drawbacks: the system is not able to adopt to any changes in primary, secondary or reference paths and the system is not able to adopt to primary disturbance changes. The first problem could be potentially addressed by some forms of gain scheduling based on some environmental parameters like temperature, which can significantly change plate response [109]. The second problem can be addressed by measurement of the noise spectrum by reference sensor and redesign of a control filter. Such redesign requires the models of the primary path and secondary paths. The knowledge about the models can be indirect, for instance the control filter can be chosen from a set of previously designed filters that are not specific to exact disturbance spectrum, e.g., using generalized disturbance [84].

The direct usage of other type of error sensors may not lead to effective emitted noise reduction. For example, when vibration sensors are used the ANVC system actually becomes AVC system, which provides good vibration reduction, but does not result in effective reduction of noise emission [47, 143, 100]. One solution is to estimate noise emission by using theoretical models. Such leads to ASAC. Precise modelling of real world setup is, however, difficult and in many cases virtually impossible. A lot of physical parameters needed for such modelling is usually not known and require estimation based on real world data. In practice it is simpler to identify paths from known signals to physical error microphones. After identification such microphones can be removed and identified models can be used to provide noise reduction close to reduction with real microphones. This idea is known as VMC (*Virtual Microphone Control*) [134]. There are many possible VMC approaches.

Fig. 6.20 shows the block diagram of a ANC system with two sets of error sensors: $\mathbf{e}_1(n)$ and $\mathbf{e}_2(n)$. To analyse possible adaptation to disturbance changes it is assumed that $\mathbf{d}(n) = \mathbf{D}\varepsilon(n)$, where $\varepsilon(n)$ is an some signal. The operation for deterministic disturbances can be analysed by assuming: $\varepsilon(n) = \delta(n)$, where $\delta(n)$ is Kronecker delta. The operation for stochastic disturbances can be analysed by assuming that $\varepsilon(n)$ s a white noise.

It is assumed that both secondary paths share some common path, $\mathbf{S}_c$. In practice all elements of secondary paths up to walls (e.g. DACs, reconstruction filters, power amplifiers, actuators, walls of the casing) are shared between those two paths. Additionally, also primary paths share some common path, $\mathbf{P}_c$, everything from primary noise source up to walls of the casing. When the active control system works using $\mathbf{e}_1(n)$ error sensors it provides some residual error:

$$\mathbf{e}_1(n) = (\mathbf{P}_1\mathbf{P}_c + \mathbf{S}_1\mathbf{S}_c\mathbf{W}\mathbf{X})\,\mathbf{d}(n). \tag{6.34}$$

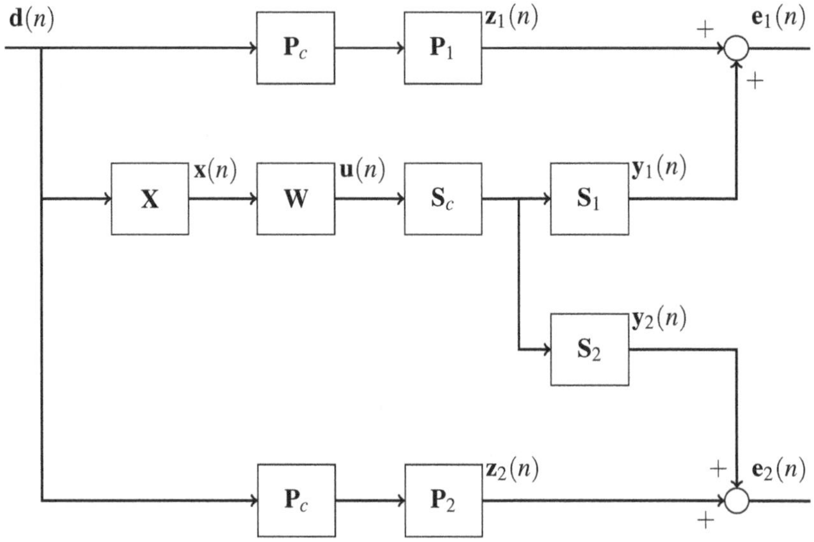

Figure 6.20 ANC system with two sets of error sensors: for real and virtual control.

The signal on the secondary sensors is equal to:

$$\mathbf{e}_2(n) = (\mathbf{P}_2\mathbf{P}_c + \mathbf{S}_2\mathbf{S}_c\mathbf{W}\mathbf{X})\,\mathbf{d}(n). \tag{6.35}$$

One possibility to create a VMC system is to ensure that the $\mathbf{e}_2(n)$ signal is the same as if a system with $\mathbf{e}_1(n)$ error sensors is running [133]. It can be obtained by identification of the path, $\mathbf{H}$, from $\mathbf{x}(n)$ to $\mathbf{e}_2(n)$ and using it as the reference model, $\hat{\mathbf{H}}$, for control. The secondary sensors outputs are equal to:

$$\mathbf{e}_2(n) = \mathbf{H}\mathbf{X}\mathbf{d}(n). \tag{6.36}$$

The $\mathbf{H}$ can be obtained by solving following equation:

$$\mathbf{H}\mathbf{X} = \mathbf{P}_1\mathbf{P}_c + \mathbf{S}_1\mathbf{S}_c\mathbf{W}\mathbf{X}. \tag{6.37}$$

If the delay in the combined $\mathbf{P}_1\mathbf{P}_c$ is smaller than the delay in $\mathbf{X}$ the $\mathbf{H}$ is non-causal. Because a causal model must be used in control system only an approximate model can be used in that case. Ideally, $\mathbf{d}(n)$ should be used directly, but that signal is unavailable for the control system, and $\mathbf{x}(n)$ must be used instead.

The goal of new control system is to minimize $\mathbf{e}_3(n) = \mathbf{e}_2(n) - \hat{\mathbf{H}}\mathbf{x}$, where $\hat{\mathbf{H}}$ is an identified model of $\mathbf{H}$ path. So, effectively the VMC system minimizes (Fig. 6.21):

$$\mathbf{e}_3(n) = \left(\mathbf{P}_2\mathbf{P}_c + \left(\mathbf{S}_2\mathbf{S}_c\mathbf{W} - \hat{\mathbf{H}}\right)\mathbf{X}\right)\mathbf{d}(n). \tag{6.38}$$

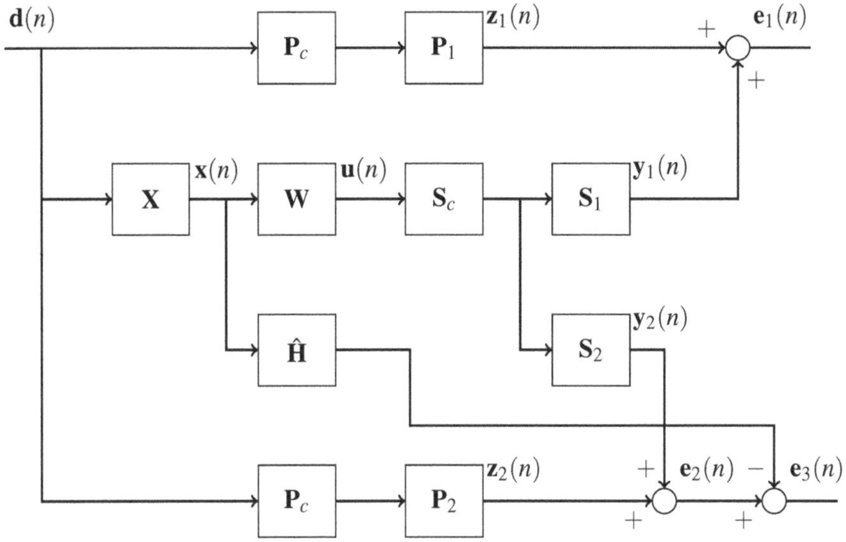

Figure 6.21 ANC system with VMC, with appropriate $\hat{\mathbf{H}}$ reduction of $\mathbf{e}_3(n)$ error signals reduces $\mathbf{e}_1(n)$ error signals.

It can be shown that this VMC system adopts, minimization of $\mathbf{e}_3(n)$ provides optimal reduction of $\mathbf{e}_1(n)$, to any changes of $\mathbf{S}_c$. To simplify equations it is assumed that the ANC system operates perfectly, $\mathbf{e}_1(n) = 0$ and also $\mathbf{e}_3(n) = 0$. It requires ideal model of $\hat{\mathbf{H}}$ (the $\mathbf{H}$ must be causal). In that case equations 6.34 and 6.38 can be rewritten as:

$$\begin{cases} \mathbf{e}_1(n) = 0 = (\mathbf{P}_1\mathbf{P}_c + \mathbf{S}_1\mathbf{S}_c\mathbf{W}\mathbf{X})\,\mathbf{d}(n) \\ \mathbf{e}_3(n) = 0 = \left(\mathbf{P}_2\mathbf{P}_c + \left(\mathbf{S}_2\mathbf{S}_c\mathbf{W} - \hat{\mathbf{H}}\right)\mathbf{X}\right)\mathbf{d}(n), \end{cases} \tag{6.39}$$

When the common secondary path $\mathbf{S}_c$ changes to $\mathbf{S}'_c$, the active control system will adopt control filter to $\mathbf{W}_s$ and provide:

$$\mathbf{e}'_3(n) = 0 = \left(\mathbf{P}_2\mathbf{P}_c + \left(\mathbf{S}_2\mathbf{S}'_c\mathbf{W}_s - \hat{\mathbf{H}}\right)\mathbf{X}\right)\mathbf{d}(n), \tag{6.40}$$

By combining equations 6.39 and 6.40, $\mathbf{S}'_c\mathbf{W}_s = \mathbf{S}_c\mathbf{W}$. Thus,

$$\mathbf{e}'_1(n) = \left(\mathbf{P}_1\mathbf{P}_c + \mathbf{S}_1\mathbf{S}'_c\mathbf{W}_s\mathbf{X}'\right)\mathbf{d}(n) = \left(\mathbf{P}_1\mathbf{P}_c + \mathbf{S}_1\mathbf{S}_c\mathbf{W}\mathbf{X}'\right)\mathbf{d}(n) = 0. \tag{6.41}$$

So with the VMC system the new control filter $\mathbf{W}_s$ provides also optimal control for $\mathbf{e}_1(n)$.

Fig. 6.22 and Table 6.5 show the performance of the VMC system applied to light-weight active noise-cancelling casing. Five walls of the casing were actively controlled (all except Bottom). Five error microphones were placed around the casing. 21 actuators and 21 accelerometers were mounted on the casing. Octahedral

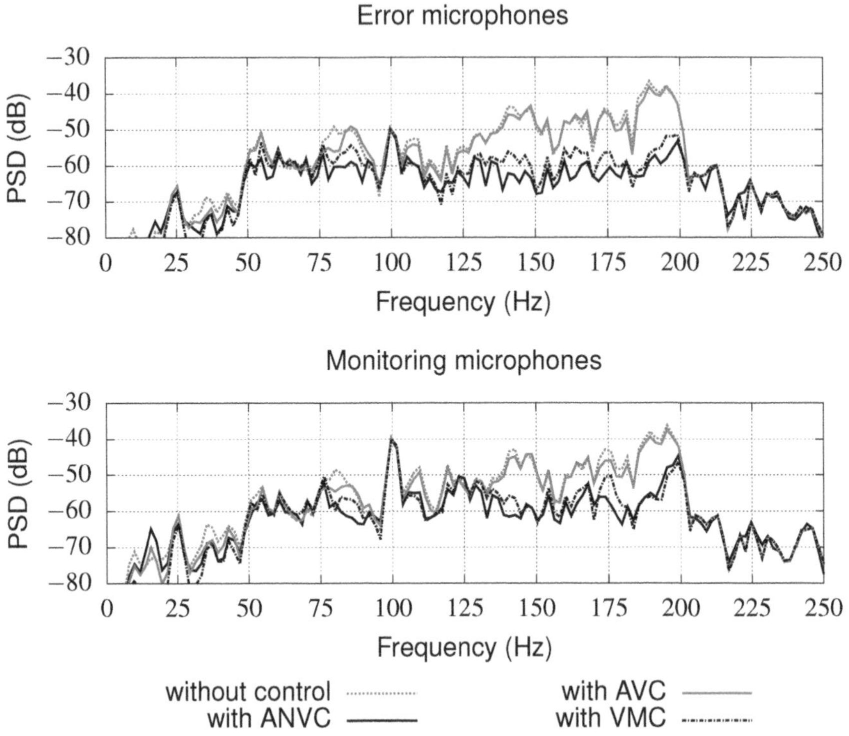

Figure 6.22 Average signal spectrums at error and monitoring microphones.

microphone arrangement were used. Three monitoring microphones were placed in the laboratory room. The multitonal noise signal, composed of sinusoids from 50 Hz to 200 Hz, with a 2 Hz step, was generated by a loudspeaker placed inside the casing.

The AVC system, except for some frequency bands, does not provide noise reduction. The ANVC system, with error microphones, provides good noise reduction for most frequencies. The VMC system, for which $\hat{\mathbf{H}}$ was identified during the operation of the ANVC system provides only a slightly worse noise reduction compared to the ANVC system and significantly better noise reduction than the AVC system.

6.4 ADAPTIVE CONTROL ALGORITHMS

6.4.1 THE LEAST MEAN SQUARE ALGORITHM

The majority of active noise and vibration systems use one of the LMS-family algorithms; therefore, it is beneficial to have some knowledge about this adaptation method. The algorithm was developed by Bernard Widrow and his doctoral student

Table 6.5
The signal levels for a multitonal noise for different control strategies (F–Front, R–Right, B–Back, L–Left, T–Top).

	F (dB)	R (dB)	B (dB)	L (dB)	T (dB)	M0 (dB)	M1 (dB)	M2 (dB)
without control	−21.8	−25.9	−29.4	−25.2	−27.6	−25.3	−22.8	−26.1
ANVC	−35.4	−37.6	−40.0	−36.8	−38.1	−32.1	−32.8	−29.3
AVC	−22.8	−26.6	−30.2	−25.7	−28.6	−26.0	−23.9	−27.0
VMC	−33.6	−35.8	−38.1	−34.9	−36.0	−31.5	−32.8	−29.3

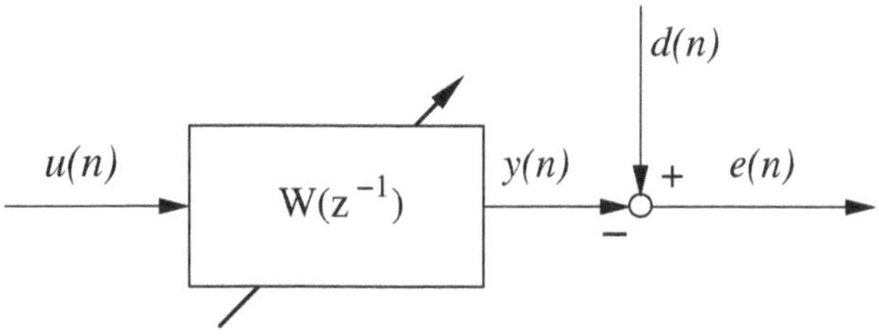

Figure 6.23 The adaptive filtering problem.

Marcian E. Hoff in 1960 [163], during their research on the steepest descent algorithm in application to the 'Adaline' device. Most probably, the algorithm was named the *Least Mean Squares* by Koford and Groner in 1966 [78]. It soon gained recognition and wide areas of applications. Also, different modifications of the LMS algorithm started to appear in the literature very soon.

LMS Algorithm Essentials

The LMS algorithm belongs to the family of stochastic gradient algorithms: it adjusts each tap of an adaptive filter in a direction of the gradient of the squared magnitude of an error with respect to the tap weights. Stochastic gradient algorithms are used to solve the adaptive filtering problem, which is depicted in Fig. 6.23. In this problem, a discrete input signal, $u(n)$, is filtered with a discrete time adaptive filter, $W(z^{-1})$. A resulting output signal, $y(n)$, is then compared with a signal with desired properties, $d(n)$. The result of this comparison, $e(n)$, is called an error. The problem of the adaptive filtering is how to adjust the filter weights to minimize a selected measure of the error.

The problem cannot be tackled without assumptions concerning the signals, the filter and the optimization criterion. Without further analysis of possible choices, we will use the assumptions enumerated below. More on the assumptions can be found in [17].

1. Both the input signals, $u(n)$ and $d(n)$, are discrete-time, real-valued, bounded or with bounded distribution.
2. The filter, for a frozen time, is a linear, discrete-time, of the finite impulse response (FIR) type.
3. The selected measure of the error to be minimized is an expected value of the mean-squared error signal.

Consider the adaptive filter with N parameters (or taps). Based on assumption 2, the filter output can be written as:

$$y(n) = \sum_{k=0}^{N-1} w_k(n)u_k(n) \in \mathbb{R}; \tag{6.42}$$

where $w_k(n) \in \mathbb{R}$ are the filter weights at the discrete time n, $u_k(n)$ are the corresponding samples of the input signal. The filter parameters can be organized into a tap-weight vector:

$$\mathbf{w}(n) = [w_0(n), w_1(n), \ldots, w_{N-1}(n)]^T \in \mathbb{R}^N. \tag{6.43}$$

In the application of the adaptive filtering problem considered in this book, the input samples come from a tapped-delay line, and can be expressed in a vector form as:

$$\mathbf{u}(n) = [u(n), u(n-1), \ldots, u(n-N+1)]^T \in \mathbb{R}^N. \tag{6.44}$$

Therefore, the output of the adaptive filter can be calculated as:

$$y(n) = \sum_{k=0}^{N-1} w_k(n)u(n-k) = \mathbf{u}^T(n)\mathbf{w}(n) = \mathbf{w}^T(n)\mathbf{u}; \tag{6.45}$$

and the error can be determined as:

$$e(n) = d(n) - \mathbf{u}^T(n)\mathbf{w}(n). \tag{6.46}$$

Based on assumption 3, the optimization criterion should be expressed as:

$$\min J(\mathbf{w}) = \min E(e^2(n)). \tag{6.47}$$

The LMS algorithm, which is the central point of this chapter, is given by:

$$\mathbf{w}(n+1) = \mathbf{w}(n) + \mu\, \mathbf{u}(n)e(n); \tag{6.48}$$

where μ is called the step size.

The algorithm thus defined has many advantages: it is simple and easy to implement, it computes fast and is surprisingly robust (the LMS algorithm robust stability was proved in 1993 [60]). Unfortunately, it also has some disadvantages, like slow convergence for some signals and the need of carefully step size selection. Due to this and other reasons, many modifications of the LMS appeared and still are being published. However, only a few of them are general enough and gained a wide recognition—those are discussed below.

Modifications of the LMS Algorithm

Normalized LMS Algorithm

Due to strong dependence of properties of the LMS algorithm on the step size, the first well-known modification of the LMS was concerned with the step size. This modification was invented independently by Nagumo and Noda [127], and Albert and Gardner [4], and it is nowadays known as the *Normalized* LMS (NLM) algorithm.

There are two ways to formulate the NLMS algorithm.

- It can be considered as the LMS algorithm with the step size dependent on the energy of the input signal $u(n)$.
- It may be obtained as a solution of the constrained optimization problem of the minimization of the Euclidean norm of a change of the filter weight vector $\mathbf{w}(n)$, subject to the constraint $\mathbf{w}^T(n+1)\mathbf{u}(n) = d(n)$ [62].

Using either of the attitudes, one obtains an update law of the NLMS algorithm as:

$$\mathbf{w}(n+1) = \mathbf{w}(n) + \frac{\tilde{\mu}}{\|\mathbf{u}(n)\|^2}\mathbf{u}(n)e(n); \tag{6.49}$$

where $\tilde{\mu}$ is called the normalized step size. The relation between the LMS and the NLMS step sizes is therefore:

$$\mu(n) = \frac{\tilde{\mu}}{\|\mathbf{u}(n)\|^2}; \tag{6.50}$$

where we have appreciated the fact that the LMS algorithm step size varies by naming it $\mu(n)$.

Leaky LMS Algorithm

The Leaky LMS (LLMS) algorithm was introduced as a response to the observation that the adaptive filter weights in many applications of the LMS algorithm tended to diverge, or drift, away from the optimal values in spite of the fact that all signals remained bounded. It was discovered that this phenomenon is due to insufficient excitation in the input signal [17]. The authors that published the algorithm first denoted it as:

$$\mathbf{w}(n+1) = \gamma\mathbf{w}(n) + \mu\,\mathbf{u}(n)e(n); \tag{6.51}$$

where $\gamma \in (0, 1]$ is the leakage factor.

Later it was discovered that the LLMS algorithm can be obtained as a result of minimization of the following cost function:

$$J(n) = e^2(n) + \alpha\|\mathbf{w}(n)\|^2 = e^2(n) + \alpha\,\mathbf{w}^T(n)\mathbf{w}(n); \tag{6.52}$$

where α is a constant. The result of minimization of this cost function with respect to the filter's weight vector $\mathbf{w}(n)$, is given by:

$$\mathbf{w}(n+1) = (1-\mu\alpha)\mathbf{w}(n) + \mu\,\mathbf{u}(n)e(n); \tag{6.53}$$

Comparing Eq. (6.51) and Eq. (6.53) results in a conclusion that both of them describe the same algorithm, with $\gamma = 1 - \mu\alpha$.

Other modifications

As mentioned above, there is a plethora of other modifications, especially concerned with techniques of step size adjustment [19]. Other, well-known modifications include:

- simplified LMS versions: sign-error, sign-regressor, and sign-sign,
- affine projection LMS,
- transform-domain LMS,
- partial update LMS.

The latter group of algorithms is covered in more details in Section 6.4.3.

6.4.2 STABILITY OF THE LMS-FAMILY ALGORITHMS

There are two best known methods to analyse the stability and convergence of the LMS algorithm. Both the methods use different sets of assumptions, with the common denominator in a form of an assumption that the input signals are wide-sense stationary. This assumption is unrealistic for a wide area of applications of the LMS algorithm.

The first method is historically older and is referred to as the *independence theory.* It assumes the input *vectors* constitute a sequence of statistically independent, identically distributed vectors (i.i.d. sequence), and that the input sequence is independent from the desired response [14, 17]. Unfortunately, this assumption is violated by each application where the input data comes from a tapped-delay line.

The second method, which is called the *small step size theory*, was developed by Butterweck and published in 1995 [22]. In this method, the main assumption is that the step size is so small that an LMS filter can be treated as a low-pass filter, with a low cut-off frequency, and that the input vector and the desired response are related by a linear multiple regression model, or that they are stationary and jointly Gaussian. These assumptions are unrealistic as well.

For the above reasons, Bismor [14] proposed a different, control-theory based method of the LMS algorithm stability analysis, which will be shorty outlined here.

Stability Analysis of the LMS Algorithm

The first step to obtain the stability conditions using control theory results is to develop the state-space model of the LMS algorithm. Substituting the error Eq. (6.46) into the LMS algorithm weight update Eq. (6.48) yields:

$$\mathbf{w}(n+1) = \mathbf{w}(n) + \mu\,\mathbf{u}(n)\left[d(n) - \mathbf{u}^T(n)\mathbf{w}(n)\right]. \tag{6.54}$$

After regrouping the terms, the above equation becomes:

$$\mathbf{w}(n+1) = \left[\mathbb{1} - \mu\,\mathbf{u}(n)\mathbf{u}^T(n)\right]\mathbf{w}(n) + \mu\,\mathbf{u}(n)d(n). \tag{6.55}$$

Equation (6.55) can be viewed as the state equation of an arbitrary nonstationary discrete-time system:

$$\tilde{\mathbf{x}}(n+1) = \mathbb{A}(n)\tilde{\mathbf{x}}(n) + \mathbb{B}(n)\tilde{\mathbf{u}}(n); \tag{6.56}$$

with the matrices and vectors:

$$\mathbb{A}(n) = \mathbb{1} - \mu\,\mathbf{u}(n)\mathbf{u}^T(n), \qquad \tilde{\mathbf{x}}(n) = \mathbf{w}(n) \tag{6.57}$$

$$\mathbb{B}(n) = \mu\,\mathbf{u}(n) \qquad \tilde{\mathbf{u}}(n) = d(n). \tag{6.58}$$

The Lyapunov stability of the system in Eq. (6.56) depends only on the state transition matrix $\mathbb{A}$, which is (fortunately) symmetric and (unfortunately) time-dependent. Based on the control theory it can be proved that the marginal stability sufficient condition for the system in Eq. 6.55 (with a symmetric state transition matrix) is that all the roots of the characteristic polynomial (or eigenvalues) of the state transition matrix $\mathbb{A}$ have absolute values less than or equal to 1. Bismor proved that the state transition matrix in Eq. (6.57) has an eigenvalue

$$\lambda_1(n) = 1 - \mu \sum_{l=0}^{N-1} u^2(n-l) = 1 - \mu\,\|\mathbf{u}(n)\|^2; \tag{6.59}$$

with the corresponding eigenvector $\mathbf{u}(n)$. The remaining eigenvalues are all 1-s. Considering the stability condition, it is possible to deduce that the LMS algorithm remains marginally stable if:

$$\forall_n \quad 0 \leq \mu < \frac{2}{\sum_{l=0}^{N-1} u^2(n-l)} = \frac{2}{\|\mathbf{u}(n)\|^2}; \tag{6.60}$$

provided $\|\mathbf{u}(n)\| \neq 0$ [14].

The above result is not unexpected as it is consistent with the literature. Observe, however, that contrary to the other literature conditions, it was achieved using only a small set of mild assumptions. Particularly, it wasn't necessary to assume neither that the step size is small (which is contradictory to fast adaptation, with the step size close to its maximum), nor that the signals are stationary. It wasn't even assumed that the signals $u(n)$ and $d(n)$ are related. Moreover, the above result is simple and easy to use, even in real-world, online applications.

Stability Analysis of the NLMS Algorithm

By comparing the right-hand side of the Eq. (6.60) with Eq. (6.50) it is possible to conclude that the NLMS algorithm stability condition is given by:

$$0 \leq \tilde{\mu} < 2. \tag{6.61}$$

This also suggest that the simplest way of application of the above stability condition is by using the normalized version of the LMS algorithm.

Stability Analysis of the LLMS Algorithm

Even if the stability analysis of the LMS algorithm with leakage is similar to the analysis of the algorithm without leakage, the results are substantially different. Therefore, the analysis will be presented below, starting with the state space model of the LLMS algorithm, which is:

$$\mathbf{w}(n+1) = \left[\gamma\mathbb{1} - \mu\,\mathbf{u}(n)\mathbf{u}^T(n)\right]\mathbf{w}(n) + \mu\,\mathbf{u}(n)d(n). \tag{6.62}$$

Thus, the state transition matrix, or the LLMS algorithm stability matrix, is equal to:

$$\mathbb{A}(n) = \gamma\mathbb{1} - \mu\,\mathbf{u}(n)\mathbf{u}^T(n). \tag{6.63}$$

It can be proved [20] that this matrix has an eigenvalue:

$$\lambda_1(n) = \gamma - \mu\sum_{l=0}^{N-1} u^2(n-l) = \gamma - \mu\,\|\mathbf{u}(n)\|^2; \tag{6.64}$$

with the corresponding eigenvector $\mathbf{u}(n)$. The remaining eigenvalues are all equal to γ, and this makes the greatest difference from the LMS algorithm due to the fact that $0 < \gamma \leq 1$ (but for $\gamma = 1$ the LLMS turns to be the LMS). Specifically, for the case of $\gamma < 1$, the LLMS can be asymptotically stable, contrary to the LMS, which can be marginally stable at the most. Based on Eq. (6.64), the asymptotical stability sufficient condition is:

$$\underset{n}{\forall} \quad |\lambda_1(n)| = \left|\gamma - \mu\sum_{l=0}^{N-1} u^2(n-l)\right| < 1; \tag{6.65}$$

which after solving for μ yields:

$$\underset{n}{\forall} \quad \frac{\gamma - 1}{\sum_{l=0}^{N-1} u^2(n-l)} < \mu < \frac{\gamma + 1}{\sum_{l=0}^{N-1} u^2(n-l)}, \tag{6.66}$$

provided $\|\mathbf{u}(n)\| \neq 0$.

The most interesting part of the inequality in Eq. (6.66) is the left-hand side numerator, which for $\gamma < 1$ is negative. This suggests that the LLMS can be asymptotically stable with a negative step size, i.e. with filter weights correction according to the gradient of the cost function. This is because of the leakage of the filter weights, which keeps the filter stable anyway. More details and many simulations supporting this theoretical result can be found in [20].

6.4.3 FAST IMPLEMENTATIONS

Structural ANC algorithms are very demanding algorithms, when it comes to computational power. Even if modern hardware offers very high clock rates and multiple cores, the multidimensional nature of structural ANC can use all these resources and demand more. One of the ways to lower this demand is to use algorithms with partial update of coefficient vectors [13]. Such algorithms are well known in signal processing, but their leaky versions, which are necessary to be used in the structural ANC applications, were rarely reported.

Partial updates (PU), or more precisely: partial parameter updates, is a technique which updates only a certain subset of all adaptive filter coefficients in each sampling period. Of course, the set of coefficients selected for the update must vary in time, and the way of selection is the main factor in which particular PU algorithms differ. There are two main groups of PU algorithms: data-independent algorithms and data-dependent algorithms [40]. The application of an algorithm belonging to the first group almost always results in a degradation of the performance. It is a designers choice how much performance should be sacrificed in order to spare a processing power, costs, etc. The application of data-dependent PU algorithms, on the other hand, may even result in an increase of the performance, compared to the full parameters update. However, these algorithms do not save as much of a processing power as the data-independent algorithms do.

Partial updates can be used with any iterative gradient-based algorithm, but in this section we will concentrate our attention on PU from the LMS family of algorithms. Moreover, we will introduce and discuss the algorithms with leakage. Therefore, our focus will be limited to the following algorithms:

- data-independent algorithms:
  - periodic partial update leaky LMS algorithm (periodic LLMS),
  - sequential partial update leaky LMS algorithm (sequential LLMS),
  - stochastic partial update leaky LMS algorithm (stochastic LLMS),
- data-dependent algorithms:
  - M-max partial update leaky LMS algorithm (M-max LLMS),
  - M-max partial update leaky normalized LMS algorithm (M-max LNLMS),
  - selective partial update leaky (normalized) LMS algorithm (selective LNLMS),
  - one tap update leaky LMS algorithm (OTU LLMS).

Periodic LLMS Algorithm

The simplest idea to lower the processing power required for an adaptive algorithm is to perform the filter weights update only once per several sampling periods. Thus, the average update cost is lowered by spreading the computations of one update over a selected number of sampling periods. The application of this idea to the LMS algorithm may be expressed as:

$$\begin{aligned} \mathbf{w}(S(n+1)) &= \gamma\mathbf{w}(Sn) + \mu\,\mathbf{u}(Sn)e(Sn); \\ \mathbf{w}(Sn+k) &= \gamma\mathbf{w}(Sn), \quad \text{for } k = 0,1,\ldots,S-1; \end{aligned} \tag{6.67}$$

where S is the period of filter weights update, i.e. the number of sampling periods used to perform one update. In between one and the next update the coefficients calculated during the previous update are used to calculate the filter output:

$$y(n) = \mathbf{w}^T(\lfloor n/S \rfloor S)\mathbf{u}(n); \tag{6.68}$$

where $\lfloor \cdot \rfloor$ denotes the floor operation.

General Notation of Other PU LLMS Algorithms

With an exception of the periodic LMS algorithm, all the other PU LLMS algorithms discussed here can be expressed by [15, 16]:

$$\mathbf{w}(n+1) = \mathbb{G}_M(n)\mathbf{w}(n) + \mu(n)\,\mathbb{I}_M(n)\mathbf{u}(n)e(n); \tag{6.69}$$

where $\mathbb{G}_M(n)$ is the leakage matrix, $\mathbb{I}_M(n)$ is the coefficient selection matrix, and where the step size $\mu(n) = \mu$ for the algorithms without the step-size normalization.

The main difference between the full update LLMS algorithm and the PU LLMS algorithms lies in the matrix $\mathbb{I}_M(n)$, which is defined as:

$$\mathbb{I}_M(n) = \begin{bmatrix} i_0(n) & 0 & \cdots & 0 \\ 0 & i_1(n) & \cdots & 0 \\ \vdots & \vdots & \ddots & \vdots \\ 0 & \cdots & 0 & i_{N-1}(n) \end{bmatrix}; \tag{6.70}$$

where

$$i_k(n) \in \{0,1\}, \quad \sum_{k=0}^{N-1} i_k(n) = M. \tag{6.71}$$

Thus, the weight selection matrix is a diagonal matrix, with the elements equal to 0 or 1, and the total number of 1s in the matrix is equal to M. The selection matrix is also subscripted with the number M, because it is the maximum number of filter weights that are *selected* for the update in each sampling period.

In each iteration, the elements on the diagonal of the selection matrix are selected as equal to 0 or 1, according to the following formula:

$$i_k(n) = \begin{cases} 1 & \text{if } k \in \mathscr{I}_{M(n)} \\ 0 & \text{otherwise} \end{cases}; \tag{6.72}$$

where $\mathscr{I}_{M(n)}$ denotes a set of the filter weights indexes that define the coefficients to be updated in the nth iteration; the number of the elements in this set is equal to, or less than M. Different variants of PU LMS algorithms define this set in a different way.

The leakage matrix $\mathbb{G}_M(n)$ is a diagonal matrix as well, and is defined as:

$$\mathbb{G}_M(n) = \begin{bmatrix} g_0(n) & 0 & \cdots & 0 \\ 0 & g_1(n) & \cdots & 0 \\ \vdots & \vdots & \ddots & \vdots \\ 0 & \cdots & 0 & g_{L-1}(n) \end{bmatrix}, \tag{6.73}$$

where

$$g_k(n) \in \{1, \gamma\}; \tag{6.74}$$

$$g_k(n) = \begin{cases} \gamma & \text{if } k \in \mathscr{I}_{M(n)} \\ 1 & \text{otherwise} \end{cases}. \tag{6.75}$$

Specifically, the matrix contains elements equal to 1 in the rows corresponding to 0-s in the $\mathbb{I}_M(n)$ matrix, and the selected value of leakage γ in the rows corresponding to 1-s in the coefficient selection matrix. In this way, the leaky version of the PU-LMS algorithm in Eq. (6.69) applies leakage in a particular iteration only to those coefficients, which are updated in this iteration. Otherwise, the advantage of processing of only a subset of parameters would be ruined.

Sequential LLMS Algorithm

A basic idea of the Sequential LLMS algorithm is to update a predefined subset of the filter weights in each iteration [43]. The number of subsets the filter is partitioned with can be calculated as:

$$B = \lceil N/M \rceil; \tag{6.76}$$

where $\lceil \cdot \rceil$ denotes the ceil operation. Only in a special case when $N \mod M$ is equal to 0, the subsets will be of equal lengths, otherwise there will be $B-1$ subsets with M coefficients and one subset with $N \mod B$ coefficients.

It is a designers choice how to partition the filter coefficients set. For example, consider a simple case with $N = 2M$, e.g. the algorithm updates half of the filter weights in each sampling period. In such case it is possible to update the parameters with even indexes in one iteration, and the parameters with odd indexes in the next iteration. The weights selection matrix (6.70) can then be written as:

$$\begin{aligned} \mathbb{I}_M(n) &= \mathrm{diag}\,(1,0,1,0,1,0\ldots) \\ \mathbb{I}_M(n+1) &= \mathrm{diag}\,(0,1,0,1,0,1\ldots). \end{aligned} \tag{6.77}$$

Another possibility is to update the first half of the weights in one iteration, and the second half in the next iteration:

$$\begin{aligned} \mathbb{I}_M(n) &= \mathrm{diag}\,(1,1,1\ldots 0,0,0\ldots) \\ \mathbb{I}_M(n+1) &= \mathrm{diag}\,(0,0,0\ldots 1,1,1\ldots). \end{aligned} \tag{6.78}$$

Other choices are also possible.

A property of the Sequential LLMS algorithm is that the subsets created by particular partitioning of the weights vector and defined by the weights selection matrix are processed in a sequence. Therefore, after B iterations all the weights are updated.

Stochastic LLMS Algorithm

It may be proved that the Sequential LMS algorithm is permanently unstable (regardless of the step size) for some input signals, e.g. for cyclostationary signals [40]. The same may apply to the Sequential LLMS algorithm (dependent of the leakage factor). To avoid this unwanted effect, the Stochastic LLMS algorithm can be applied. This algorithm selects the subsets of the weights to be updated on a random basis. The random selection should be organized in a way that selects each of the subsets with equal probability. Thus, the set of filter weight indexes from Eq. (6.72) is defined as:

$$\mathscr{P}\big(M(n)=k\big)=\pi_k,\quad k=1,\dots B,\quad \sum_{k=1}^{B}\pi_k=1; \tag{6.79}$$

where $\mathscr{P}$ denotes the probability density function of an independent random process $M(n)$, and $B=\lceil N/M\rceil$.

Please note that it is necessary to randomize the subset selection rather than the decision if to update every individual parameter. This assures that the complexity reduction is attained during each sampling period. Please note also that the computational cost of the Stochastic LLMS algorithm is slightly higher than in case of the Sequential LLMS, as the random selection mechanism requires more time than the sequential selection. For further details refer to [54].

M-max LLMS Algorithm

Contrary to the previously described PU algorithms, the M-max LLMS algorithm [42] is a data-dependent algorithm. The idea of the M-max LLMS is based on selecting M those entries of the input vector $\mathbf{u}(n)$, that result in the largest magnitude changes of the filter weights. Therefore, the weights selection matrix (6.70) entries can be defined as:

$$i_k(n)=\begin{cases}1 & \text{if } |u(n-k+1)|\in \max\limits_{1\le l\le N}\left(|u(n-l+1)|,M\right)\\ 0 & \text{otherwise}\end{cases}; \tag{6.80}$$

where $\max_l(u_l,M)$ denotes a set of M maxima of the elements u_l [40]. This algorithm does not use normalization, therefore the step size is constant: $\mu(n)=\mu$.

The M-max LLMS algorithm requires ranking (sorting) of the input vector elements, based on their absolute value. Therefore, the time savings this algorithm offers are smaller than in case of the data-independent algorithms. On the other hand, computationally efficient sorting algorithms are now available [40]. Especially in cases when the input vector is constructed from a tapped-delay line (as in Eq. (6.44)), where only one sample is exchanged in each sampling period, the shift structure of the input vector can be exploited to sort very efficiently [144].

M-max LNLMS Algorithm

The idea of M-max partial updates can also be applied to the normalized version of the LLMS algorithm. In this case the weights selection matrix is defined identically as in the case of M-max LLMS algorithm (see Eq. 6.80). However, this algorithm performs the normalization of the step size:

$$\mu(n)=\frac{\tilde{\mu}}{\mathbf{u}^T(n)\mathbf{u}(n)}. \tag{6.81}$$

Selective LLMS Algorithm

The M-max LNLMS algorithm has a subtle drawback: it uses the power of the whole input vector to normalize the step size in spite of the fact that only some of its elements are used for the actual weight update. To correct this issue, it is possible to calculate the normalized step size as:

$$\mu(n) = \frac{\tilde{\mu}}{\mathbf{u}^T(n)\mathbb{I}_M(n)\mathbf{u}(n)}. \tag{6.82}$$

This results in the algorithm known as the selective LLMS algorithm [2, 99].

It is also possible to *derive* the Selective LLMS algorithm as a result of instantaneous approximation of the Newton's method, or as an application of the minimum disturbance principle [40]. Both the derivations result in the update Eq. (6.69) and the step-size normalization given by Eq. (6.82).

Although the difference between the Selective LLMS and the M-max LNLMS algorithms is small, it results in a substantial differences in performance of the algorithms, specifically in the rate of convergence. Examples are given in below.

One Tap Update LLMS Algorithm

Consider a case of the Selective LLMS algorithm with the extreme choice of $M = 1$, i.e. the algorithm that updates only one tap during each sampling period. By combining Eqs. (6.82) and (6.69) and using $M = 1$ we notice that the algorithm is given by:

$$\mathbf{w}(n+1) = \gamma\mathbf{w}(n) + \frac{\tilde{\mu}}{u_{\max}^2(n)}\mathbb{I}_1(n)\mathbf{u}(n)e(n), \tag{6.83}$$

where $u_{\max}(n) = \max|\mathbf{u}(n)|$ denotes the maximum absolute value in the input vector at a discrete time n. The maximum absolute value $u_{\max}(n)$ is assumed to be unique, and if it is not, a single element corresponding to one of the maximum absolute values in the input vector is selected for the update at a random basis.

Remembering that $w_k(n)$ denotes the $k-th$ filter coefficient, the algorithm may be expressed as:

$$w_k(n+1) = \begin{cases} \gamma w_k(n) + \dfrac{\tilde{\mu}e(n)}{u(n-k)}, & \text{if } |u(n-k)| = \max|\mathbf{u}(n)| \\ w_k(n), & \text{otherwise} \end{cases}. \tag{6.84}$$

Thus, this simple update algorithm developed by Douglas [41], updates only one coefficient—the coefficient that corresponds to the input sample with the maximum absolute value. This algorithm will be referred to as the one tap update LLMS (OTU LLMS) algorithm.

Considering the shift structure of the input vector, the *maxline* algorithm [144] can be used for finding its maximum absolute value. The worst case computational complexity of this algorithm is $N+1$ comparisons, and only one multiplication, one division and one addition per iteration. However, it has been shown in [144] that the average number of comparisons, in case of a random input data with the uniform distribution, is approximately equal to 3 and does not increase with the filter length.

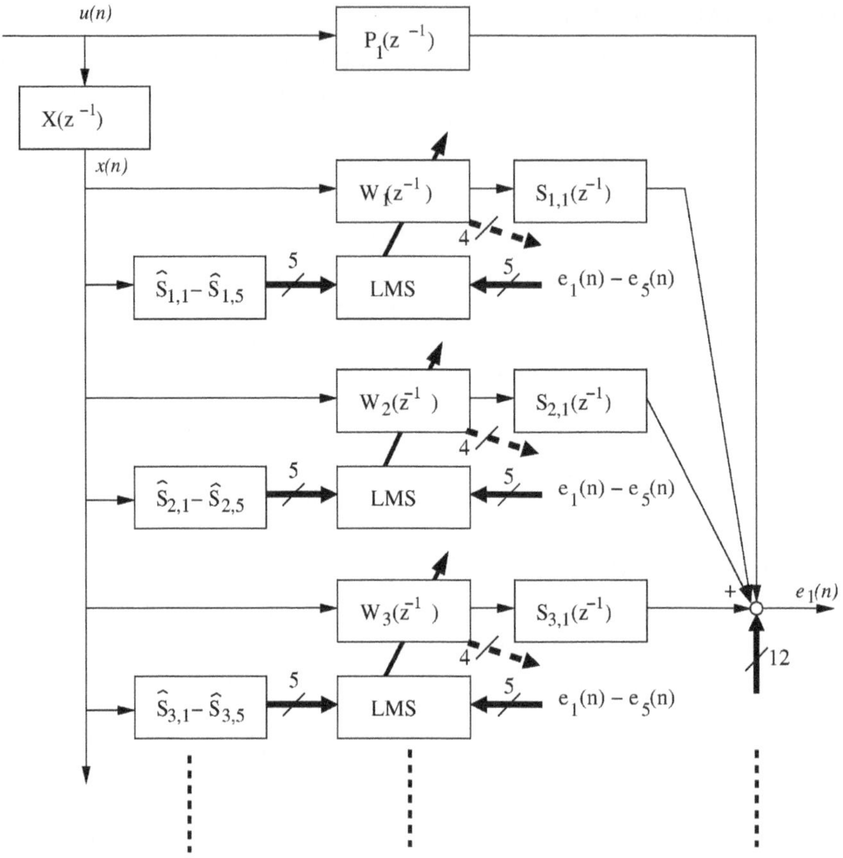

Figure 6.24 Block diagram of the Multiple-Error Filtered-x LMS algorithm.

EXPERIMENTAL SETUP

All the models used in the research described in this chapter were obtained by identification experiments performed on the laboratory setup consisting of a cubic casing with rigid corners and 1 mm thick aluminium plates. The casing was positioned in a laboratory at least 70 cm from the walls; therefore five plates were controlled, while the bottom was passively isolated. Each plate, 420×420 mm, was equipped with three electrodynamic Monacor EX-1 5 Watt actuators, mounted in carefully selected positions to improve controllability of the system. An error microphone was positioned in front of each wall. More details on the setup can be found in [103].

This setup resulted in 75 different secondary paths (15 actuators times 5 error microphones), which were identified in a form of FIR filters with 256 parameters. The identification experiments are described in details in [18].

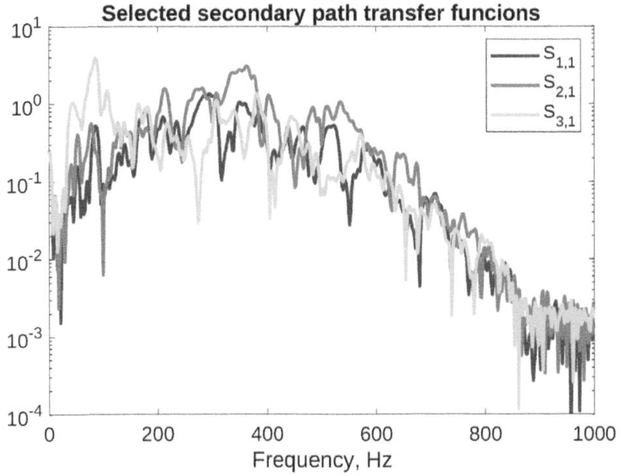

Figure 6.25 Secondary path transfer functions from the front plate actuators to the front plate error microphone.

A part of the block diagram of the simulated control system, associated with one plate (front) and one error signal, is presented in Fig. 6.24. The primary signal, $u(n)$, is routed through the reference path transfer function, $X(z^{-1})$, to produce the reference signal, $x(n)$. This reference signal forms the input to the three control filters, associated with the front plate. Each control filter output signal is then filtered with five different secondary path transfer functions. Of these transfer functions, only one for each filter is visible in the figure (i.e. $S_{1,1}(z^{-1})$, $S_{2,1}(z^{-1})$, and $S_{3,1}(z^{-1})$), while the remaining four belong to the other error signal paths. In total,, each error signal is a sum of 16 signals: one primary and 15 secondary, coming from the 15 actuators. All the 5 error signals are used by each of the adaptation algorithms (denoted as LMS), as well as the five reference signals (each filtered through a different secondary path transfer function estimate). Exemplary secondary path transfer functions are illustrated in Fig. 6.25. Parameters of the simulated system are summarized in Table 6.6.

The step sizes were experimentally adjusted with a great care to achieve as fast as possible convergence speed, but to avoid unwanted effects that frequently appear in ANC systems with too large step sizes [11, 12]. Different leakage factors were used for different control filters, with values from 0.99999 to 0.99. The procedure to adjust the leakage factors was also experimental and depended on the observation of the filter output signals. If a selected output signal was of a high power or of still increasing values, the leakage factor was lowered until the effect was prevented.

Simulation results

Many simulations were performed, using different primary signals, always resulting in a stable system, with different levels of attenuation. Below the results of one of the experiments, using a single frequency (tonal) signal of 112 Hz and a white noise

Table 6.6
Parameters of the simulated system.

Parameter	Value
Sampling frequency	2 kHz
Filter length	128
Step size	2×10^{-4}
Normalized step size	5×10^{-6}
Leakage	0.99...0.99999

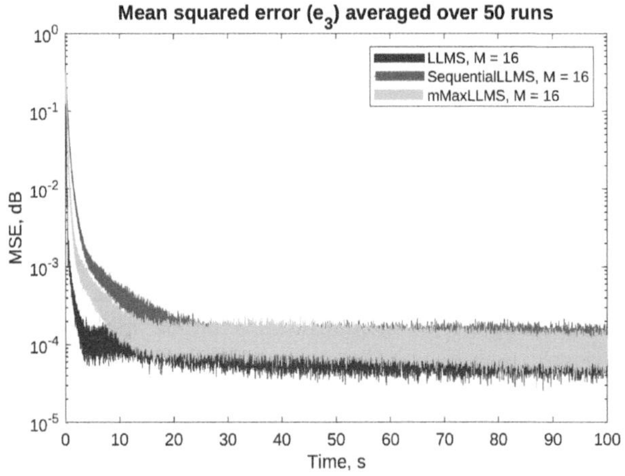

Figure 6.26 Simulation results for algorithms without step size normalization, $M = 16$.

of small variance (0.0001), are presented and discussed.

Fig. 6.26 presents the MSE (averaged over 50 runs) for the third error microphone signal during simulation with the leaky sequential PU-LMS and leaky M-max PU-LMS algorithms, where the results of the full-update LLMS algorithm are showed for comparison. The number of parameters updated in each iteration was selected as 16 (out of 128). As can be noticed (and as expected), the full-update LLMS algorithm is the fastest to converge, the Sequential LLMS is the slowest, and the M-max PU LLMS is between. This order (LMS fastest, Sequential PU-LMS slowest) was the same for all error microphones.

Fig. 6.27 presents analogous plots for the algorithms with the normalized step sizes, and contains the results of the full-update Leaky NLMS algorithm for comparison. In this case, the initial speed of convergence is comparable for all the algorithms, but the final attenuation levels differ significantly. Both the NLMS (purple)

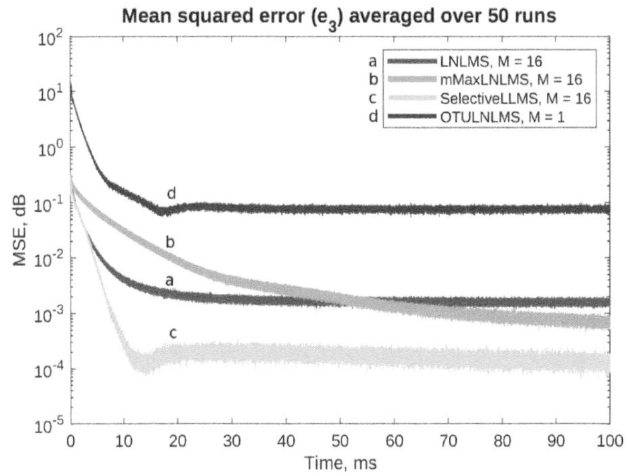

Figure 6.27 Simulation results for algorithms with step size normalization, $M = 16$.

and M-max PU-NLMS (green) algorithms result in around 25 dB of attenuation, while the Selective PU-LMS reaches more than 40 dB. On the other hand, the One Tap Update algorithms does not perform impressively and reaches only around 10 dB of attenuation. Very similar results were obtained for the other error microphones.

The observed behaviour is typical to the PU algorithms, among which the data-independent algorithms, while offering the best computational power saves, suffer from decrease of convergence speed, proportional to the number of parameters which are not updated in each iteration. The data-dependent algorithms, on the other hand, usually perform comparably with their non-partial versions, but in some cases can even outperform the full-update algorithms

Figs. 6.28 and 6.29 contain the results for the number of parameters updated in each iteration equal to 8, for the first error signal. The results are similar except for the fact that some of the algorithms exhibit a small increase of the MSE after initial convergence. Ultimately the MSE starts to decrease around the middle of the experiment. The attenuation levels after 100 s of the simulation is lower than in case of $M = 16$.

As expected, the worst results were obtained for $M = 4$—see Figs. 6.30 and 6.31. The attenuation levels obtained during the 100 s of the simulation are significantly worse: only around 20 dB for the Selective PU-NLMS algorithms, and around 10 dB for the others (except for the full-update algorithms, which are shown for comparison). From this fact one can conclude that 4 parameters for update is a number too small in this complex application.

To conclude, the family of PU-LMS algorithms offers substantial computational power savings without significant degradation of performance: in fact, the performance can even be improved when compared to full update algorithms. PU-LMS can be used with the leakage and can be applied to a structural ANC system of high

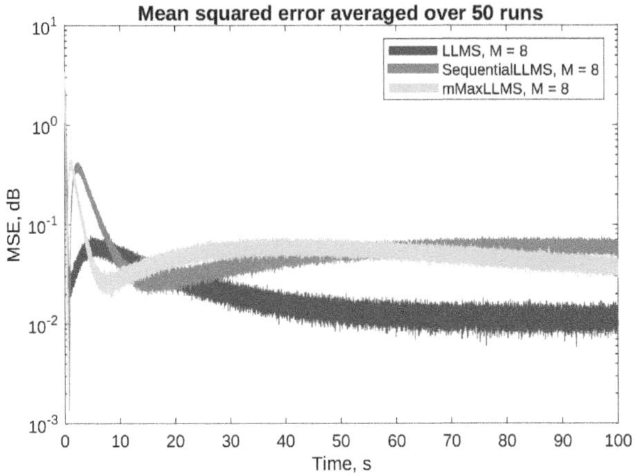

Figure 6.28 Simulation results for algorithms without step size normalization, $M = 8$.

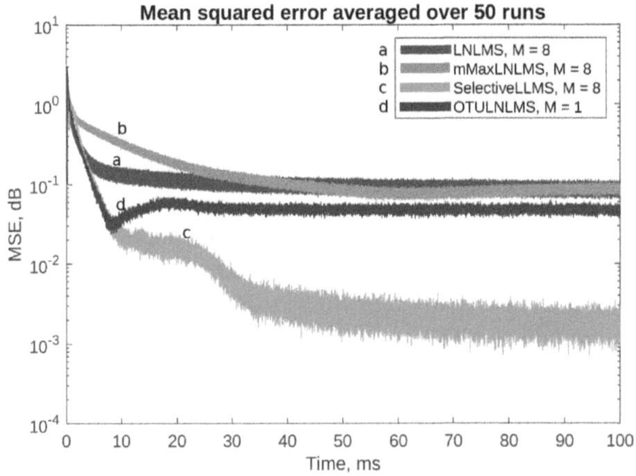

Figure 6.29 Simulation results for algorithms with step size normalization, $M = 8$.

dimensionality: 5 error sensors, 15 actuators and 75 secondary paths in total. Partial update LMS algorithms perform well even when only one parameter out of 16 is selected for update in each iteration.

6.4.4 MULTICHANNEL IMPLEMENTATIONS

Multichannel adaptive filters

Adaptive control algorithms have two significant advantages, when applied to active noise-cancelling casings. First, due to delays in practical active noise-cancelling

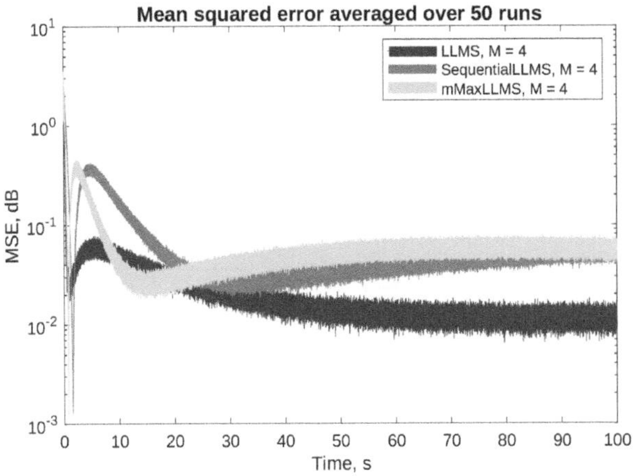

Figure 6.30 Simulation results for algorithms without step size normalization, $M = 4$.

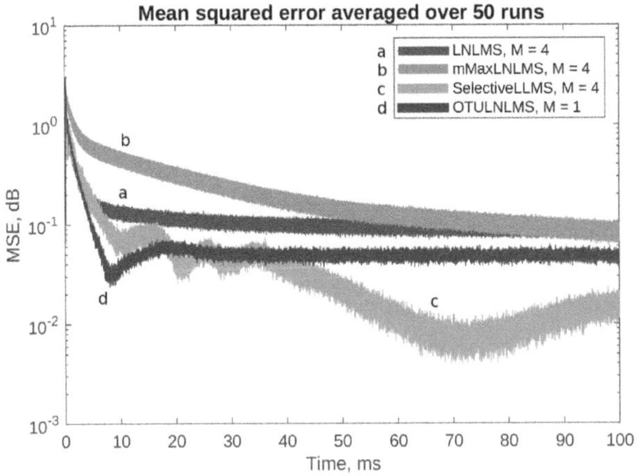

Figure 6.31 Simulation results for algorithms without step size normalization, $M = 4$.

casings the optimal control filter is non-causal and in practice the filter must predict future noise. Such prediction may be very successful for deterministic noise signals, e.g. tonal or multitonal, frequently generated by devices with rotating parts. However, when the noise is nonstationary, for instance the rotational speed may change in time, the control filter must be tuned for new noise. Adaptive control does exactly that. Second, the active casing itself is non-stationary. For instance temperature change might significantly change its frequency response [109].

The basic filter adaptation algorithms usually assume that the output of the filter is directly added to the error $\mathbf{e}'(n)$, without secondary path (Fig. 6.32). Additionally, the

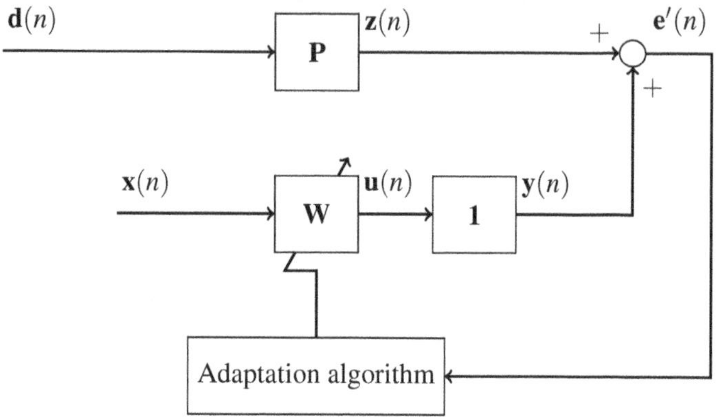

Figure 6.32 Adaptive ANC system without secondary path block diagram.

source of the reference signal is not important, both feedforward and IMC structures can be used.

As stated in Section 6.2.2 the goal of control system is to reduce noise emission. Because the noise usually contains random components the expected value of the noise emission should be minimized:

$$J_3 = \mathrm{E}\{J_2(n)\} = \mathrm{E}\left\{\sum_{i=0}^{M_M} m_i^2(n)\right\}. \tag{6.85}$$

In adaptive systems the expected value must be estimated using the currently available data. When the expected value operator is replaced with the arithmetic mean based on currently available data, from sample 0 to sample n, the cost function depends on time and becomes:

$$J_3 \approx J_4(n) = \frac{1}{n+1}\sum_{k=0}^{n}\sum_{i=0}^{M_M} m_i^2(n-k), \tag{6.86}$$

and this cost function is minimized by the RLS (*Recursive Least Squares*) algorithm [61]. The classical RLS algorithm has very high computational complexity and in active control applications Fast RLS algorithms are usually used [24, 31, 85]. Due to infinite time window, for which the expected value is calculated, such algorithm is not suitable for non-stationary systems. Thus, for active control applications usually RLS with exponential window (WRLS) is used [27]:

$$J_3 \approx J_5(n) = \frac{1-\lambda}{1-\lambda^{n+1}}\sum_{k=0}^{n}\left(\lambda^k \sum_{i=0}^{M_M} m_i^2(n-k)\right), \tag{6.87}$$

where $\lambda \in (0,1]$ is the exponential weighting factor. $\frac{1-\lambda}{1-\lambda^{n+1}}$ factor is needed only to correctly estimate the expected value, when this constant factor is dropped the result of minimization will be the same.

The algorithm can be simplified and the computational load reduced can be decreased by using the simplest estimator of expected value, the current value:

$$J_2(n) = \sum_{i=0}^{M_M} m_i^2(n). \tag{6.88}$$

This cost function can be minimized using MELMS (*Multiple Error Least Mean Squares*) algorithm [45]. This algorithm and its variants due to simplicity, robustness and speed are very popular in active control. The convergence is, however, slower than for RLS. The basic weight update equation, without secondary path, for MELMS with leakage is:

$$w_{k,c,o}(n+1) = \alpha_{k,c,o}(n) w_{k,c,o}(n) - \sum_{l'=0}^{M_E} \mu_{k,c,l',o}(n) x_k(n-o) e'_{l'}(n), \tag{6.89}$$

where $\mu_{k,c,l',o}(n)$ is the step size, and $0 \ll \alpha_{k,c,o}(n) \leq 1$ is the leak coefficient.

The third popular algorithm for active control is the AP (*Affine Projection*) algorithm. The AP algorithm uses an average of K past samples as an estimator of expected value:

$$J_6(n) = \frac{1}{K} \sum_{k=0}^{K-1} J_2(n-k) = \frac{1}{K} \sum_{k=0}^{K-1} \sum_{i=0}^{M_M} m_i^2(n-k). \tag{6.90}$$

The AP algorithm has better convergence rate than LMS (*Least Mean Squares*), but worse than RLS. For $K = 1$ the AP algorithm is equivalent to LMS algorithm. The basic AP algorithm requires solving system of K linear equations with K variables. Fast variants provide LMS like computational complexity [52]

Adaptive filters with a secondary path

Fig. 6.33 shows the structure of an adaptive system with a secondary path. In systems with secondary path the error on output of the adaptive filter, needed for weight update, is not available. There are two basic structures that neutralize the effect of secondary path: filtered reference (FX) and filtered error (FE). The FX structure is suitable for systems for which $W_{k,c}$ and $S_{c,l}$ operators commute: $W_{k,c}S_{c,l} = S_{c,l}W_{k,c}$. In that case the order of control filter and secondary path can be reversed (Fig. 6.34).

$$y_l(n) = \sum_{c=0}^{M_U} S_{c,l} \sum_{k=0}^{M_X} W_{k,c} x_k(n) = \sum_{c=0}^{M_U} \sum_{k=0}^{M_X} S_{c,l} W_{k,c} x_k(n) = \sum_{c=0}^{M_U} \sum_{k=0}^{M_X} W_{k,c} S_{c,l} x_k(n). \tag{6.91}$$

The system with inverted order of operators is not a physical representation of the system and it can be only be simulated, and in simulation only a model of the secondary path, $\hat{\mathbf{S}}$, can be used. Thus, the $W_{k,c}S_{c,l}$ is replaced by $W_{k,c}\hat{S}_{c,l}$. In such system

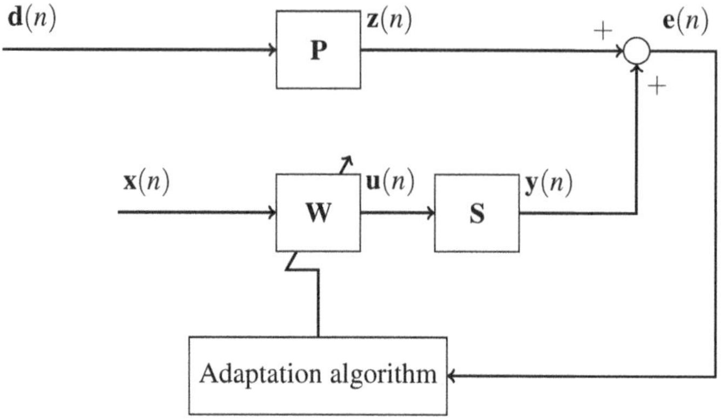

Figure 6.33 Adaptive ANC system with secondary path block diagram.

the output of adaptive filter $W_{k,c}$ is directly added to error **e** and algorithms without secondary path can be used.

Unfortunately, $\sum_{c=0}^{M_U}\sum_{k=0}^{M_X}\hat{S}_{c,l}x_k(n)$ cannot be easily expressed as a vector, a matrix multiplied by a vector of reference signals. Instead, we assume that the product of this operator, $\hat{\mathbf{S}}'$, is a signal represented with 3-dimensional array $\mathbf{R}(n)$, with values defined as:

$$r_{k,c,l}(n) = \hat{S}_{c,l}x_k(n). \tag{6.92}$$

The calculation of full $\mathbf{R}(n)$ array in each sample can be very expensive, it requires $N_X N_U N_E$ filtrations, each by filters with $M_S + A_S$ parameters.

When the Filtered-reference structure is used the Eq. 6.89 becomes:

$$w_{k,c,o}(n+1) = \alpha_{c,k,o}(n)w_{c,k,o}(n) - \sum_{l=0}^{M_E}\mu_{k,c,l,o}(n)r_{k,c,l}(n-o)e_l(n), \tag{6.93}$$

The second solution is to use FE (*Filtered Error*) structure. To make the secondary path 'invisible' to the adaptation algorithm the error can be filtered by an inverse of appropriate secondary path:

$$S_{c,l}S_{c,l}^{-1} = 1. \tag{6.94}$$

Unfortunately, in practice the inverse of secondary path is non-causal. However, the effect of secondary path and its 'inverse' can be equal to a simple delay:

$$S_{c,l}S_{c,l}^{\prime -1} = q^D. \tag{6.95}$$

The same delay must be added to reference signals.

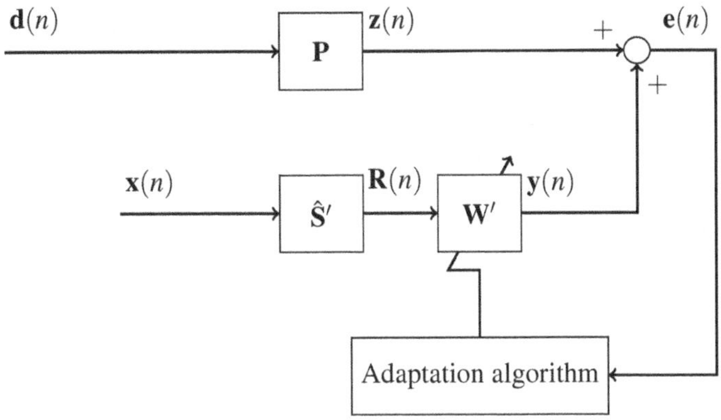

Figure 6.34 Adaptive ANC system with control filter and secondary path swapped.

Variable step-size algorithms

The stability and convergence rate of the LMS algorithm depends on the step size. The maximal step size depends on the reference signal power. A commonly used variant of the LMS algorithm that removes this dependence is the NLMS (*Normalized Least Mean Squares*) algorithm with step-size defined as:

$$\mu_{k,c,l,o}(n) = (P_{k,c,l}(n) + \zeta)^{-1} \mu_n, \tag{6.96}$$

where μ_n is the normalized step size, and ζ is an optional small constant added to avoid division by zero. In classic NLMS $\zeta = 0$, $\alpha_{c,k,o}(n) = 1$ and for stable operation without the secondary paths $0 < \mu_n < 2$ with optimal value for fastest convergence $\mu_n = 1$. The $P_{k,c,l}(n)$ is equal to:

$$P_{k,c,l}(n) = \sum_{k'=0}^{M_X} \sum_{c'=0}^{M_U} \sum_{l'=0}^{M_E} \sum_{o=0}^{M_W} (r_{k',c',l'}(n-o))^2. \tag{6.97}$$

Switched error LMS

For large systems, the MEFXLMS (*Multiple Error Filtered-Reference Least Mean Squares*) algorithm has very high computational requirements. The calculation of output has $O(N_X N_U N_W)$ asymptotical computational complexity, but Multichannel FXLMS (*Filtered-Reference Least Mean Squares*) has $O(N_X N_U N_E (N_W + N_S))$. Additionally, the number of operations per coefficient is larger, thus scale proportionality constant factor is larger. In many applications N_S, related to filtration of reference signals by secondary path models, is smaller or comparable to N_W, and the additional dependence on N_S is not a problem. However, the dependency on the number of error signals N_E leads to significant increase of computational requirements. This dependence can be eliminated by using only a constant number of error signals, one in the simplest case, in each iteration [113]. So, in each iteration only a subset of errors is active. As long as the step-size is small enough, and the switching is faster than convergence rate, the switching has virtually no effect on steady-state performance. In that case, the adaptation algorithm is not fast enough, to minimize only currently selected errors. Effectively, the adaptation algorithm acts as low-pass filter and high-frequency switching is attenuated. The equation of step-size for SEFXLMS (*Switched Error Filtered-Reference Least Mean Squares*) becomes:

$$\mu_{k,c,l,o}(n) = (P_{k,c,l}(n) + \zeta)^{-1} g_l \mu_n, \tag{6.98}$$

where $g_l(n)$ is equal to 1 if the jth error is active in nth iteration, it is equal to 0 otherwise. As long as $\sum_{l'=0}^{M_E} g_l < C$, the asymptotical computational complexity is reduced to: $O(N_X N_U (N_W + N_E N_S))$. The computational complexity can be further reduced to: $O(N_X N_U (N_W + N_S))$, if the number the number of filtered reference signals does not depend on N_E, thus not all $\mathbf{R}(n)$ signals are calculated in each iteration. The error is not active, $g_l = 0$, $\mathbf{R}(n)$ is not needed in Eq. (6.96), however it is still needed in classical NLMS normalization. Thus, to effectively use SEFXLMS another variable-step size algorithm must be used. However, It is possible to approximate NLMS behaviour without using unneeded for basic adaptation equation data. This can be done, by using Eq. (6.98), but approximating $P_{k,c,l}(n)$.

The variable step-size algorithm should not introduce unintentional weighting of error signals, thus should satisfy the following condition [102]:

$$\forall_{l_1,l_2 \in \{0,1,\ldots,M_E\}} \left(E\left\{\mu_{k,c,l_1,o}(n)\right\} = E\left\{\mu_{k,c,l_2,o}(n)\right\}\right). \tag{6.99}$$

If the switching pattern and the reference signal are independent, this condition can be satisfied if two following conditions are satisfied:

$$\forall_{l_1,l_2 \in \{0,1,\ldots,M_E\}} \left(E\left\{g_{l_1}(n)\right\} = E\left\{g_{l_2}(n)\right\}\right) \tag{6.100}$$

and:

$$\forall_{l_1,l_2 \in \{0,1,\ldots,M_E\}} \left(E\left\{P_{k,c,l_1}(n)\right\} = E\left\{P_{k,c,l_2}(n)\right\}\right). \tag{6.101}$$

Eq. (6.100) is satisfied with any correctly selected switching pattern [117]. The independence of switching pattern and reference signal can be satisfied by using random switching. Further details will be presented in Section 6.4.4.

Reference signal power estimation

In many cases noise is stationary or its statistical properties change slowly compared to the switching period. In that case, it can be assumed that reference signals are stationary, and the expected value of filtered reference signals does not depend on time:

$$E\left\{\sum_{o=0}^{M_W} \left(r_{k,c,l}(n-o)\right)\right\} = E\left\{\sum_{o=0}^{M_W} \left(r_{k,c,l}(n-o-d_{k',c',l'}(n))\right)\right\}, \tag{6.102}$$

where $d_{k',c',l'}(n) > 0$ are arbitrary integers used shift time indexes to values for which $r_{k,c,l}(n)$ is known. So the $P_{k,c,l}$ can be approximated by $P^R_{k,c,l}$ (variant "G")[102]:

$$P_{k,c,l} \approx P^R_{k,c,l}(n) = \sum_{k'=0}^{M_X} \sum_{c'=0}^{M_U} \sum_{l'=0}^{M_E} \sum_{o=0}^{M_W} \left(r_{k',c',l'}(n-o-d_{k',c',l'}(n))\right)^2. \tag{6.103}$$

For stationary signals the window used to estimate signal power does not need to be rectangular, as in NLMS. Arbitrary window can be used [102]:

$$E\left\{\sum_{o=0}^{M_W} \left(r_{k,c,l}(n-o)\right)^2\right\} = N_W E\left\{\sum_{o=0}^{\infty} h_i \left(r_{k,c,l}(n-o)\right)^2\right\}, \tag{6.104}$$

where h_o are window coefficients, and $\sum_{o=0}^{\infty} h_i = 1$. When an exponential window is used, $h_i = \beta(1-\beta)^i$, where $0 < \beta < 1$, the power can be estimated by following equation [102]:

$$P_{k,c,l} \approx P^E_{k,c,l}(n) = \sum_{l=0}^{M_E} P_l(n). \tag{6.105}$$

where:

$$P_l(n) = \begin{cases} (1-\beta)P_l(n-1) + \beta N_W \sum_{k'=0}^{M_X} \sum_{c'=0}^{M_U} \left(r_{k',c',l}(n)\right)^2 & \text{if } g_l(n) = 1 \\ P_l(n-1) & \text{if } g_l(n) = 0 \end{cases}. \tag{6.106}$$

To approximate rectangular window with N_W length the β coefficient should be equal to $\frac{1}{N_W}$.

The performance of SEFXLMS with NLMS step-size normalization and both approximations has been tested in simulation on models and noise samples from a real washing machine [120, 121]. The testing was performed using simulation to provide exactly the same conditions for all variants. As expected, the performance of both variants for stationary noise of the washing machine spinning at 1200 rpm is indistinguishable from the performance with full NLMS normalization (Fig. 6.35). The average step-size is very close to NLMS (Fig. 6.36). Even for non-stationary change of spinning speed from 1000 rpm to 1200 rpm no significant changes we observed (Fig. 6.37).

Switching

The set of active error signals can be described by the $\mathbf{g}(n)$ vector:

$$\mathbf{g}(n) = [g_0(n), g_1(n), \ldots, g_{M_E}(n)]^T. \tag{6.107}$$

For MEFXLMS all error signals are always enabled, $\mathbf{g}(n) = [1, 1, \ldots, 1]^T$. For SEFXLMS the set of active errors is selected from all used possibilities:

$$\mathbf{g}(n) = \mathbf{g}_{\mathbf{a}(n)}, \tag{6.108}$$

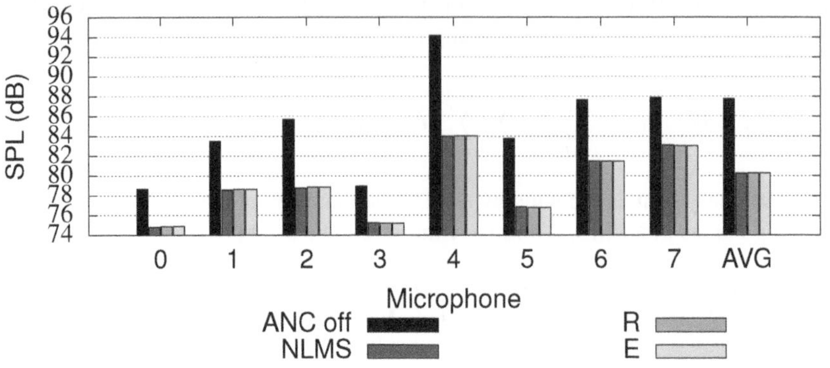

Figure 6.35 The power on all error microphones for final phase of spinning at 1200 rpm for a different normalization algorithms.

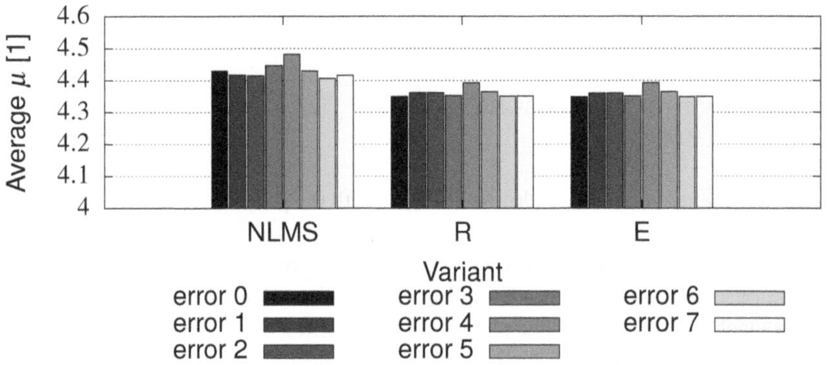

Figure 6.36 Average step-size for 1200 rpm spinning noise for different microphones.

where $\mathbf{a}(n) \in \{0, 1, \ldots, M_Q\}$ is the switching sequence that selects active error set, where $N_Q = M_Q + 1$ is the number of possibilities. All of used possibilities can be written as the switching matrix:

$$\mathbf{G} = [\mathbf{g}_0, \mathbf{g}_1, \ldots, \mathbf{g}_{M_Q}] = \begin{bmatrix} g_{0,0} & g_{0,1} & \cdots & g_{0,M_Q} \\ g_{1,0} & g_{1,1} & \cdots & g_{1,M_Q} \\ \vdots & \vdots & \ddots & \vdots \\ g_{M_E,0} & g_{M_E,1} & \cdots & g_{M_E,M_Q} \end{bmatrix}. \tag{6.109}$$

The switching is described by the switching matrix $\mathbf{G}$ and the switching sequence $\mathbf{a}(n)$. The switching must satisfy condition 6.100. It can be satisfied when each

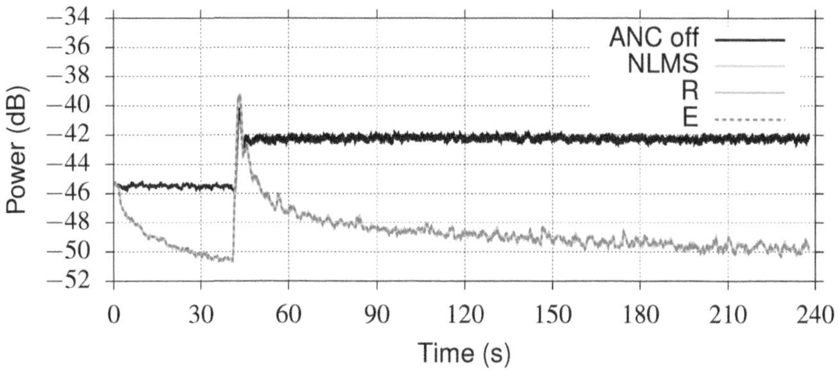

Figure 6.37 The average power on error microphones for an excerpt of washing machine noise for a different normalization algorithms.

possible set of active error signals is selected with the same probability (frequency):

$$\forall_{i_1,i_2 \in \{0,1,\ldots,M_Q\}} \forall_{n_1,n_2 \in \mathbb{Z}} \left(P\{\mathbf{a}(n_1) = i_1\} = P\{\mathbf{a}(n_2) = i_2\}\right), \tag{6.110}$$

and each error is selected with the same frequency in rows of the switching matrix:

$$\forall_{i_1,i_2 \in \{0,1,\ldots,M_E\}} \left(\sum_{i_3=0}^{M_Q} \left(g_{i_1,i_3}\right) = \sum_{i_4=0}^{M_Q} \left(g_{i_2,i_4}\right) \right), \tag{6.111}$$

To avoid any bias caused by switching correlated with noise, the switching sequence $\mathbf{a}(n)$ should be random. However, in practice round-robin switching can be used:

$$\mathbf{a}(n) = \left\lfloor \frac{n}{N_B} \right\rfloor \mod N_Q, \tag{6.112}$$

where N_B is the number of samples for which selected set of active errors is enabled. This is related to filtration of reference signals, by reference paths. When the error is activated, to be able to perform control filter weight update also N_S previous filtered reference signal samples are needed. So the filtered-reference filtering must be enabled earlier, before error signal is activated. Thus, to avoid filtering too many signals N_B should be larger than N_S. In that case, filtered reference signals for at most $2N_Q$ error signals must be calculated. Alternatively, after switching errors for N_S adaptation can be disabled. In that case, filtered reference signals for at N_Q error signals are calculated. This variant will be called "slow".

Different SEFXLMS variants were tested on a models of a lightweight casing and a washing machine. The lightweight casing uses 21 actuators and 5 error microphones. The washing machine uses 13 actuators and 8 error microphones.

The simplest variant uses only one active error. For lightweight casing the switching matrix is equal to:

$$\mathbf{G}_{L1} = \begin{bmatrix} 1 & 0 & 0 & 0 & 0 \\ 0 & 1 & 0 & 0 & 0 \\ 0 & 0 & 1 & 0 & 0 \\ 0 & 0 & 0 & 1 & 0 \\ 0 & 0 & 0 & 0 & 1 \end{bmatrix}. \tag{6.113}$$

Other possibilities with 5 columns, such as:

$$\mathbf{G}_{L1B} = \begin{bmatrix} 1 & 0 & 0 & 0 & 0 \\ 0 & 0 & 1 & 0 & 0 \\ 0 & 1 & 0 & 0 & 0 \\ 0 & 0 & 0 & 1 & 0 \\ 0 & 0 & 0 & 0 & 1 \end{bmatrix}, \tag{6.114}$$

are not really different when the numbering of physical microphones is arbitrary. Variants with duplicated columns:

$$\mathbf{G}_{L1C} = \begin{bmatrix} 1 & 0 & 0 & 0 & 0 & 1 & 0 & 0 & 0 & 0 \\ 0 & 1 & 0 & 0 & 0 & 0 & 1 & 0 & 0 & 0 \\ 0 & 0 & 1 & 0 & 0 & 0 & 0 & 1 & 0 & 0 \\ 0 & 0 & 0 & 1 & 0 & 0 & 0 & 0 & 1 & 0 \\ 0 & 0 & 0 & 0 & 1 & 0 & 0 & 0 & 0 & 1 \end{bmatrix}, \tag{6.115}$$

and

$$\mathbf{G}_{L1D} = \begin{bmatrix} 1 & 0 & 0 & 0 & 0 & 0 & 0 & 0 & 0 & 1 \\ 0 & 1 & 0 & 0 & 0 & 0 & 0 & 0 & 1 & 0 \\ 0 & 0 & 1 & 0 & 0 & 0 & 0 & 1 & 0 & 0 \\ 0 & 0 & 0 & 1 & 0 & 0 & 1 & 0 & 0 & 0 \\ 0 & 0 & 0 & 0 & 1 & 1 & 0 & 0 & 0 & 0 \end{bmatrix} \tag{6.116}$$

are different, but the same effect can be obtained by choosing a different $\mathbf{a}(n)$ function.

Fig. 6.38 shows the error signal power on all microphones around the lightweight casing for 150 Hz tone and time needed to obtain 20 dB attenuation for a different normalized step-size μ_n. For "slow" variant, $N_B = 2N_S$ and after active error change for N_S samples adaptation is disabled. For "fast" variant, $N_B = N_S$. Fig. 6.39 shows the difference between both variants.

The "slow" variant is roughly 2 times slower than "fast" variant, due to disabled adaptation phase. This comes at cost of increased computational load, calculation of $N_X N_U$ secondary paths, related to next active error, is needed. Both variants are slower than MEFXLMS. When, the convergence speed is important the "fast" variant is recommended.

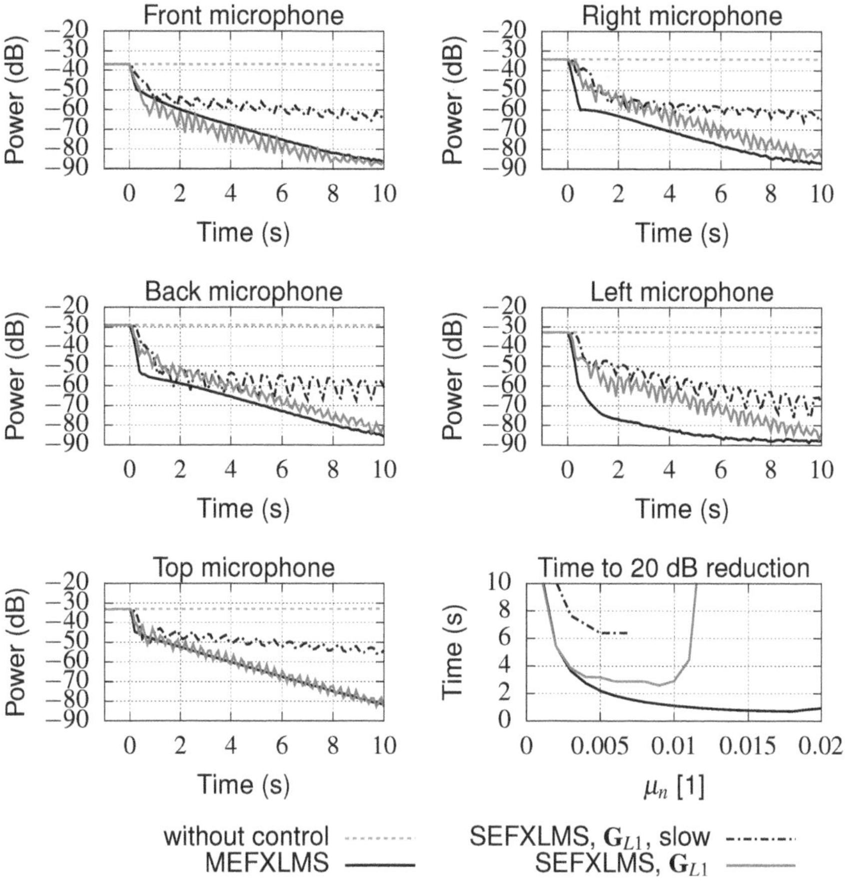

Figure 6.38 Error microphone signals power and time needed to obtain 20 dB noise reduction for different control algorithms (150 Hz tone, $\mu_n = 0.005$, lightweight casing).

When two errors are active at the same time, two basic possibilities are possible:

$$\mathbf{G}_{L2A} = \begin{bmatrix} 1 & 0 & 0 & 0 & 1 \\ 1 & 1 & 0 & 0 & 0 \\ 0 & 1 & 1 & 0 & 0 \\ 0 & 0 & 1 & 1 & 0 \\ 0 & 0 & 0 & 1 & 1 \end{bmatrix} \tag{6.117}$$

"fast" variant

$\frac{n}{N_S}$	0	1	2	3	4	5	6	7	8	...
$\mathbf{a}(n)$	0	1	2	3	4	0	1	2	3	...
error 0	+	–	–	–	FX	+	–	–	–	...
error 1	FX	+	–	–	–	FX	+	–	–	...
error 2	–	FX	+	–	–	–	FX	+	–	...
error 3	–	–	FX	+	–	–	–	FX	+	...
error 4	–	–	–	FX	+	–	–	–	FX	...

"slow" variant

$\frac{n}{N_S}$	0	1	2	3	4	5	6	7	8	...
$\mathbf{a}(n)$	0	0	1	1	2	2	3	3	4	...
error 0	FX	+	–	–	–	–	–	–	–	...
error 1	–	–	FX	+	–	–	–	–	–	...
error 2	–	–	–	–	FX	+	–	–	–	...
error 3	–	–	–	–	–	–	FX	+	–	...
error 4	–	–	–	–	–	–	–	–	FX	...

Figure 6.39 Comparison of "fast" and "slow" variants of SEFXLMS ("+"—enabled error, "FX"—disabled adaptation, enabled filtered-reference; "–"—disabled adaptation).

and

$$\mathbf{G}_{L2B} = \begin{bmatrix} 1 & 0 & 1 & 0 & 0 \\ 1 & 0 & 0 & 1 & 0 \\ 0 & 1 & 0 & 1 & 0 \\ 0 & 1 & 0 & 0 & 1 \\ 0 & 0 & 1 & 0 & 1 \end{bmatrix}. \tag{6.118}$$

The $\mathbf{G}_{L2B}$ variant provides faster convergence (Fig. 6.40),

especially at higher μ_n at the cost of slightly increased computational load, $4N_XN_U$ filtered reference signals need to be calculated instead of $3N_XN_U$.

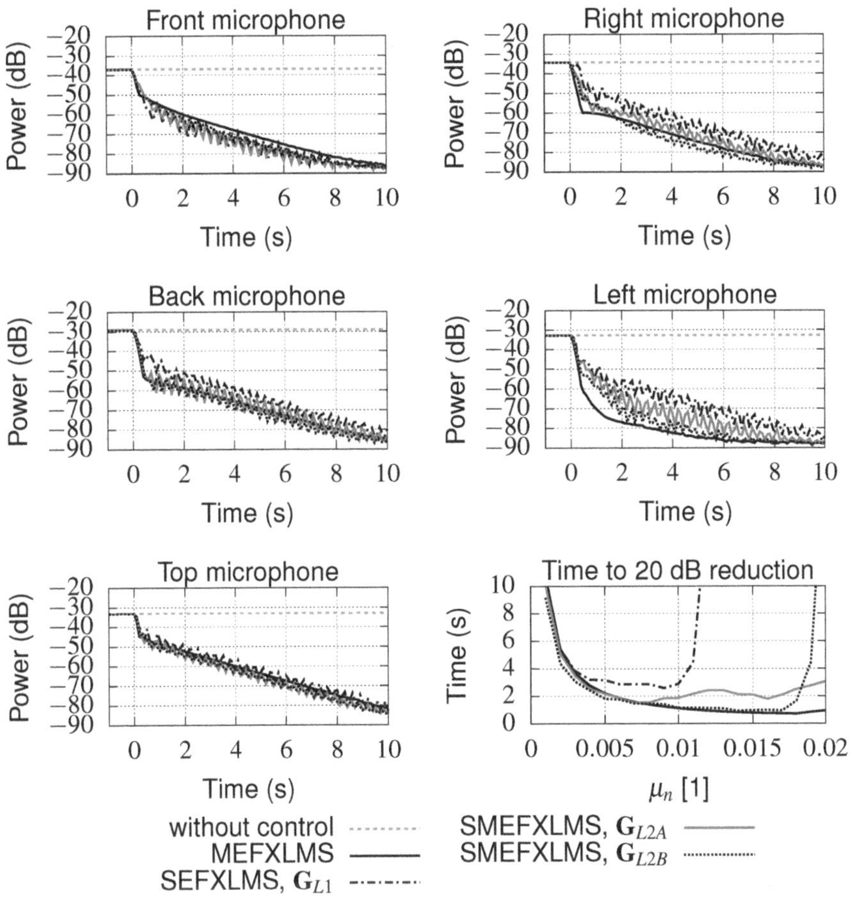

Figure 6.40 Error microphone signals power and time needed to obtain 20 dB noise reduction for different control algorithms (150 Hz tone, $\mu_n = 0.005$, lightweight casing).

For a washing machine, the following switching matrices has been tested:

$$\mathbf{G}_{W1} = \begin{bmatrix} 1 & 0 & 0 & 0 & 0 & 0 & 0 & 0 \\ 0 & 1 & 0 & 0 & 0 & 0 & 0 & 0 \\ 0 & 0 & 1 & 0 & 0 & 0 & 0 & 0 \\ 0 & 0 & 0 & 1 & 0 & 0 & 0 & 0 \\ 0 & 0 & 0 & 0 & 1 & 0 & 0 & 0 \\ 0 & 0 & 0 & 0 & 0 & 1 & 0 & 0 \\ 0 & 0 & 0 & 0 & 0 & 0 & 1 & 0 \\ 0 & 0 & 0 & 0 & 0 & 0 & 0 & 1 \end{bmatrix}, \tag{6.119}$$

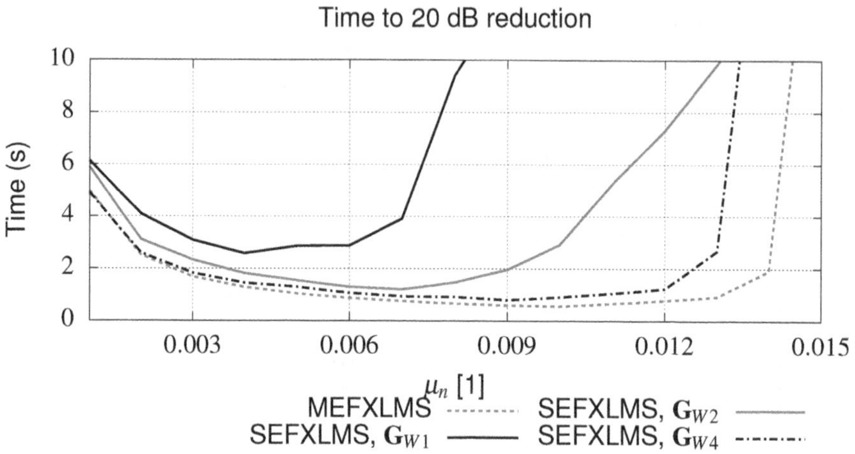

Figure 6.41 Time needed to obtain 10 dB noise reduction for different control algorithms (114 Hz tone, $\mu_n = 0.003$, washing machine).

$$\mathbf{G}_{W2} = \begin{bmatrix} 1 & 0 & 0 & 0 \\ 1 & 0 & 0 & 0 \\ 0 & 1 & 0 & 0 \\ 0 & 1 & 0 & 0 \\ 0 & 0 & 1 & 0 \\ 0 & 0 & 1 & 0 \\ 0 & 0 & 0 & 1 \\ 0 & 0 & 0 & 1 \end{bmatrix}, \tag{6.120}$$

and

$$\mathbf{G}_{W4} = \begin{bmatrix} 1 & 0 \\ 1 & 0 \\ 1 & 0 \\ 1 & 0 \\ 0 & 1 \\ 0 & 1 \\ 0 & 1 \\ 0 & 1 \end{bmatrix}. \tag{6.121}$$

The results are presented in Fig. 6.41. A larger number of error microphones allows for higher step-size and provides faster convergence. However, for small step-sizes, the differences are small.

Selection of step-size and leak factor

The performance of SEFXLMS system depends on the normalized step-size μ_n and the leak factor $\alpha_{k,c,o}(n)$. Higher leaks require higher step-size to achieve the same

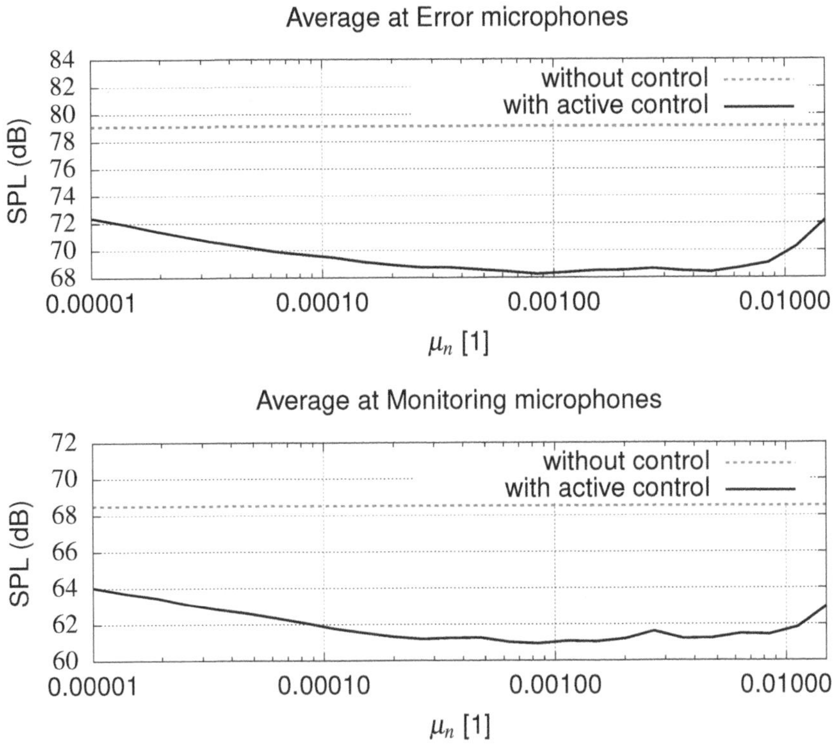

Figure 6.42 Average SPL at error and monitor microphones for reproduced spinning noise at 1200 rpm for a different μ_n after 15 minutes, $\gamma = 0.0001$.

noise reduction. It is convenient to introduce another coefficient γ and assume:

$$\alpha_{k,c,o}(n) = 1 - \mu_n \gamma. \tag{6.122}$$

Fig. 6.42 shows the average SPL at error and monitoring microphones for a washing machine casing with loudspeaker placed inside drum for a different μ_n and the same γ. The SPL is measured 15 minutes after enabling the active control. When the step-size is too low the system is to slow to achieve full noise reduction after 15 minutes. When the step-size is too high, the noise reduction is degraded.

Fig. 6.43 shows the average SPL at error and monitoring microphones for a different γ and the same μ_n. The degraded performance when the leak is high is expected. However, the best performance should be achieved without leak, but small leak, improves noise reduction. This might be related with a fact that leak reduces control signal powers and the system is underdetermined without the leak. The system has 13 actuators and only 4 error signals and there are infinite number of different control signals that will provide the same noise reduction on error microphones. A small leak will give preference to the combination with the smallest power.

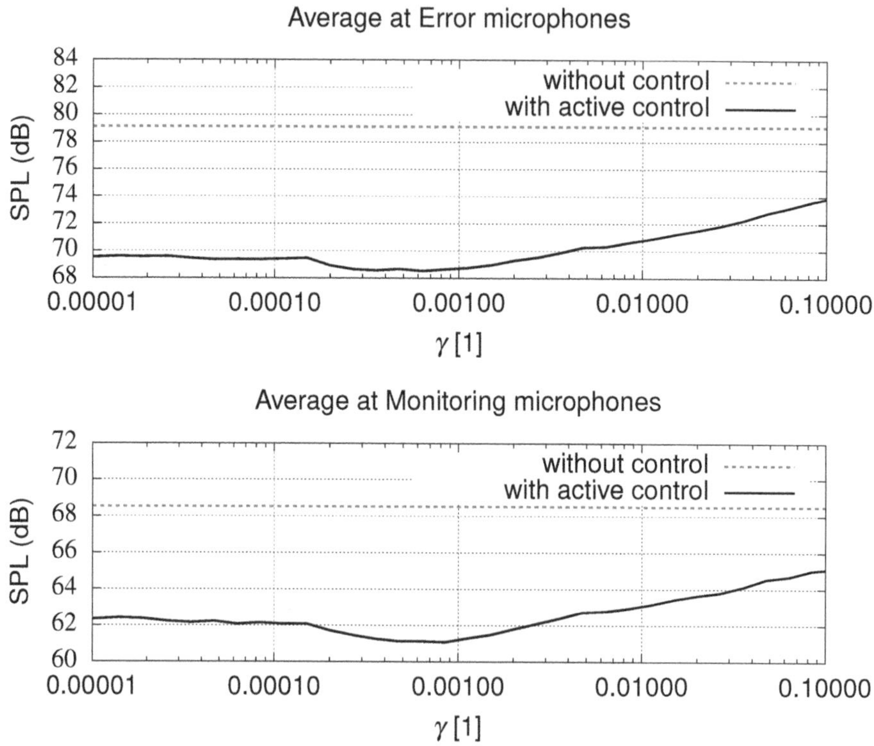

Figure 6.43 Average SPL at error and monitor microphones for reproduced spinning noise at 1200 rpm for a different γ after 15 minutes, $\mu_n = 0.001$.

6.5 SUMMARY

This chapter was devoted to active control solutions, which add energy to the system and rely on external energy power supply, but offer best noise reduction performance.

First, the necessary control hardware was described. This includes selection and position optimization of sensors and actuators. It is a very important step, as the final performance of the system depends on the controllability and observability measures of the system. Both actuators and microphones are considered from this point of view. The employed control unit and related solutions are also appropriately presented.

Second, different control structures are discussed and compared. Practical solutions including feedback cancellation and virtual microphone control are also presented in details. Finally, the control algorithms themselves are discussed. The FxLMS algorithm is considered as a basis, providing essentials and stability analysis. Then, appropriate modification are considered, including fast implementations and multichannel architecture.

It follows from the experiments performed on the active casings that active control is am effective approach for noise reduction. It can respond to time-varying operating conditions. The complexity and cost of implementation can be high. However, the active casing approach is much cheaper than classical active control, and noise reduction can be obtained globally.

7 Summary

In this book a consistent approach to global reduction of noise generated by different kinds of devices has been proposed. The idea is to control vibration of the device casing in order to reduce its noise radiation. The main benefit of that is the autonomous character of the solution, what means that neither measurement components, nor loudspeakers, nor any additional apparatus mounted outside the casing is required. Three different types of casings, with different complexities of the phenomena taking place, were considered, proposing solutions with varying degrees of complexity, while providing the opportunity to tailor them to the specific problem or noise-generating device.

A range of specific control strategies for noise-controlling casings are proposed in this book, which can be divided into three categories followed from their complexity and energy demand: passive, semi-active and active methods.

Passive solutions are characterized by a complete lack of the need for external energy supply. One approach proposed by the authors is to optimally shape the vibroacoustic properties of the controlled structure (by shifting the natural frequencies) basing on a previously prepared theoretical or numerical model. These modifications can be made by adding mass, a damper, or increasing the stiffness of a selected structure component in an appropriate manner, provided, e.g. by an optimization algorithm. Functionally graded materials can be applied for the casing walls to increase their acoustic isolation. Passive solutions can also include systems originally based on semi-active or even active solutions, but powered by energy harvesting within the system, which eliminates the need for external energy supply. Even 8 dB noise reduction was achievable with the passive solutions for a real-world noise. This is possible with a very low or even marginal cost of implementation, because many of these solutions can be applied during manufacturing process of the casing.

The second category are semi-active solutions that require little power, but external energy is only used to favourably change the properties of the system, therefore it is not directly transferred into the system. Examples of such solutions are switched shunt circuits connected to piezoelectric or electro-dynamic actuators. In such systems, the mechanical energy of vibrations is converted to electrical energy and then dissipated or used to resist the vibrations. As a result, additional damping is introduced in a selected frequency band or bands, however, systems of this type do not induce themselves additional vibrations of the structure or in the acoustic medium. Authors' concepts are elements that favourably change the properties of the structure, such as switchable link (coupling or decoupling selected components) or adjustable additional masses. They operate like the previously described passive modifications, but in the semi-active variant they can be adapted in real time to the requirements, e.g. to the current noise spectrum. The energy is then used to switch or maintain a given state of the semi-active actuator. Semi-active solutions resulted even in 12 dB noise reduction. The cost of implementation and operation is really small because of

DOI: 10.1201/9781003273806-7

low complexity of the electronic circuit, and using energy for the process of changing parameters of the system and response of the structure instead of compensating for the noise itself.

The third category are active solutions. They have a greater energy demand but achieve the highest levels of noise reduction exceeding 20 dB. They consist of actuators, sensors and a controller which, basing on the measurement signals, calculates the control signals in accordance with the selected algorithm. The actuators and sensors may be both acoustic (e.g. speakers and microphones) and structural (e.g. electro-dynamic exciters and accelerometers). These systems introduce additional energy directly into the system, but they do not have to completely eliminate system vibration. Often, it is enough to alter the vibration distribution appropriately to enhance the effective insulation of the casing. Virtual sensing can be used to control acoustic pressures estimated using other measurements. The proposed solutions, contrary to classical approach to noise reduction are generally much less complicated, require less energy and do not interfere with arrangement of the space where the device is located, providing global noise reduction at the same time.

The range of proposed innovative noise-reducing solutions offer a feasible alternative to classical approaches, covering a wide range of potential applications, even including cabins of vehicles and aircrafts. The authors believe that proposed noise-controlling casings have potential to be widespread and hence contribute to the struggle against excessive noise.

The presented idea can be further developed and improved by applying new achievements in measurement and actuation technology, signal processing, control, and new materials development.

A Stiffness matrix elements

The elements of submatrices of $\mathbf{K}_p$, $\mathbf{K}_b$ and $\mathbf{K}_r$ defined in (3.34) are given in this Appendix.

The elements of $\mathbf{K}_p$ can be derived as:

$$K_{pcc,ij} = \int_0^1\int_0^1 \left\{ \frac{b}{a}\kappa_x h G_{xz}\left(\frac{\partial\phi_i}{\partial\xi}\frac{\partial\phi_j}{\partial\xi}\right) + \frac{a}{b}\kappa_y h G_{yz}\left(\frac{\partial\phi_i}{\partial\eta}\frac{\partial\phi_j}{\partial\eta}\right)\right\} d\xi d\eta\,, \tag{A.1a}$$

$$K_{pcd,ij} = \int_0^1\int_0^1 \left\{ b\kappa_x h G_{xz}\left(\frac{\partial\phi_i}{\partial\xi}\psi_{x,j}\right)\right\} d\xi d\eta\,, \tag{A.1b}$$

$$K_{pce,ij} = \int_0^1\int_0^1 \left\{ a\kappa_y h G_{yz}\left(\frac{\partial\phi_i}{\partial\eta}\psi_{y,j}\right)\right\} d\xi d\eta\,, \tag{A.1c}$$

$$K_{pdd,ij} = \int_0^1\int_0^1 \left\{ \frac{b}{a}D_x\left(\frac{\partial\psi_{x,i}}{\partial\xi}\frac{\partial\psi_{x,j}}{\partial\xi}\right) + \frac{a}{b}D_{xy}\left(\frac{\partial\psi_{x,i}}{\partial\eta}\frac{\partial\psi_{x,j}}{\partial\eta}\right) + ab\kappa_x h G_{xz}\left(\psi_{x,i}\psi_{x,j}\right)\right\} d\xi d\eta\,, \tag{A.1d}$$

$$K_{pde,ij} = \int_0^1\int_0^1 \left\{ \frac{1}{2}\left(D_x\nu_y + D_y\nu_x\right)\left(\frac{\partial\psi_{x,i}}{\partial\xi}\frac{\partial\psi_{y,j}}{\partial\eta}\right) + D_{xy}\left(\frac{\partial\psi_{x,i}}{\partial\eta}\frac{\partial\psi_{y,j}}{\partial\xi}\right)\right\} d\xi d\eta\,, \tag{A.1e}$$

$$K_{pee,ij} = \int_0^1\int_0^1 \left\{ \frac{a}{b}D_y\left(\frac{\partial\psi_{y,i}}{\partial\eta}\frac{\partial\psi_{y,j}}{\partial\eta}\right) + \frac{b}{a}D_{xy}\left(\frac{\partial\psi_{y,i}}{\partial\xi}\frac{\partial\psi_{y,j}}{\partial\xi}\right) + ab\kappa_y h G_{yz}\left(\psi_{y,i}\psi_{y,j}\right)\right\} d\xi d\eta\,, \tag{A.1f}$$

$i = 1,2,...,N, \qquad j = 1,2,...,N.$

The elements of $\mathbf{K}_b$ can be derived as:

$$K_{bcc,ij} = b\int_0^1 \left\{ k_{tx0}\left(\phi_i\phi_j\right)\Big|_{\xi=0} + k_{tx1}\left(\phi_i\phi_j\right)\Big|_{\xi=1}\right\} d\eta + a\int_0^1 \left\{ k_{ty0}\left(\phi_i\phi_j\right)\Big|_{\eta=0} + k_{ty1}\left(\phi_i\phi_j\right)\Big|_{\eta=1}\right\} d\xi\,, \tag{A.2a}$$

$$K_{bdd,ij} = b\int_0^1 \left\{ k_{rx0}\left(\psi_{x,i}\psi_{x,j}\right)\Big|_{\xi=0} + k_{rx1}\left(\psi_{x,i}\psi_{x,j}\right)\Big|_{\xi=1}\right\} d\eta \tag{A.2b}$$

$$K_{bee,ij} = a\int_0^1 \left\{ k_{ry0}\left(\psi_{y,i}\psi_{y,j}\right)\Big|_{\eta=0} + k_{ry1}\left(\psi_{y,i}\psi_{y,j}\right)\Big|_{\eta=1} \right\} d\xi \tag{A.2c}$$

$$i = 1,2,...,N, \qquad j = 1,2,...,N.$$

The elements of $\mathbf{K}_r$ can be derived as:

$$K_{rcc,ij} = \sum_{r=0}^{N_r} \left\{ \frac{G_{r,i}A_{r,i}}{\beta_{r,i}} \int_{\xi_{r0,i}}^{\xi_{r1,i}} \left[\frac{\cos\alpha_i}{a}\left(\frac{\partial\phi_i}{\partial\xi}\frac{\partial\phi_j}{\partial\xi}\right) + \frac{a\sin^2\alpha_i}{b^2\cos\alpha_i}\left(\frac{\partial\phi_i}{\partial\eta}\frac{\partial\phi_j}{\partial\eta}\right) + \frac{\sin\alpha_i}{b}\left(\frac{\partial\phi_i}{\partial\xi}\frac{\partial\phi_j}{\partial\eta} + \frac{\partial\phi_i}{\partial\eta}\frac{\partial\phi_j}{\partial\xi}\right)\right]\Bigg|_{\eta=g_{r,i}(\xi)} d\xi \right\} \tag{A.3a}$$

$$K_{rcd,ij} = \sum_{r=0}^{N_r} \left\{ \frac{G_{r,i}A_{r,i}}{\beta_{r,i}} \int_{\xi_{r0,i}}^{\xi_{r1,i}} \left[\cos\alpha_i\left(\frac{\partial\phi_i}{\partial\xi}\psi_{x,j}\right) + \frac{a\sin\alpha_i}{b}\left(\frac{\partial\phi_i}{\partial\eta}\psi_{x,j}\right)\right]\Bigg|_{\eta=g_{r,i}(\xi)} d\xi \right\} \tag{A.3b}$$

$$K_{rce,ij} = \sum_{r=0}^{N_r} \left\{ \frac{G_{r,i}A_{r,i}}{\beta_{r,i}} \int_{\xi_{r0,i}}^{\xi_{r1,i}} \left[\sin\alpha_i\left(\frac{\partial\phi_i}{\partial\xi}\psi_{y,j}\right) + \frac{a\sin^2\alpha_i}{b\cos\alpha_i}\left(\frac{\partial\phi_i}{\partial\eta}\psi_{y,j}\right)\right]\Bigg|_{\eta=g_{r,i}(\xi)} d\xi \right\} \tag{A.3c}$$

$$\begin{aligned} K_{rdd,ij} = \sum_{r=0}^{N_r} \Bigg\{ & \frac{G_{r,i}A_{r,i}}{\beta_{r,i}} \int_{\xi_{r0,i}}^{\xi_{r1,i}} \left[a\cos\alpha_i\left(\psi_{x,i}\psi_{x,j}\right)\right]\Big|_{\eta=g_{r,i}(\xi)} d\xi \\ & + E_{r,i}I_{r,i}\int_{\xi_{r0,i}}^{\xi_{r1,i}} \left[\frac{\cos^3\alpha_i}{a}\left(\frac{\partial\psi_{x,i}}{\partial\xi}\frac{\partial\psi_{x,j}}{\partial\xi}\right) + \frac{a\sin^2\alpha_i\cos\alpha_i}{b^2}\left(\frac{\partial\psi_{x,i}}{\partial\eta}\frac{\partial\psi_{x,j}}{\partial\eta}\right) \right. \\ & \left. + \frac{\cos^2\alpha_i\sin\alpha_i}{b}\left(\frac{\partial\psi_{x,i}}{\partial\xi}\frac{\partial\psi_{x,j}}{\partial\eta} + \frac{\partial\psi_{x,i}}{\partial\eta}\frac{\partial\psi_{x,j}}{\partial\xi}\right)\right]\Bigg|_{\eta=g_{r,i}(\xi)} d\xi \\ & + G_{r,i}J_{r,i}\int_{\xi_{r0,i}}^{\xi_{r1,i}} \left[\frac{\sin^2\alpha_i\cos\alpha_i}{a}\left(\frac{\partial\psi_{x,i}}{\partial\xi}\frac{\partial\psi_{x,j}}{\partial\xi}\right) + \frac{a\sin^4\alpha_i}{b^2\cos\alpha_i}\left(\frac{\partial\psi_{x,i}}{\partial\eta}\frac{\partial\psi_{x,j}}{\partial\eta}\right) \right. \\ & \left. + \frac{\sin^3\alpha_i}{b}\left(\frac{\partial\psi_{x,i}}{\partial\xi}\frac{\partial\psi_{x,j}}{\partial\eta} + \frac{\partial\psi_{x,i}}{\partial\eta}\frac{\partial\psi_{x,j}}{\partial\xi}\right)\right]\Bigg|_{\eta=g_{r,i}(\xi)} d\xi \Bigg\} \end{aligned} \tag{A.3d}$$

$$K_{rde,ij} = \sum_{r=0}^{N_r} \left\{ \frac{G_{r,i}A_{r,i}}{\beta_{r,i}} \int_{\xi_{r0,i}}^{\xi_{r1,i}} \left[a \sin\alpha_i \left(\psi_{x,i}\psi_{y,j}\right)\right]\Big|_{\eta=g_{r,i}(\xi)} d\xi \right.$$

$$+ \left(E_{r,i}I_{r,i} - G_{r,i}J_{r,i}\right) \int_{\xi_{r0,i}}^{\xi_{r1,i}} \left[\frac{\cos^2\alpha_i \sin\alpha_i}{a} \left(\frac{\partial\psi_{x,i}}{\partial\xi}\frac{\partial\psi_{y,j}}{\partial\xi} \right) + \frac{a\sin^3\alpha_i}{b^2} \left(\frac{\partial\psi_{x,i}}{\partial\eta}\frac{\partial\psi_{y,j}}{\partial\eta} \right) \right.$$

$$\left. \left. + \frac{\sin^2\alpha_i\cos\alpha_i}{b} \left(\frac{\partial\psi_{x,i}}{\partial\xi}\frac{\partial\psi_{y,j}}{\partial\eta} + \frac{\partial\psi_{x,i}}{\partial\eta}\frac{\partial\psi_{y,j}}{\partial\xi} \right) \right] \Big|_{\eta=g_{r,i}(\xi)} d\xi \right\} \tag{A.3e}$$

$$K_{ree,ij} = \sum_{r=0}^{N_r} \left\{ \frac{G_{r,i}A_{r,i}}{\beta_{r,i}} \int_{\xi_{r0,i}}^{\xi_{r1,i}} \left[\frac{a\sin^2\alpha_i}{\cos\alpha_i} \left(\psi_{y,i}\psi_{y,j}\right) \right]\Big|_{\eta=g_{r,i}(\xi)} d\xi \right.$$

$$+ E_{r,i}I_{r,i} \int_{\xi_{r0,i}}^{\xi_{r1,i}} \left[\frac{\sin^2\alpha_i\cos\alpha_i}{a} \left(\frac{\partial\psi_{y,i}}{\partial\xi}\frac{\partial\psi_{y,j}}{\partial\xi} \right) + \frac{a\sin^4\alpha_i}{b^2\cos\alpha_i} \left(\frac{\partial\psi_{y,i}}{\partial\eta}\frac{\partial\psi_{y,j}}{\partial\eta} \right) \right.$$

$$\left. + \frac{\sin^3\alpha_i}{b} \left(\frac{\partial\psi_{y,i}}{\partial\xi}\frac{\partial\psi_{y,j}}{\partial\eta} + \frac{\partial\psi_{y,i}}{\partial\eta}\frac{\partial\psi_{y,j}}{\partial\xi} \right) \right] \Big|_{\eta=g_{r,i}(\xi)} d\xi$$

$$+ G_{r,i}J_{r,i} \int_{\xi_{r0,i}}^{\xi_{r1,i}} \left[\frac{\cos^3\alpha_i}{a} \left(\frac{\partial\psi_{y,i}}{\partial\xi}\frac{\partial\psi_{y,j}}{\partial\xi} \right) + \frac{a\sin^2\alpha_i\cos\alpha_i}{b^2} \left(\frac{\partial\psi_{y,i}}{\partial\eta}\frac{\partial\psi_{y,j}}{\partial\eta} \right) \right.$$

$$\left. \left. + \frac{\cos^2\alpha_i\sin\alpha_i}{b} \left(\frac{\partial\psi_{y,i}}{\partial\xi}\frac{\partial\psi_{y,j}}{\partial\eta} + \frac{\partial\psi_{y,i}}{\partial\eta}\frac{\partial\psi_{y,j}}{\partial\xi} \right) \right] \Big|_{\eta=g_{r,i}(\xi)} d\xi \right\} \tag{A.3f}$$

$i = 1, 2, ..., N, \qquad j = 1, 2, ..., N.$

B Mass matrix elements

The elements of submatrices of $\mathbf{M}_p$, $\mathbf{M}_m$ and $\mathbf{M}_r$ defined in (3.36) are given in this Appendix.

The elements of $\mathbf{M}_p$ can be derived as:

$$M_{pcc,ij} = \int_0^1 \int_0^1 \left\{ ab\rho_p h \left(\phi_i \phi_j \right) \right\} d\xi d\eta \,, \tag{B.1a}$$

$$M_{pdd,ij} = \int_0^1 \int_0^1 \left\{ \frac{1}{12} ab\rho_p h^3 \left(\psi_{x,i} \psi_{x,j} \right) \right\} d\xi d\eta \,, \tag{B.1b}$$

$$M_{pee,ij} = \int_0^1 \int_0^1 \left\{ \frac{1}{12} ab\rho_p h^3 \left(\psi_{y,i} \psi_{y,j} \right) \right\} d\xi d\eta \,, \tag{B.1c}$$

$$i = 1,2,...,N, \qquad j = 1,2,...,N.$$

The elements of $\mathbf{M}_m$ can be derived as:

$$M_{mcc,ij} = \sum_{k=0}^{N_a} \left\{ m_{a,k} \left(\phi_i \phi_j \right) \right\} \Big|_{\substack{\xi=\xi_{a,k} \\ \eta=\eta_{a,k}}} + \sum_{k=0}^{N_s} \left\{ m_{s,k} \left(\phi_i \phi_j \right) \right\} \Big|_{\substack{\xi=\xi_{s,k} \\ \eta=\eta_{s,k}}}$$
$$+ \sum_{k=0}^{N_m} \left\{ m_{m,k} \left(\phi_i \phi_j \right) \right\} \Big|_{\substack{\xi=\xi_{m,k} \\ \eta=\eta_{m,k}}} , \tag{B.2a}$$

$$M_{mdd,ij} = \sum_{k=0}^{N_a} \left\{ I_{ax,k} \left(\psi_{x,i} \psi_{x,j} \right) \right\} \Big|_{\substack{\xi=\xi_{a,k} \\ \eta=\eta_{a,k}}} + \sum_{k=0}^{N_s} \left\{ I_{sx,k} \left(\psi_{x,i} \psi_{x,j} \right) \right\} \Big|_{\substack{\xi=\xi_{s,k} \\ \eta=\eta_{s,k}}}$$
$$+ \sum_{k=0}^{N_m} \left\{ I_{mx,k} \left(\psi_{x,i} \psi_{x,j} \right) \right\} \Big|_{\substack{\xi=\xi_{m,k} \\ \eta=\eta_{m,k}}} , \tag{B.2b}$$

$$M_{mee,ij} = \sum_{k=0}^{N_a} \left\{ I_{ay,k} \left(\psi_{y,i} \psi_{y,j} \right) \right\} \Big|_{\substack{\xi=\xi_{a,k} \\ \eta=\eta_{a,k}}} + \sum_{k=0}^{N_s} \left\{ I_{sy,k} \left(\psi_{y,i} \psi_{y,j} \right) \right\} \Big|_{\substack{\xi=\xi_{s,k} \\ \eta=\eta_{s,k}}}$$
$$+ \sum_{k=0}^{N_m} \left\{ I_{my,k} \left(\psi_{y,i} \psi_{y,j} \right) \right\} \Big|_{\substack{\xi=\xi_{m,k} \\ \eta=\eta_{m,k}}} , \tag{B.2c}$$

$$i = 1,2,...,N, \qquad j = 1,2,...,N.$$

DOI: 10.1201/9781003273806-B

The elements of $\mathbf{M}_r$ can be derived as:

$$M_{rcc,ij} = \sum_{r=0}^{N_r} \left\{ \frac{A_{r,i}\rho_{r,i}a}{\cos\alpha_i} \int_{\xi_{r0,i}}^{\xi_{r1,i}} (\phi_i\phi_j)\bigg|_{\eta=g_{r,i}(\xi)} d\xi \right\} \tag{B.3a}$$

$$M_{rdd,ij} = \sum_{r=0}^{N_r} \left\{ \frac{A_{r,i}\rho_{r,i}k_{r,i}a}{\cos\alpha_i} \int_{\xi_{r0,i}}^{\xi_{r1,i}} (\psi_{x,i}\psi_{x,j})\bigg|_{\eta=g_{r,i}(\xi)} d\xi \right\} \tag{B.3b}$$

$$M_{ree,ij} = \sum_{r=0}^{N_r} \left\{ \frac{A_{r,i}\rho_{r,i}k_{r,i}a}{\cos\alpha_i} \int_{\xi_{r0,i}}^{\xi_{r1,i}} (\psi_{y,i}\psi_{y,j})\bigg|_{\eta=g_{r,i}(\xi)} d\xi \right\} \tag{B.3c}$$

$$i = 1,2,...,N, \qquad j = 1,2,...,N.$$

References

1. O. Abdeljaber, O. Avci, and D. J. Inman. Active vibration control of flexible cantilever plates using piezoelectric materials and artificial neural networks. *Journal of Sound and Vibration*, 363:33–53, 2016.
2. T. Aboulnasr and K. Mayyas. Complexity reduction of the NLMS algorithm via selective coefficient update. *IEEE Transactions on Signal Processing*, 47(5):1421–1424, May 1999.
3. J. H. Affdl and J. Kardos. The Halpin-Tsai equations: a review. *Polymer Engineering & Science*, 16(5):344–352, 1976.
4. A. E. Albert and L. S. Gardner. *Stochastic Approximation and Nonlinear Regression.* MIT Press, Cambridge, MA, 1967.
5. H. Amirinezhad, A. Tarkashvand, and R. Talebitooti. Acoustic wave transmission through a polymeric foam plate using the mathematical model of functionally graded viscoelastic (FGV) material. *Thin-Walled Structures*, 148:106466, 2020.
6. B. D. Anderson and J. B. Moore. *Optimal Control: Linear Quadratic Methods.* Courier Corporation, 2007.
7. K. Balachandran, D. Park, and P. Manimegalai. Controllability of second-order integrodifferential evolution systems in Banach spaces. *Computers & Mathematics with Applications*, 49(11–12):1623–1642, 2005.
8. E. A. Bender. *An Introduction to Mathematical Modeling.* Courier Corporation, 2000.
9. J.-P. Berenger. A perfectly matched layer for the absorption of electromagnetic waves. *Journal of Computational Physics*, 114(2):185–200, 1994.
10. K. Billon, N. Montcoudiol, A. Aubry, R. Pascual, F. Mosca, F. Jean, C. Pezerat, C. Bricault, and S. Chesné. Vibration isolation and damping using a piezoelectric flextensional suspension with a negative capacitance shunt. *Mechanical Systems and Signal Processing*, 140:106696, 2020.
11. D. Bismor. Comments on "A new feedforward hybrid ANC system". *IEEE Signal Processing Letters*, 21(5):635–637, May 2014.
12. D. Bismor. Comments on "A new feedforward hybrid ANC system"—an addendum. *IEEE Signal Processing Letters*, 21(5):642–642, May 2014.
13. D. Bismor. Partial update LMS algorithms in active noise control. In *Proceedings of Forum Acusticum 2014*, pages 1–7, Cracow, 2014.
14. D. Bismor. Extension of LMS stability condition over a wide set of signals. *International Journal of Adaptive Control and Signal Processing*, 29(5):653–670, 2015.
15. D. Bismor. Variable step size partial update LMS algorithm for fast convergence. In M. J. Crocker, M. Pawelczyk, F. Pedrielli, E. Carletti, and S. Luzzi, editors, *Proceedings of the 22st International Congress on Sound and Vibration*, pages 1–8, Florence, Italy, 2015. International Institute of Acoustics and Vibration.
16. D. Bismor. Simulations of partial update LMS algorithms in application to active noise control. *Procedia Computer Science*, 80:1180–1190, 2016. International Conference on Computational Science 2016, ICCS 2016, 6–8 June 2016, San Diego, California, USA.
17. D. Bismor. *Stability of a Class of Least Mean Square Algorithms.* Problemy współczesnej automatyki i robotyki. AOW EXIT, Warszawa, 2016.

18. D. Bismor. *Postepy akustyki*, chapter Leaky Partial Updates in Application to Structural Active Noise Control, pages 44–53. Polskie Towarzystwo Akustyczne, Oddział w Krakowie, Cracow, 2021.
19. D. Bismor, K. Czyz, and Z. Ogonowski. Review and comparison of variable step-size LMS algorithms. *International Journal of Acoustics and Vibration*, 21(1):24–39, March 2016.
20. D. Bismor and M. Pawelczyk. Stability conditions for the leaky LMS algorithm based on control theory analysis. *Archives of Acoustics*, 41(4):731–740, 2016.
21. G. E. Box, A. Luceño, and M. del Carmen Paniagua-Quinones. *Statistical Control by Monitoring and Adjustment*, volume 700. John Wiley & Sons, 2011.
22. H. J. Butterweck. A steady-state analysis of the LMS adaptive algorithm without use of the independence assumption. *Proceedings of ICASSP*, pages 1404–1407, 1995.
23. H. Caldwell and D. Cabrera. Comparison of acoustic retroreflection from corner cube arrays using FDTD simulation. In *Proceedings of ACOUSTICS*, volume 7, 2018.
24. G. Carayannis, D. Manolakis, and N. Kalouptsidis. A fast sequential algorithm for least-squares filtering and prediction. *IEEE Transactions on Acoustics, Speech, and Signal Processing*, 31(6):1394–1402, 1983.
25. S. Chemishkian et al. H-optimal mapping of actuators and sensors in flexible structures. In *Proceedings of the 1998 37th IEEE Conference on Decision and Control (CDC)*, pages 821–826, 1998.
26. D. Chhabra, G. Bhushan, and P. Chandna. Optimal placement of piezoelectric actuators on plate structures for active vibration control via modified control matrix and singular value decomposition approach using modified heuristic genetic algorithm. *Mechanics of Advanced Materials and Structures*, 23(3):272–280, 2016.
27. B.-Y. Choi and Z. Bien. Sliding-windowed weighted recursive least-squares method for parameter estimation. *Electronics Letters*, 25(20):1381–1382, 1989.
28. A. Chraponska, S. Wrona, J. Rzepecki, K. Mazur, and M. Pawelczyk. Active structural acoustic control of an active casing placed in a corner. *Applied Sciences*, 9(6):1059, 2019.
29. M. D. Christie, S. Sun, L. Deng, D. Ning, H. Du, S. Zhang, and W. Li. The variable resonance magnetorheological pendulum tuned mass damper: mathematical modelling and seismic experimental studies. *Journal of Intelligent Material Systems and Structures*, 31(2):263–276, 2020.
30. L.-L. Chung, Y.-A. Lai, C.-S. W. Yang, K.-H. Lien, and L.-Y. Wu. Semi-active tuned mass dampers with phase control. *Journal of Sound and Vibration*, 332(15):3610–3625, 2013.
31. J. Cioffi and T. Kailath. Fast, recursive-least-squares transversal filters for adaptive filtering. *IEEE Transactions on Acoustics, Speech, and Signal Processing*, 32(2):304–337, 1984.
32. C. Claeys, E. Deckers, B. Pluymers, and W. Desmet. A lightweight vibro-acoustic metamaterial demonstrator: numerical and experimental investigation. *Mechanical Systems and Signal Processing*, 70:853–880, 2016.
33. W. W. Clark. Vibration control with state-switched piezoelectric materials. *Journal of Intelligent Material Systems and Structures*, 11(4):263–271, 2000.
34. R. R. Craig and A. J. Kurdila. *Fundamentals of Structural Dynamics*. John Wiley & Sons, 2006.
35. M. J. Crocker. *Handbook of Noise and Vibration Control*. John Wiley & Sons, 2007.

36. L. Dal Bo, P. Gardonio, D. Casagrande, and S. Saggini. Smart panel with sweeping and switching piezoelectric patch vibration absorbers: Experimental results. *Mechanical Systems and Signal Processing*, 120:308–325, 2019.
37. R. Darleux, B. Lossouarn, and J.-F. Deü. Passive self-tuning inductor for piezoelectric shunt damping considering temperature variations. *Journal of Sound and Vibration*, 432:105–118, 2018.
38. N. de Melo Filho, C. Claeys, E. Deckers, and W. Desmet. Metamaterial foam core sandwich panel designed to attenuate the mass-spring-mass resonance sound transmission loss dip. *Mechanical Systems and Signal Processing*, 139:106624, 2020.
39. N. de Melo Filho, L. Van Belle, C. Claeys, E. Deckers, and W. Desmet. Dynamic mass based sound transmission loss prediction of vibro-acoustic metamaterial double panels applied to the mass-air-mass resonance. *Journal of Sound and Vibration*, 442:28–44, 2019.
40. K. Doğançay. *Partial-Update Adaptive Signal Processing. Design, Analysis and Implementation*. Academic Press, Oxford, 2008.
41. S. Douglas. A family of normalized LMS algorithms. *IEEE Signal Processing Letters*, 1(3):49–51, March 1994.
42. S. Douglas. Analysis and implementation of the max-NLMS adaptive filter. In *Conference Record of the Twenty-Ninth Asilomar Conference on Signals, Systems and Computers, 1995*, volume 1, pages 659–663 vol.1, Oct 1995.
43. S. C. Douglas. Simplified stochastic gradient adaptive filters using partial updating. In *1994 Sixth IEEE Digital Signal Processing Workshop*, pages 265–268, 1994.
44. T. Ehrig, M. Dannemann, R. Luft, C. Adams, N. Modler, and P. Kostka. Sound transmission loss of a sandwich plate with adjustable core layer thickness. *Materials*, 13(18):4160, 2020.
45. S. E. Elliott, I. M. Stothers, and P. A. Nelson. A multiple error LMS algorithm and its application to the active control of sound and vibration. *IEEE Transactions on Acoustics, Speech and Signal Processing*, ASSP-35(10):1423–1434, 1987.
46. F. Engesser. Die knickfestigkeit gerader stäbe. *Zentralblatt der Bauverwaltung*, 11:483–486, 1891.
47. F. Fahy and P. Gardonio. *Sound and Structural Vibration*. Elsevier, Oxford, second edition, 2007.
48. F. J. Fahy and P. Gardonio. *Sound and Structural Vibration: Radiation, Transmission and Response*. Academic Press, 2007.
49. G. Ferrari and M. Amabili. Active vibration control of a sandwich plate by non-collocated positive position feedback. *Journal of Sound and Vibration*, 342:44–56, 2015.
50. M. Ferrer, M. de Diego, G. Piñero, and A. Gonzalez. Active noise control over adaptive distributed networks. *Signal Processing*, 107:82–95, 2015.
51. P. Gardonio, M. Zientek, and L. Dal Bo. Panel with self-tuning shunted piezoelectric patches for broadband flexural vibration control. *Mechanical Systems and Signal Processing*, 134:106299, 2019.
52. S. L. Gay. The fast affine projection algorithm. In *Acoustic Signal Processing for Telecommunication*, pages 23–45. Springer, 2000.
53. S. Ghinet, N. Atalla, and H. Osman. Diffuse field transmission into infinite sandwich composite and laminate composite cylinders. *Journal of Sound and Vibration*, 289(4–5):745–778, 2006.
54. M. Godavarti and A. O. Hero. Partial update LMS algorithms. *IEEE Transactions on Signal Processing*, 53(7):2382–2399, 2005.

55. N. Hagood and A. von Flotow. Damping of structural vibrations with piezoelectric materials and passive electrical networks. *Journal of Sound and Vibration*, 146(2):243–268, 1991.
56. J. Hale and A. Daraji. Optimal placement of sensors and actuators for active vibration reduction of a flexible structure using a genetic algorithm based on modified H_∞. *Journal of Physics: Conference Series*, 382(1):012036, 2012.
57. C. Hansen. Fundamentals of acoustics. *American Journal of Physics*, 19, 01 1951.
58. C. Hansen, S. Snyder, X. Qiu, L. Brooks, and D. Moreau. *Active Control of Noise and Vibration*. CRC Press, 2012.
59. R. H. Hardin and N. J. A. Sloane. McLaren's improved snub cube and other new spherical designs in three dimensions. *Discrete & Computational Geometry*, 15(4):429–441, Apr 1996.
60. B. Hassibi, A. H. Sayed, and T. Kailath. LMS is H^∞ optimal. In *Proceedings of the 32nd IEEE Conference on Decision and Control*, pages 74–79 vol.1, 1993.
61. M. H. Hayes. *Statistical Digital Signal Processing and Modeling*. John Wiley & Sons, 2009.
62. S. Haykin. *Adaptive Filter Theory, Fourth Edition*. Prentice Hall, New York, 2002.
63. C. W. Isaac, M. Pawelczyk, and S. Wrona. Comparative study of sound transmission losses of sandwich composite double panel walls. *Applied Sciences*, 10(4):1543, 2020.
64. C. W. Isaac, S. Wrona, M. Pawelczyk, and N. Roozen. Numerical investigation of the vibro-acoustic response of functionally graded lightweight square panel at low and mid-frequency regions. *Composite Structures*, 259:113460, 2021.
65. T. Kaczorek. Fractional positive continuous-time linear systems and their reachability. *International Journal of Applied Mathematical Computation Science (AMCS)*, 18(2):223–228, 2008.
66. T. Kaczorek. Positive fractional 2d continuous-discrete time linear systems. *Bulletin of the Polish Academy of Sciences, Technical Sciences*, 59(4):575–579, 2011.
67. O. Kaiser, S. Pietrzko, and M. Morari. Feedback control of sound transmission through a double glazed window. *Journal of Sound and Vibration*, 263(4):775–795, 2003.
68. J. Kang. *Urban Sound Environment*. CRC Press, 2006.
69. C. Kelley and J. Kauffman. Adaptive synchronized switch damping on an inductor: A self-tuning switching law. *Smart Materials and Structures*, 26, 02 2017.
70. K. Kim, B.-H. Kim, T.-M. Choi, and D.-S. Cho. Free vibration analysis of rectangular plate with arbitrary edge constraints using characteristic orthogonal polynomials in assumed mode method. *International Journal of Naval Architecture and Ocean Engineering*, 4(3):267–280, 2012.
71. J. Klamka. *Controllability of Dynamical Systems*. Kluwer Academic Publishers. Dordrecht. The Netherlands, 1991.
72. J. Klamka. Controllability of dynamical systems—a survey. *Archives of Control Sciences*, 2 (XXXVIII)(3–4):281–307, 1993.
73. J. Klamka. Constrained controllability of nonlinear systems. *Journal of Mathematical Analysis and Applications*, 201(2):365–374, 1996.
74. J. Klamka. Constrained exact controllability of semilinear systems. *Systems & Control Letters*, 47(2):139–147, 2002.
75. J. Klamka. Controllability of dynamical systems. a survey. *Bulletin of the Polish Academy of Sciences: Technical Sciences*, 61(2):335–342, 2013.
76. J. Klamka. *Controllability and Minimum Energy Control, Monograph in Series Studies in Systems, Decision and Control*, volume 162. Springer Verlag, Berlin, 2018.

77. J. Klamka and J. Wyrwał. Controllability of second-order infinite-dimensional systems. *Systems & Control Letters*, 57(5):386–391, 2008.
78. J. Koford and G. Groner. The use of an adaptive threshold element to design a linear optimal pattern classifier. *IEEE Transactions on Information Theory*, 12(1):42–50, Jan 1966.
79. N. Kournoutos and J. Cheer. A system for controlling the directivity of sound radiated from a structure. *The Journal of the Acoustical Society of America*, 147(1):231–241, 2020.
80. J. Kowal, J. Pluta, J. Konieczny, and A. Kot. Energy recovering in active vibration isolation system results of experimental research. *Journal of Vibration and Control*, 14(7):1075–1088, 2008.
81. K. R. Kumar and S. Narayanan. The optimal location of piezoelectric actuators and sensors for vibration control of plates. *Smart Materials and Structures*, 16(6):2680, 2007.
82. B. Lam, S. Elliott, J. Cheer, and W.-S. Gan. Physical limits on the performance of active noise control through open windows. *Applied Acoustics*, 137, 2018.
83. F. Langfeldt, H. Hoppen, and W. Gleine. Broadband low-frequency sound transmission loss improvement of double walls with Helmholtz resonators. *Journal of Sound and Vibration*, page 115309, 2020.
84. M. Latos and M. Pawelczyk. Feedforward vs. feedback fixed-parameter h2 control of non-stationary noise. *Archives of Acoustics*, 34(4):521–535, 2009.
85. D. Lee, M. Morf, and B. Friedlander. Recursive least squares ladder estimation algorithms. *IEEE Transactions on Acoustics, Speech, and Signal Processing*, 29(3):627–641, 1981.
86. A. W. Leissa. Vibration of plates. Technical report, DTIC Document, 1969.
87. S. Leleu, H. Abou-Kandil, and Y. Bonnassieux. Piezoelectric actuators and sensors location for active control of flexible structures. In *Proceedings of the 17th IEEE Instrumentation and Measurement Technology Conference*, volume 2, pages 818–823. IEEE, 2000.
88. S. Leleu, H. Abou-Kandil, and Y. Bonnassieux. Piezoelectric actuators and sensors location for active control of flexible structures. *Instrumentation and Measurement, IEEE Transactions on*, 50(6):1577–1582, 2001.
89. X. Li, K. Yu, R. Zhao, J. Han, and H. Song. Sound transmission loss of composite and sandwich panels in thermal environment. *Composites Part B: Engineering*, 133:1–14, 2018.
90. K. Liew, Y. Xiang, S. Kitipornchai, and M. Lim. Vibration of rectangular Mindlin plates with intermediate stiffeners. *Journal of Vibration and Acoustics*, 116(4):529–535, 1994.
91. H. Lissek, R. Boulandet, and R. Fleury. Electroacoustic absorbers: bridging the gap between shunt loudspeakers and active sound absorption. *The Journal of the Acoustical Society of America*, 129(5):2968–2978, 2011.
92. W. Liu, Z. Hou, and M. A. Demetriou. A computational scheme for the optimal sensor/actuator placement of flexible structures using spatial H_2 measures. *Mechanical Systems and Signal Processing*, 20(4):881–895, 2006.
93. X. Liu, G. Cai, F. Peng, and H. Zhang. Piezoelectric actuator placement optimization and active vibration control of a membrane structure. *Acta Mechanica Solida Sinica*, 31(1):66–79, 2018.

94. A. E. H. Love. The small free vibrations and deformation of a thin elastic shell. *Philosophical Transactions of the Royal Society of London*, 179:491–546, 1888.
95. X. Ma, K. Chen, and J. Xu. Active control of sound transmission through orthogonally rib stiffened double-panel structure: Mechanism analysis. *Applied Sciences*, 9(16):3286, 2019.
96. Q. Mao. Improvement on sound transmission loss through a double-plate structure by using electromagnetic shunt damper. *Applied Acoustics*, 158:107075, 2020.
97. Q. Mao and S. Pietrzko. Experimental study for control of sound transmission through double glazed window using optimally tuned Helmholtz resonators. *Applied Acoustics*, 71(1):32–38, 2010.
98. Q. Mao and S. Pietrzko. *Control of Noise and Structural Vibration*. Springer, 2013.
99. K. Mayyas. A variable step-size selective partial update LMS algorithm. *Digital Signal Processing*, 23(1):75–85, 2013.
100. K. Mazur. *Active Control of Sound with a Vibrating Plate*. PhD thesis, Silesian University of Technology, Gliwice, Poland, 2013.
101. K. Mazur. Free dSPACE DS1104 drivers, 2014 [Online].
102. K. Mazur, A. Chraponska, S. Wrona, J. Rzepecki, and M. Pawelczyk. Switched-error FXLMS step-size normalizations for active noise-reducing casings. In *ICSV*, 2021.
103. K. Mazur and M. Pawelczyk. Active control of noise emitted from a device casing. In *Proceedings of the 22nd International Congress on Sound and Vibration*, Florence, Italy, 2015.
104. K. Mazur, J. Rzepecki, A. Pietruszewska, S. Wrona, and M. Pawelczyk. Vibroacoustical performance analysis of a rigid device casing with piezoelectric shunt damping. *Sensors (Basel, Switzerland)*, 21:2517, 2021.
105. K. Mazur, S. Wrona, A. Chraponska, J. Rzepecki, and M. Pawelczyk. Synchronized switch damping on inductor for noise-reducing casing. In *Proceedings of 26th International Congress on Sound and Vibration*, Montreal, Canada, 7–11 July, 2019.
106. K. Mazur, S. Wrona, and M. Pawelczyk. Design and implementation of multichannel global active structural acoustic control for a device casing. *Mechanical Systems and Signal Processing*, 98C:877–889, 2018.
107. K. Mazur, S. Wrona, and M. Pawelczyk. Active noise control for a washing machine. *Applied Acoustics*, 146:89–95, 2019.
108. K. Mazur, S. Wrona, A. Pietruszewska, J. Rzepecki, and M. Pawelczyk. Synchronized switch damping on inductor for noise-reducing casing. In *26th International Congress on Sound and Vibration*, Montreal, Canada, 2019.
109. K. J. Mazur and M. Pawelczyk. Active noise-vibration control using the filtered-reference LMS algorithm with compensation of vibrating plate temperature variation. *Archives of Acoustics*, 36(1):65–76, 2011.
110. K. J. Mazur and M. Pawelczyk. Nonlinear active noise control of sound transmitted through a plate. *Archives of Acoustics*, 37(3):381–382, 2012.
111. K. J. Mazur and M. Pawelczyk. Active noise control with a single nonlinear control filter for a vibrating plate with multiple actuators. *Archives of Acoustics*, 38(4):537–545, 2013.
112. K. J. Mazur and M. Pawelczyk. Hammerstein nonlinear active noise control with the filtered-error LMS algorithm. *Archives of Acoustics*, 38(2):197–203, 2013.
113. K. J. Mazur and M. Pawelczyk. Multiple-error adaptive control of an active noise-reducing casing. In *Postêpy akustyki 2015: Progress of Acoustics 2015*, pages 701–712. 2015.

114. K. J. Mazur and M. Pawelczyk. Internal model control for a light-weight active noise-reducing casing. *Archives of Acoustics*, 41(2):315–322, 2016.
115. K. J. Mazur and M. Pawelczyk. Virtual microphone control for a light-weight active noise-reducing casing. In *23rd International Congress on Sound and Vibration : ICSV 2016, Athens Greece, 10–14 July 2016*, pages 1–8. 2016.
116. K. J. Mazur and M. Pawelczyk. Virtual microphone control for an active noise-cancelling casing. In *Active Noise and Vibration Control: Selected, Peer Reviewed Papers from the 12th Conference on Active Noise and Vibration Control Methods (MARDiH'2015), June 8–11, 2015, Cracow, Poland*, pages 57–66. 2016.
117. K. J. Mazur, S. Wrona, A. Chrapońska, J. Rzepecki, and M. Pawelczyk. FXLMS with multiple error switching for active noise-cancelling casings. *Archives of Acoustics*, 44(4):775–782, 2019.
118. K. J. Mazur, S. Wrona, and M. Pawelczyk. Design and implementation of multichannel global active structural acoustic control for a device casing. *Mechanical Systems and Signal Processing*, 98:877–889, 2018.
119. K. J. Mazur, S. Wrona, and M. Pawelczyk. Placement of microphones for an active noise-reducing casing. In *25th International Congress on Sound and Vibration 2018 (ICSV 25), Hiroshima, Japan, 8–12 July 2018*, pages 944–951. 2018.
120. K. J. Mazur, S. Wrona, and M. Pawelczyk. Active noise control for a washing machine. *Applied Acoustics*, 146:89–95, 2019.
121. K. J. Mazur, S. Wrona, and M. Pawelczyk. Performance evaluation of active noise control for a real device casing. *Applied Sciences*, 10(1):1–13, 2020.
122. J. Milton, J. Cheer, and S. Daley. Active structural acoustic control using an experimentally identified radiation resistance matrix. *The Journal of the Acoustical Society of America*, 147(3):1459–1468, 2020.
123. R. D. Mindlin. Influence of rotary inertia and shear on flexural motions of isotropic elastic plates. *Journal of Applied Mechanics*, 18:31–38, 1951.
124. M. Misol. Full-scale experiments on the reduction of propeller-induced aircraft interior noise with active trim panels. *Applied Acoustics*, 159:107086, 2020.
125. M. Misol, S. Algermissen, M. Rose, and H. P. Monner. Aircraft lining panels with low-cost hardware for active noise reduction. In *2018 Joint Conference-Acoustics*, pages 1–6. IEEE, 2018.
126. L. Morzyński and G. Szczepański. Double panel structure for active control of noise transmission. *Archives of Acoustics*, 43(4):689–696, 2018.
127. J. Nagumo and A. Noda. A learning method for system identification. *Automatic Control, IEEE Transactions on*, 12(3):282–287, June 1967.
128. J. Nalepa and M. Kawulok. Adaptive memetic algorithm enhanced with data geometry analysis to select training data for SVMS. *Neurocomputing*, 185:113–132, 2016.
129. P. A. Nelson and S. J. Elliott. *Active Control of Sound*. Academic press, 1993.
130. F. Neri, C. Cotta, and P. Moscato. *Handbook of Memetic Algorithms*, volume 379. Springer, 2012.
131. J. Nicholson and L. Bergman. Vibration of thick plates carrying concentrated masses. *Journal of Sound and Vibration*, 103(3):357–369, 1985.
132. A. N. Norris and D. M. Photiadis. Thermoelastic relaxation in elastic structures, with applications to thin plates. *The Quarterly Journal of Mechanics and Applied Mathematics*, 58(1):143–163, 2005.
133. M. Pawelczyk. Adaptive noise control algorithms for active headrest system. *Control Engineering Practice*, 12:1101–1112, 2004.

134. M. Pawelczyk. Feedback control of acoustic noise at desired locations. *Silesian University of Technology, Gliwice*, 2005.
135. M. Pawelczyk. *Application-Oriented Design of Active Noise Control Systems*. Academic Publishing House Exit, 2013.
136. M. Pawelczyk, J. Rzepecki, and S. Wrona. Semi-active electromagnetic element for damping of transverse vibration of planar structures, PL Patent 426875, 2018.
137. M. Pawelczyk, S. Wrona, and K. Mazur. Metody redukcji hałasu urządzeń (in English: *Methods of Device Noise Control*). In P. Kulczycki, J. Korbicz, and J. Kacprzyk, editors, *Automatyka, robotyka i przetwarzanie informacji*, pages 741–762. PWN Scientific Publisher; and Automatic Control and Robotics Committee of the Polish Academy of Sciences, Warsaw, 2019.
138. M. Pawelczyk, S. Wrona, and K. Mazur. Methods of device noise control. In P. Kulczycki, J. Korbicz, and J. Kacprzyk, editors, *Automatic Control, Robotics, and Information Processing, Part of the Studies in Systems, Decision and Control book series (SSDC, volume 296)*, pages 821–843. Springer; and Automatic Control and Robotics Committee of the Polish Academy of Sciences, 2020.
139. PCI Special Interest Group. *PCI Local Bus Specification Revision 3.0*, 2004.
140. A. Pietruszewska, J. Rzepecki, C. Isaac, K. Mazur, and M. Pawelczyk. Spectral analysis of macro-fiber composites measured vibration of double-panel structure coupled with solenoids. *Sensors*, 20:3505, 06 2020.
141. A. Pietruszewska, J. Rzepecki, K. Mazur, S. Wrona, and M. Pawelczyk. Influence of double-panel structure modification on vibroacoustical properties of a rigid device casing. *Archives of Acoustics*, 45:119–127, 02 2020.
142. S. Pietrzko. *Contributions to Noise and Vibration Control Technology*. Akademia Górniczo-Hutnicza, 2009.
143. S. J. Pietrzko. *Contributions to Noise and Vibration Control Technology*. AGH—University of Science and Technology Press, Cracow, 2009.
144. I. Pitas. Fast algorithms for running ordering and max/min calculation. *IEEE Transactions on Circuits and Systems*, 36(6):795–804, Jun 1989.
145. S. S. Rao. *Vibration of Continuous Systems*. John Wiley & Sons, 2007.
146. W. Rdzanek. *Structural Vibroacoustics of Surface Elements [in Polish: Wibroakustyka strukturalna elementów powierzchniowych]*. Rzeszów University of Technology Publishing House, 2011.
147. J. N. Reddy. *Theory and Analysis of Elastic Plates and Shells*. CRC Press, 2006.
148. K. Renji, P. Nair, and S. Narayanan. Modal density of composite honeycomb sandwich panels. *Journal of Sound and Vibration*, 195(5):687–699, 1996.
149. C. Richard, D. Guyomar, D. Audigier, and H. Bassaler. Enhanced semi-passive damping using continuous switching of a piezoelectric device on an inductor. In *Smart Structures*, 2000.
150. J. Rzepecki, A. Chraponska, K. Mazur, S. Wrona, and M. Pawelczyk. Semi-active reduction of device casing vibration using a set of piezoelectric elements. In *2019 20th International Carpathian Control Conference (ICCC)*, Wieliczka, Poland, 2019.
151. J. Rzepecki, A. Pietruszewska, S. Budzan, C. W. Isaac, K. Mazur, and M. Pawelczyk. Chladni figures in modal analysis of a double-panel structure. *Sensors*, 20(15):4084, 2020.
152. B. Sikora and J. Klamka. Cone-type constrained relative controllability of semilinear fractional systems with delays. *Kybernetika*, 53(2).

153. B. Sikora and J. Klamka. Constrained controllability of fractional linear systems with delays in control. *Systems and Control Letters*, 106(1):9–15, 2017.
154. O. J. M. Smith. A controller to overcome dead time. *ISA J.*, 6:28–33, 1959.
155. R. M. Stallman. *Using the GNU Compiler Collection for GCC version 4.7.3*. GNU Press, 2003.
156. S. P. Timoshenko and S. Woinowsky-Krieger. *Theory of Plates and Shells*. McGraw-Hill, 1959.
157. F. Tornabene. Free vibration analysis of functionally graded conical, cylindrical shell and annular plate structures with a four-parameter power-law distribution. *Computer Methods in Applied Mechanics and Engineering*, 198(37–40):2911–2935, 2009.
158. E. Turco and P. Gardonio. Sweeping shunted electro-magnetic tuneable vibration absorber: Design and implementation. *Journal of Sound and Vibration*, 407:82–105, 2017.
159. S. V. Vaseghi. *Advanced Digital Signal Processing and Noise Reduction*. John Wiley & Sons, 2008.
160. M. Vatavu, V. Năstăsescu, F. Turcu, and I. Burda. Voltage-controlled synthetic inductors for resonant piezoelectric shunt damping: a comparative analysis. *Applied Sciences*, 9:4777, 11 2019.
161. K. Velten. *Mathematical Modeling and Simulation: Introduction for Scientists and Engineers*. John Wiley & Sons, 2009.
162. T. Wei, B. Wang, J. Cao, and B. Bao. Experimental comparisons of two detection methods for semi-passive piezoelectric structural damping. *Journal of Vibration Engineering & Technologies*, 5:367–379, 09 2017.
163. B. Widrow and M. E. Hoff. Adaptive switching circuits. In *1960 IRE WESCON Convention Record, part 4*, pages 96–140, 1960. Available online at http://isl-www.stanford.edu/ widrow/papers/c1960adaptiveswitching.pdf.
164. S. Wrona. *Modelling and Control of Device Casing Vibrations for Active Reduction of Acoustic Noise*. PhD thesis, Silesian University of Technology, Gliwice, Poland, 2016.
165. S. Wrona, M. de Diego, and M. Pawelczyk. Shaping zones of quiet in a large enclosure generated by an active noise control system. *Control Engineering Practice*, 80:1–16, 2018.
166. S. Wrona, K. Mazur, M. Pawelczyk, and J. Klamka. Optimal placement of actuators for active control of a washing machine casing. In *Proceedings of 13th Conference on Active Noise and Vibration Control Methods*, Kazimierz Dolny, Poland, 12–14 June, 2017.
167. S. Wrona and M. Pawelczyk. Application of an memetic algorithm to placement of sensors for active noise-vibration control. *Mechanics and Control*, 32(3):122–128, 2013.
168. S. Wrona and M. Pawelczyk. Active reduction of device narrowband noise by controlling vibration of its casing based on structural sensors. In *Proceedings of 22nd International Congress on Sound and Vibration*, Florence, Italy, 12–16 July, 2015.
169. S. Wrona and M. Pawelczyk. Identification of elastic boundary conditions of light-weight device casing walls using experimental data. In *Proceedings of 21st International Conference On Methods and Models in Automation and Robotics (MMAR), IEEE*, Międzyzdroje, Poland, 29 August–1 September, 2016.
170. S. Wrona and M. Pawelczyk. Optimal placement of actuators for active structural acoustic control of a light-weight device casing. In *Proceedings of 23rd International Congress on Sound and Vibration*, Athens, Greece, 10–14 July, 2016.

171. S. Wrona and M. Pawelczyk. Shaping frequency response of a vibrating plate for passive and active control applications by simultaneous optimization of arrangement of additional masses and ribs. Part I: Modeling. *Mechanical Systems and Signal Processing*, 70–71:682–698, 2016.
172. S. Wrona and M. Pawelczyk. Shaping frequency response of a vibrating plate for passive and active control applications by simultaneous optimization of arrangement of additional masses and ribs. Part II: Optimization. *Mechanical Systems and Signal Processing*, 70–71:699–713, 2016.
173. S. Wrona, M. Pawelczyk, and J. Cheer. Acoustic radiation-based optimization of the placement of actuators for active control of noise transmitted through plates. *Mechanical Systems and Signal Processing*, 147:107009, 2021.
174. S. Wrona, M. Pawelczyk, and L. Cheng. A novel semi-active actuator with tunable mass moment of inertia for noise control applications. *Journal of Sound and Vibration*, 509:116244, 2021.
175. S. Wrona, M. Pawelczyk, and L. Cheng. Semi-active links in double-panel noise barriers. *Mechanical Systems and Signal Processing*, 154:107542, 2021.
176. S. Wrona, M. Pawelczyk, and X. Qiu. Shaping the acoustic radiation of a vibrating plate. *Journal of Sound and Vibration*, 476:115285, 2020.
177. J. Wyrwał. Simplified conditions of initial observability for infinite-dimensional second-order damped dynamical systems. *Journal of Mathematical Analysis and Applications*, 478(1):33–57, 2019.
178. J. Wyrwal, M. Pawelczyk, L. Liu, and Z. Rao. Double-panel active noise reducing casing with noise source enclosed inside–modelling and simulation study. *Mechanical Systems and Signal Processing*, 152:107371, 2021.
179. J. Wyrwal, R. Zawiski, M. Pawelczyk, and J. Klamka. Modelling of coupled vibro-acoustic interactions in an active casing for the purpose of control. *Applied Mathematical Modelling*, 50:219–236, 2017.
180. D. Young. Vibration of rectangular plates by the Ritz method. *Journal of Applied Mechanics-Transactions of the ASME*, 17(4):448–453, 1950.
181. Y. Yu, Q. Cai, and S. Guo. A new construction method of common embedded cross compiler tool based on newlib. In *2010 IEEE International Conference on Intelligent Computing and Intelligent Systems*, volume 3, pages 817–819, 2010.
182. X. Zhao, Y. Lee, and K. M. Liew. Free vibration analysis of functionally graded plates using the element-free kp-Ritz method. *Journal of Sound and Vibration*, 319(3-5):918–939, 2009.
183. W. Zheng, Y. Lei, Q. Huang, and S. Li. Improving low frequency performance of double-wall structure using piezoelectric transducer/loudspeaker shunt damping technologies. *Journal of Low Frequency Noise, Vibration and Active Control*, 31(3):175–192, 2012.

Index